Mode of action of antifungal agents

Sites of action of some antifungal agents

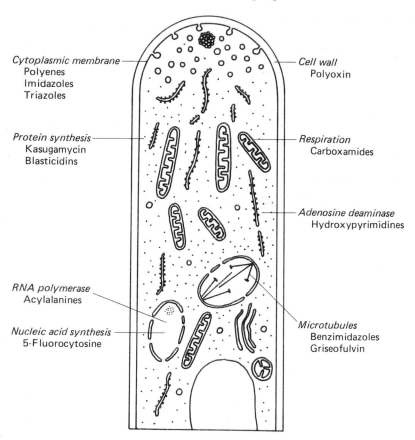

Cytoplasmic membrane
  Polyenes
  Imidazoles
  Triazoles

Cell wall
  Polyoxin

Protein synthesis
  Kasugamycin
  Blasticidins

Respiration
  Carboxamides

Adenosine deaminase
  Hydroxypyrimidines

RNA polymerase
  Acylalanines

Nucleic acid synthesis
  5-Fluorocytosine

Microtubules
  Benzimidazoles
  Griseofulvin

Mode of action of antifungal agents

SYMPOSIUM OF
THE BRITISH MYCOLOGICAL SOCIETY
HELD AT THE UNIVERSITY OF MANCHESTER
MANCHESTER, SEPTEMBER 1983

EDITED BY

A.P.J.TRINCI & J.F.RYLEY

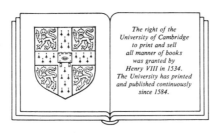

The right of the
University of Cambridge
to print and sell
all manner of books
was granted by
Henry VIII in 1534.
The University has printed
and published continuously
since 1584.

CAMBRIDGE UNIVERSITY PRESS
CAMBRIDGE
LONDON NEW YORK NEW ROCHELLE
MELBOURNE SYDNEY

Published by the Press Syndicate of the University of Cambridge
The Pitt Building, Trumpington Street, Cambridge CB2 1RP
32 East 57th Street, New York, NY 10022, USA
296 Beaconsfield Parade, Middle Park, Melbourne 3206, Australia

First published 1984

Printed in Great Britain by The Pitman Press, Bath

Library of Congress catalogue card number: 84–4279

British Library Cataloguing in Publication Data

Mode of action in antifungal agents. –
   (British Mycological Society Symposia)

   1. Antifungal agents
   I. Trinci, A.P.J. II. Ryley, J.F.
   III. Series
   616.9′69061   RM410

ISBN 0 521 26171 6

Contents

Contents

Contributors

B. C. Baldwin, *ICI Plant Protection Division, Jealott's Hill Research Station, Bracknell, Berkshire, RG12 6EY, UK*

K. J. Brent, *Long Ashton Research Station, Long Ashton, Bristol, BS18 9AF, UK*

T. G. Burland, *McArdle Laboratory, University of Wisconsin, Madison, WI 53706, USA*

R. Y. Cartwright, *Public Health Laboratory, St Luke's Hospital, Guildford, GU1 3NT, UK*

M. L. Chance, *Department of Parasitology, Liverpool School of Tropical Medicine, Pembroke Place, Liverpool, L3 5QA, UK*

W. Cools, *Department of Comparative Biochemistry, Janssen Pharmaceutica, B-2340 Beerse, Belgium*

L. C. Davidse, *Department of Phytopathology, Agricultural University Wageningen, PO Box 8025, 6700 EE Wageningen, The Netherlands*

J. Dekker, *Department of Phytopathology, Agricultural University, Wageningen, PO Box 8025, 6700 EE Wageningen, The Netherlands*

R. Fritz, *Institut National de la Recherche Agronomique, Laboratoire de Phytopharmacie, Etoile de Choisy, 78000 Versailles (Yvelines), France*

K. Gull, *Biological Laboratory, University of Kent, Canterbury, Kent, CT2 7NJ, UK*

D. W. Hollomon, *Insecticides and Fungicides Department, Rothamsted Experimental Station, Harpenden, Hertfordshire, AL5 2JQ, UK*

D. Kerridge, *Department of Biochemistry, University of Cambridge, Tennis Court Road, Cambridge, CB2 1QW, UK*

P. Kuhn, *Shell Research Ltd, Sittingbourne Research Centre, Sittingbourne, Kent, ME9 8AG, UK*

W. Lauwers, *Analytical Department, Janssen Pharmaceutica, B-2340 Beerse, Belgium*

P. Leroux, *Institut National de la Recherche Agronomique, Laboratoire de Phytopharmacie, Etoile de Choisy, 78000 Versailles (Yvelines), France*

P. Marichal, *Department of Comparative Biochemistry, Janssen Pharmaceutica, B-2340 Beerse, Belgium*

R. R. C. New, *Department of Biochemistry, University of Liverpool, PO Box 147, Liverpool, L69 3BX, UK*

N. N. Ragsdale, *Cooperative State Research Service, USDA, Washington, DC 20250, USA*

W. G. Rathmell, *ICI Plant Protection Division, Jealott's Hill Research Station, Bracknell, Berkshire, RG12 6EY, UK*

J. F. Ryley, *ICI Pharmaceuticals Division, Mereside Alderley Park, Macclesfield, Cheshire, SK10 4TG, UK*

A. Kaars Sijpesteijn, *Netherlands Organization for Applied Scientific Research, Institute of Applied Chemistry, PO Box 5009, 3502 JA Utrecht, The Netherlands*

H. D. Sisler, *Department of Botany, University of Maryland, College Park, Maryland, 20742, USA*

A. P. J. Trinci, *Department of Botany, University of Manchester, Manchester, M13 9PL, UK*

H. Van den Bossche, *Department of Comparative Biochemistry, Janssen Pharmaceutica, B-2340 Beerse, Belgium*

M. Wade, *Shell Research Ltd, Sittingbourne Research Centre, Sittingbourne, Kent, ME9 8AG, UK*

W. L. Whelan, *Department of Biochemistry, University of Cambridge, Tennis Court Road, Cambridge, CB2 1QW, UK*

G. Willemsens, *Department of Comparative Biochemistry, Janssen Pharmaceutica, B-2340 Beerse, Belgium*

Preface

Fungi are responsible for a variety of diseases of plants and animals and cause considerable economic loss; for example, they account for 70–80% of the damage to crops caused by disease (Chapter 1). For many years industrial mycologists have been seeking antifungal agents to combat disease in plants and animals, while academic mycologists have complemented this work with more basic mode-of-action studies. Seldom, however, have those interested in combating plant mycopathogens and those concerned with human or animal mycoses come together. This situation was remedied at a symposium organised by the British Mycological Society which was held at the University of Manchester during September 1983; this volume is the outcome of that symposium.

Surveys by Brent (Chapter 1) and by Cartwright (Chapter 2) emphasise the great diversity in pathogenic effects produced by fungi in plants and in man, and draw attention to the need to control the whole disease process rather than the fungal pathogen alone. Potential targets for the selective inhibition of fungal growth are discussed by Baldwin (Chapter 3), and consideration is given in various chapters to the development by some mycopathogens of resistance to site-specific fungicides or antifungal drugs. In contrast, resistance has not been a problem in the case of the older, multi-site fungicides or antifungal drugs such as amphotericin B (although amphotericin-B-resistance does occasionally occur).

Several chapters consider the modes of action of antifungal compounds at the cellular or biochemical level. These antifungal compounds have provided cell and molecular biologists with specific cellular probes which they can use to investigate eukaryotic cells. For example, benzimidazole carbamates bind to β-tubulins and interfere with microtubule activities (Chapter 15), metalaxyl inhibits the activity of a chromatin-bound RNA polymerase (Chapter 11), and griseofulvin is the only compound known to bind to microtubule-associated proteins. Similarly, studies on the mode of action of amphotericin B have provided valuable information about the structure and function of fungal membranes (Chapter 16). However, we still do not know the precise modes of action of many antifungal agents, including those of some of the older fungicides (Chapters 7 and 10). It is worth noting here that only with the azoles – imidazoles (Chapter 14) and triazoles (Chapter 12) – which inhibit ergosterol biosynthesis, do we find similar compounds effective in both areas of concern; this observation emphasises the importance of ergosterol in the fungal membrane. However, considering the diversity of habitat adopted by a wide range of mycopathogens, it is hardly surprising that different types of chemicals are usually effective in the different disease areas.

Not only does an antifungal agent have to interact with the parasite, causing its death, or at least preventing its growth and development, it also has to survive a whole range of obstacles in the plant or animal to reach the fungal lesion and has to have a variety of physicochemical and pharmacokinetic properties relevant to the intended form of application (Chapter 17). No wonder then that we have not yet reached the stage of rational drug design in which we first identify a susceptible intracellular (or extracellular) process, and then synthesise a drug to antagonise it!

Because the approach to antifungal discovery has to be empirical – or at the best rationally empirical – vast programmes of chemical synthesis have to be supported; some of the high-throughput, simple screens established to look for activity in a very large number of chemicals are described in Chapter 4. Just as we were glad to bring together agricultural and medical mycologists to broaden their concepts of the fungal disease process, so too we hope that those from academia will have realised something of the scale on which industrial research has to be carried out if there is to be any glimmer of hope of success. Modern antifungal agents are far from perfect, and the spectre of resistance is always in the background. May ideas generated and exchanged in this symposium help us all in our endeavours for effective control of fungal disease.

October 1983

A. P. J. Trinci
Botany Department, University of Manchester
J. F. Ryley
ICI, Alderley Park

1

Fungal diseases of plants and the effectiveness of present methods of control

K. J. BRENT

Long Ashton Research Station, University of Bristol, Long Ashton, Bristol BS18 9AF, UK

Introduction

This chapter aims to give a practical context for subsequent chapters on the modes of action of agricultural fungicides, particularly for those who are not familiar with the problems of controlling crop diseases. Inevitably this is a very general and simplified account. Special consideration is given to chemical treatment which, in most situations, remains the predominant method of coping with plant disease. However, I must emphasise the importance of integrating chemical applications with other crop-protection measures, especially since these will receive less attention in this volume than they really deserve.

The damage to crops and plants caused by fungi

Virtually all crops throughout the world, and also most ornamental plants and trees, are subject to attack by pathogenic organisms. These pathogens can be fungi, bacteria or viruses, and members of each of these groups can cause severe damage. However, in contrast to the position in human or animal disease, it is primarily the fungi that cause the greatest recognisable damage to plant health and to crop productivity. Often fungal infections remain small, or occur too late in the season to have serious effects, but when conditions favour invasion by the pathogen and when the timing of infection coincides with a critical stage of plant growth, then the fungi can cause severe or even total losses in yield or in the quality of the plant or its products.

The damage resulting from plant diseases can take many forms, but four main types can be distinguished.

(i) Acute damage to the plants, leading to severe injury or death within a few weeks. Examples include: potato late blight (caused by *Phytophthora infestans*), grape downy mildew (*Plasmopara viticola*), wheat leaf rust (*Puccinia recondita*) and tomato wilt (*Fusarium oxysporum* f.sp. *lycopersici*).

(ii) Chronic infections which gradually debilitate the plant, and decrease its productivity. The plant or a large part of it may eventually be killed. Examples include: apple canker (caused by *Nectria galligena*), barley powdery mildew (*Erysiphe graminis* f.sp. *hordei*), many virus diseases.

(iii) Direct damage to produce, which lowers or destroys its value. Even localised or superficial damage can greatly decrease the quality of certain produce. Examples include: apple scab (caused by *Venturia inaequalis*), coffee berry disease (*Colletotrichum coffeanum*) and potato scurf (*Rhizoctonia solani*).

(iv) Post-harvest damage to produce. This can arise from pathogens which were already present on the produce before harvest, and which may have penetrated the tissue to a limited extent, or from pathogens which arrive during post-harvest handling or storage. Examples include: green and blue mould on citrus (caused by *Penicillium* spp.), brown rot of pome and stone fruit (*Sclerotinia* spp.) and potato dry rot (*Fusarium caeruleum*).

This classification of types of damage cannot be applied rigidly in practice since some pathogens may act in more than one way or have intermediate effects. For example, after a severe epidemic of blight on the leaves of potato plants (type (i)) the pathogen may be washed by heavy rain into the soil, reach and invade the tubers and destroy directly their palatability and nutritional value (type (iii)). Apple scab, which normally affects fruit finish (type (iii)), may develop to such an extent in the leaves and twigs that it affects tree growth (type (ii)).

The economic losses on a world scale caused by diseases cannot be determined at all accurately, because of the enormous job of collecting data from crops over huge areas in different countries, because damage varies greatly from year to year, and because changes in the amounts and cultivars of crops planted, in methods of husbandry, and in the use of control measures produce a continually changing picture. Even to determine the degree of crop loss caused by disease in one field or orchard in one season is difficult; this is especially the case when soil-borne pathogens are involved, whose effects are often very hard to assess quantitatively, or mixed infections caused by two or more fungi, which occur commonly. Techniques of crop-loss appraisal are discussed fully in the manual edited by Chiarappa (1971–81), and are reviewed by James (1974).

The most comprehensive attempt to draw together world-wide data on crop losses from pests and diseases remains that of Cramer (1967), who used statistics collected by the Food and Agriculture Organization (FAO), and other organisations, for all major crops. Fortunately this work is now being brought up-to-date. Losses due to diseases, insect pests and weeds were estimated conservatively at about 12%, 14% and 9% respectively of potential world agricultural production (i.e. production if the effects of disease, pests and weeds were removed). The estimates took account of the benefits of control measures in use at the time, and if these were excluded the figures would have been larger. Deterioration during storage of produce was not considered; this would have increased substantially the figures for both diseases and pests. The relative contributions of fungi, bacteria and viruses were not estimated by Cramer; examination of his data for different crop–country combinations suggests that fungi accounted for most (70–80%) of the assessed damage from disease, although their relative importance varied greatly from crop to crop. The estimates were made by integrating the effects of many different pathogens, and no one disease or group of diseases predominated; rice blast (caused by *Pyricularia oryzae*) was, however, singled out for its special world-wide importance, especially in Asia.

Plant pathogenic fungi and their behaviour

Hundreds of fungal species of all kinds cause plant disease. A few leading examples from the main taxonomic groups are given in Table 1.1. The Ascomycotina (ascomycetes) appear to predominate overall as plant pathogens, particularly as many members of the Deuteromycotina (Fungi Imperfecti) have now been transferred to this sub-division. Oomycetes (a group in the Mastigomycotina) and members of the Basidiomycotina (basidiomycetes) are also well represented, but few members of the Zygomycetes (a group in the Zygomycotina) or the lower fungi appear to cause significant problems.

Most of these fungi are facultative parasites; they can grow on living plant tissue or on non-living substrates such as agar media. Some are obligate parasites which either cannot be grown *in vitro* or which grow only to a limited extent on special media. The facultative parasites are necrotrophic; they kill the plant tissue as they grow through it, by

Table 1.1. *Some important fungal pathogens of plants*

Plasmodiophoromycetes	*Plasmodiophora brassicae*	Club root of brassica crops
Oomycetes	*Plasmopara viticola*	Grape downy mildew
	Phytophthora infestans	Potato late blight
	Pythium spp.	Seedling damping-off
Zygomycetes	*Rhizopus stolonifer*	Fruit storage rots
Ascomycotina	*Botrytis cinerea*	Grey mould of fruits and vegetables
	Colletotrichum coffeanum	Coffee berry disease
	Diplocarpon rosae	Rose black spot
	Erysiphe graminis f.sp. *hordei*	Barley powdery mildew
	Fusarium oxysporum f.sp. *vasinfectum*	Cotton wilt
	Nectria galligena	Apple canker
	Venturia inaequalis	Apple scab
Basidiomycotina	*Armillaria mellea*	Honey fungus on trees and bushes
	Puccinia recondita	Wheat leaf rust
	Tilletia caries	Wheat bunt
Deuteromycotina	*Pyricularia oryzae*	Rice blast

producing catabolic enzymes or toxins, and grow on the degradation products. The obligate parasites are biotrophic: the infected host plant stays alive for a considerable period while the fungus feeds on it, often through special absorbing organs (haustoria); ultimately the host becomes injured or dies. Some fungal species, e.g. the rusts and powdery mildews, grow only on one host species, and these fungi may have races which grow only on certain cultivars. Other fungi, e.g. *Botrytis cinerea*, grow on many different kinds of host plants. The underlying mechanisms of these widely differing selectivities are not understood.

The pathogen arrives on its potential host plant in many forms. It may travel as wind-borne spores, or spread from plant to plant in water-splashes during rain showers. Most spores require the presence of free water on the plant surface in order to germinate and then invade; most plant diseases tend, therefore, to thrive in wet climates or seasons. Some pathogens, notably the powdery mildews, can infect in the absence of free water, although high humidity is needed. Upon germination, generally a germ-tube grows out and produces an appressorium (a terminal swelling) from which a penetration hypha emerges and enters the plant epidermis. There are many variations of this behaviour; some pathogens penetrate through a stoma; others penetrate only the cuticle and grow between the cuticle and the epidermis; soil-borne pathogens and wood-rotting fungi may enter through wounds. The powdery mildew fungi grow over the surface of plants, producing at intervals haustoria which penetrate only the epidermal cells.

After a period of biotrophic or necrotrophic growth in or on the host tissue, most pathogens ultimately produce large numbers of spores at the plant surface. The 'generation time' from arrival of inoculum to sporulation is often 10–14 d but may be considerably longer (for example several months in the cereal bunts and smuts). If conditions are favourable, several generations may occur during one season so that the amount of disease in the crop increases exponentially. In some diseases, the associated loss of yield is roughly proportional to the amount of actively metabolising tissue that is visibly diseased. In others, however, more complex relationships obtain; for example, if a pathogen releases mobile phytotoxins the amount of damage may be disproportionately large relative to the directly infected area. The minimum amount of a particular crop disease for which specific treatment is considered cost-effective is known as the 'threshold value'.

Methods of disease control

Five different approaches to the control of plant diseases can be distinguished, and these are described briefly below. Often, and wisely, two or more of these different methods are used together.

(1) *Cultural practices*. The actions which a grower can take to avoid or decrease plant diseases are many and varied. They include: the rotation of crops; adjustment of sowing date; cutting out diseased parts of trees and bushes; disinfecting pruning knives; killing weeds which may harbour diseases; burning diseased plant debris; and altering the wetness status, pH or nutritional status of the soil. Such practices can greatly decrease risks of infection, although they are seldom sufficient by themselves to give optimum yields. Conversely, some cultural methods, such as intensive use of nitrogen fertilisers, monoculture, or direct-drilling or minimal cultivation, can enhance the development of some diseases; their use may well be justified on overall agronomic or economic grounds, but warrant more intensive use of crop-protection measures and close monitoring for signs of disease.

(2) *Quarantine, notification, eradication and stock-certification schemes*. These are statutory schemes which are organised and regulated on a national scale. They are of great value in preventing the upsurge of 'new' diseases in particular countries or regions, or the resurgence of diseases that have been eliminated or decreased to acceptable levels.

(3) *Plant breeding*. For centuries humans have tended to select only the most vigorous and healthy-looking plants for taking seed or planting material for future crops, and so have fostered some degree of disease resistance. However, in the last 50 years much larger advances have been made in the production of disease-resistant cultivars of many crops (Russell, 1978). Some cultivars have given virtually complete and long-lasting freedom from formerly devastating disease in certain crops; an example is the planting some 30 years ago of the Lacatan cultivar of banana, which is resistant to wilt (*Fusarium oxysporum* f.sp. *cubense*), to replace the susceptible Gros Michel. However, resistance is often partial, needing to be complemented by chemical treatment. Frequently it is short-lived, because the pathogen adapts by producing new strains that overcome the resistance. For some diseases, such as *Botrytis* infections in many crops or diseases of fruit trees or grape vines, it has not yet proved possible to develop commercially acceptable cultivars with any substantial degree of resistance. In the next few decades major advances in incorporating potent and durable disease resistance into

crops are expected. These will result from progress in molecular genetics ('genetic engineering'), in tissue and protoplast culture, in physiological and biochemical investigation of plant-disease resistance and in conventional plant-breeding research.

(4) *Biological agents.* The application of antagonistic organisms to control certain insect and mite pests has become effective and well established in situations of intensive or protected cropping. However, instances of the successful biological control of fungi in crops are few and specialised. Well-known examples, concerning diseases that are difficult to control by chemical treatment, are: the application of *Phlebia gigantea* (=*Peniophora gigantea*) to control the proliferation on pine stumps of the tree pathogen *Heterobasidion annosum* (= *Fomes annosus*) (Rishbeth, 1963), and the application of *Trichoderma viride* to protect plum trees against silver leaf disease caused by *Chondrostereum purpureum* (Corke, 1980).

It is hard to say whether biological agents will acquire a greater importance in the control of plant pathogenic fungi in the future. Until recently the potential of biological agents seemed limited, but progress in genetic engineering may allow in time the 'alteration' of many potential microbial agents so that they work more effectively. Each biological agent will probably work only against one or a few specific diseases, and complementation with chemicals or other treatments will be necessary.

(5) *Chemical control.* At present, the battle against most fungal diseases depends primarily on the application of chemicals. For convenience and in line with common practice these will all be called 'fungicides', although few of the chemicals in commercial use have been shown unequivocally to be fungicides *sensu stricto*. Some are known to exert a fungistatic action, to work only after conversion to an active form, to affect fungal morphogenesis, or to interfere with mechanisms of pathogenesis, and one (fosetyl-Al) is considered by some workers to act primarily by enhancing the defence system of the host plant.

The use of various potions to combat blights and mildews has been recorded down the ages, but the first effective use of fungicides on a large scale began around 1860 with the application of sulphur dust to control powdery mildews in orchards and vineyards. Over the past century the extent of fungicide use has increased enormously, in tonnage, monetary value, area treated and diversity. In 1900, two types of chemical were used commercially, sulphur and copper fungicides; in 1940, six – sulphur, copper, mercurials, thiram, salicylanilide, quintozene; and in 1960 and 1980 about 20 and 80 respectively. In 1982, world fungicide sales at farmer price were of the order of £2000 million (£2000 m), and accounted for about 22% of the world pesticide market (Woodburn, 1983).

An account of the methods of discovery of fungicides is given by Ryley & Rathmell in Chapter 4, and of their biochemical modes of action by Baldwin in Chapter 3. The next sections in this chapter will focus on their practical use in modern agriculture and horticulture.

The range of fungicides in commercial use

Some of the more important fungicides in use in 1983 are listed in Table 1.2, together with the approximate date of their first commercial application, and an indication of the pathogens which they affect.

Table 1.2 illustrates the large number of different groups of fungicides that are in use, and these have widely differing biological properties and biochemical modes of action. Even so, the list is very selective and many useful fungicides have been omitted. It is notable that the older fungicides, such as sulphur, copper and the dithiocarbamates, are still in use; in some situations their application is now actually increasing because of resistance or other problems encountered with some of the newer compounds. They are relatively cheap and generally give good disease control, provided they are well and regularly applied.

The surface fungicides form superficial deposits on the plant which can be redistributed by rain or dew, or by vapour action; they inhibit the germination of fungal spores already lying on the surface or arriving later. They penetrate readily into the spores, to high concentrations, but not into plants; it is this differential uptake that accounts for their selectivity. There can be a little uptake by certain plant species, and signs of phytotoxicity (e.g. 'scorching' or russetting) can occur. The newer fungicides are more potent and less likely to damage the crop, but all these surface fungicides, classical and modern, share some serious limitations. Sprays must be applied repeatedly, generally at 7–14 d intervals, both to replace material lost by weathering and to protect the newly formed leaves, flowers or fruit. Ten or more successive sprays are often applied on vegetable and fruit crops in certain countries. The surface fungicides cannot eradicate established infections which are mostly internal. Also they cannot move down to plant roots to protect against soil-borne pathogens.

For many years plant pathologists realised that these limitations could

Table 1.2. *Range of fungicides in commercial use*

Group and approximate date of introduction	Examples	Main current uses
Surface fungicides		
Elemental sulphur (ancient)		Mainly powdery mildews
Copper compounds (1885)	Cuprous oxide, copper oxychloride, Bordeaux mixture	Many fungi, bacteria
Mercurials (1920)	Phenylmercuric chloride	Cereal seed treatments, pruning wounds
	Mercurous chloride	Club root
Dithiocarbamates (1935)	Maneb, zineb, mancozeb, thiram	Many fungi/crops
Dinitrophenols (1950)	Dinocap, binapacryl	Powdery mildews
Phthalimides (1956)	Captan, captafol, dichlofluanid	Various fungi
Isophthalonitriles (1975)	Chlorothalonil	Many fungi
Dicarboximides (1976)	Iprodione, procymidone, vinclozolin	*Botrytis*, etc.
Systemic fungicides		
Antibiotics (1966)	Kasugamycin, polyoxin	Rice diseases
Benzimidazoles (1969)	Benomyl, carbendazim, thiophanate-methyl, thiabendazole	Many fungi/crops (not oomycetes)
2-Aminopyrimidines (1969)	Ethirimol, bupirimate	Powdery mildews
Carboxamides (1969)	Carboxin, oxycarboxin	Basidiomycotina
Morpholines (1969)	Tridemorph, fenpropimorph	Various fungi (powdery mildews, rusts)
Phosphorothiolates (1972)	IBP, edifenphos	Rice blast
Triazoles (1976)	Triadimefon, triadimenol, propiconazole, prochloraz, diclobutrazol	Various fungi (not oomycetes)
Pyrimidine carbinols (1977)	Fenarimol, nuarimol	Similar to triazoles
Acylalanines (1978)	Metalaxyl, ofurace	Oomycetes
Alkylphosphonates (1978)	Fosetyl-Al	Oomycetes

Table 1.3. *Development of fungicides for apple powdery mildew*

Inorganic fungicides (e.g. sulphur)	→	Organic surface fungicides (e.g. dinocap, binapacryl)	→	Systemic fungicides (e.g. bupirimate, fenarimol, triadimefon)
5000 ppm[a]		250 ppm		50–100 ppm

[a] Typical high-volume spray concentration

Table 1.4. *Characteristics of surface and systemic fungicides*

'Surface fungicides' (also called 'residual' or 'contact' fungicides)	'Systemic fungicides'
Non-penetrant	Penetrant
Redistributed over plant surfaces	Redistributed over and within plants
Protectant only	Protectant and eradicant (curative)
Multisite inhibitors	Specific-site inhibitors
Broad spectrum	Selective
Low-risk for resistance	High-risk for resistance

be overcome by fungicides which would readily penetrate into and move in plants. Research in the 1940s and 1950s produced interesting experimental fungicides with systemic properties, for example griseofulvin, but none achieved agricultural importance. About 15 years ago the situation changed dramatically when five major groups of effective systemic fungicides emerged almost simultaneously, as may be seen in Table 1.2.

These systemic fungicides are both selective and potent, reaching the target organisms without damaging the host plants. Their potency is reflected by the doses recommended. For example, in Table 1.3, we can see an evolutionary line in fungicides for apple powdery mildew.

Almost all of the systemic fungicides move upwards from root to shoot, and from leaf bases towards the tips. They move in the transpiration stream of the xylem. They also have a translaminar action, moving across from one leaf surface to the other. Generally they do not move in the phloem system, from treated leaves into younger leaves or down to the roots. The lack of phloem movement has been a major limitation in fungicides for many years; however, fosetyl-Al and metalaxyl do show this pattern of movement to some extent.

It is customary to divide fungicides into 'protectant' and 'systemic', although I prefer the term 'surface' to 'protectant' since systemic fungicides are themselves good protectants. The two groupings of properties given in Table 1.4 are broadly valid, but there are some exceptions. For example, dinocap, a protectant, acts specifically on powdery mildews and mites, whereas some systemics have a fairly broad spectrum of action. Also some surface fungicides, e.g. mercurials, dodine, do penetrate certain plants sufficiently to exert a locally curative action.

Practical achievements

On a world scale how much potential loss of yield is saved by fungicide treatment? This is very hard to judge. Experience suggests 1:4 as a rough-and-ready average cost:benefit ratio in chemical crop protection; on this crude assumption, current world fungicide sales of about £2000 m might be saving about £8000 m in crop value compared with the estimated loss of £100 000 m. All these figures are 'guess-timates', but they do suggest that whilst the contribution of fungicides is substantial, there is much scope for further increases in productivity by wider and more rational usage, and by the introduction of better treatments.

Year-to-year fluctuations are usual in all plant diseases; in situations where these are large and irregular, then the traditional prophylactic treatments ('insurance policies') may not be worthwhile, and 'fire-brigade' treatments with eradicant fungicides may be needed. Some of the systemic fungicides can be used in this way, and some useful 'fire-alarm' systems based either on continual observations for the attainment of threshold disease levels in the crop or on short-term disease-forecasting systems have been developed. However, there can be problems in applying unplanned treatments at short notice, and in high-risk situations most farmers opt to pay for preventive treatments.

The initial introduction of sulphur, copper and mercury must have made very great differences to crop yields and quality. Properly applied, they can have transforming effects in the face of disease epidemics. The further impact of the more recent treatments is less spectacular and rather harder to evaluate. Amongst the different crop diseases, we can see three main trends in chemical control over the last 30 years.

(i) *Little or no change.* Classical fungicides which still pre-dominate; e.g. grape powdery mildew (sulphur), coffee rust (copper), cereal seed treatment (mercury in some countries).

(ii) *Major or total replacement of old products.* Examples include potato blight (copper replaced by organic surface fungicides – e.g. mancozeb, and by systemics – e.g. metalaxyl).

(iii) *Novel uses for fungicides.* For example: control of shoot diseases of cereals (mainly by systemics), control of loose smut of cereals (carboxin, benomyl).

I would guess that the world position in 1940 was (i) = 100%, and has now reached a position such that (i) = 20%, (ii) = 60%, (iii) = 20%.

In group (ii), the organic surface fungicides have taken the lion's share, giving economic improvements in disease control and yield. Systemics are being increasingly adopted, sometimes giving better results, or requiring fewer sprays, or less critical timing, but still being used mainly in schedules of repeated sprays in order to protect new growth. Examples of the fuller use of systemic properties to confer long-term protection or to reduce costs are disappointingly few at present. One interesting example is the use of granules of certain systemic fungicides (e.g. isoprothiolane, IBP) to combat rice blast disease in Japan. They are thrown into the paddy water and are transported via the roots to the leaves and heads where they control the disease. One treatment replaces two or three tedious spray or dust applications, often with better results. Another interesting example is the application of dimethirimol to the base of glasshouse cucumber plants to protect them against powdery mildew; one soil drench can replace and out-perform several sprays which, under glasshouse conditions, are difficult and unpleasant to apply.

'Novel uses' have occurred mainly in wheat and barley. Spray trials against wheat rust were done in the USA in 1891, with little success (Large, 1940). Nearly 80 years passed before trials in Europe with the newly found systemic fungicides tridemorph and ethirimol proved the value of controlling barley mildew by chemical treatment. Now the majority of wheat and barley crops in Europe receive several fungicide treatments, and substantial yield increases are usually obtained. Sometimes these cannot be ascribed to the control of any obvious disease, so that unexplained 'tonic' effects are frequently reported. The precision and convenience of fungicide spraying in cereals has been aided by the use of 'tramlines' – pathways left in the crop to guide tractor passage for spraying and fertilising operations; this practice has become almost universal in European cereal crops over the last 5 years.

A further novel use of systemic fungicides is in seed treatments to prevent loose smut disease of cereals. The pathogen is deep-seated within the seed, and is not affected by surface treatments (e.g. with the mercurials). The systemic fungicides carboxin and benomyl penetrate the seedlings and eradicate the fungus, and seed treatments containing these fungicides are now used to eliminate infections in cereal seed crops.

Acquired resistance

This subject is dealt with by Dekker in Chapter 5, but a brief mention should be made here. The appearance of insecticide-resistant forms has been a major problem in insect control. Until about 1970, this phenomenon of acquired resistance had caused very little trouble with fungicides, although a few local isolated cases had been reported after many years of use. However, various fungi produced strains highly resistant to the systemic fungicide benomyl surprisingly soon after it came into commercial use. This resistance arose in many different countries and crops, and was coupled with obvious failure of disease control. The resistant forms were cross-resistant to several other closely related fungicides. A major resistance problem also arose with another systemic fungicide, dimethirimol, shortly after its introduction as a treatment to control cucumber mildew. Since then resistance has caused difficulties in the use of a number of fungicides, e.g. acylalanines (see Chapter 11), dicarboximides (see Chapter 10).

It is becoming clear that the degree of persistence and spread of resistant populations depend greatly on the intensity and continuity of fungicide usage, on the population dynamics of the pathogen and on environmental conditions. It is important to avoid unnecessary applications wherever possible and to integrate treatment with non-chemical control measures. Break periods have been introduced to avoid prolonged contact between the fungicide and large populations of the pathogen, and the use either of mixed fungicide treatments or of programmes based on more than one fungicide with a different mode of action is probably advantageous. The more potent and selective modern fungicides do seem to carry a greater risk of encountering resistance (probably because of their more specific modes of action) and older, more durable, fungicides such as mancozeb are finding increasing use as 'companion compounds'. Much research is needed on this subject, and a close watch must be kept on standard of performance and on the response of pathogens over the years.

Prospects

There is little doubt that the discovery of new fungicides will continue for some time. Many companies are actively searching for new products with novel properties or different modes of action, although these are becoming more difficult to find. The newer materials will continue to replace the classical fungicides and some of the earlier organic fungicides. Overall increases in areas treated and in expenditure with fungicides can also be expected. However, the tonnage applied will probably decline because increases in potency and persistence of action and increased attention to risk analysis and forecasting will lead to fewer and smaller doses.

Standards of disease control will increase still further, and correspondingly the amounts and regularity of yields and crop quality will also continue to rise. The biggest opportunity for improved disease control probably lies with the soil-borne pathogens that attack the roots and vascular systems of plants. At present these are very difficult to treat, and fungicides which move down to the roots could be one answer, particularly if they had a broad spectrum of action. Better systemic treatments for bacterial diseases are badly needed. Chemicals giving direct control of plant virus infections are very desirable; they are also very hard to find. Improvement in formulations and greater precision in application methods, with an increasing trend towards low-volume and ultra-low-volume sprays, are also likely to come.

Our gradually increasing background understanding of fungicidal action, of diseased-plant physiology and of epidemiology, our experience of integrated pest-management systems and our awareness of the risks of acquired resistance should all lead to the more intelligent and cost-effective use of our existing fungicides and of those yet to be discovered.

References

Chiarappa, L. (1971–81). *Crop Loss Assessment Methods*. FAO manual on the evaluation and prevention of losses by pests, diseases and weeds (also supplements 1, 2 and 3). Farnham Royal: Commonwealth Agricultural Bureaux.

Corke, A. T. K. (1980). Biological control of tree diseases. *Long Ashton Research Station Report 1979*, pp. 190–8.

Cramer, H. H. (1967). Plant protection and world crop production. *Pflanzenschutz-Nachrichten Bayer*, **20**, 5–160.

James, W. C. (1974). Assessment of plant diseases and losses. *Annual Review of Phytopathology*, **12**, 27–48.

Large, E. C. (1940). *The Advance of the Fungi*. London: Jonathan Cape.

Rishbeth, J. (1963). Stump protection against *Fomes annosus*. III. Inoculation with *Peniophora gigantea*. *Annals of Applied Biology*, **52**, 63–77.

Russell, G. E. (1978). *Plant Breeding for Pest and Disease Resistance*. London: Butterworths.

Woodburn, A. (1983). The pesticide world market, current trends and development of new products. *Gifap Bulletin*, **(9)**, 1–6.

2

Fungal diseases of man and the effectiveness of present methods of control

R. Y. CARTWRIGHT

Public Health Laboratory, St Luke's Hospital, Guildford GU1 3NT, UK

Introduction

Fungal diseases of man were recognised in some of the earliest medical literature. Hippocrates, in his third book of Aphorisms, describes 'new-born infants that suffer from aphthae, vomiting, cough, insomnia, nightmares, inflammation of the umbilicus and discharging ears'. The aphthae are most likely the lesions of oral thrush or candidosis, a condition common in sickly infants.

Langenbeck, in 1839, described the fungus in buccal aphthae in a case of typhus, but incorrectly deduced it to be the causative organism of the underlying disease. Its association with vaginal infection was reported by Wilkinson (1849) ten years later in a paper in the *Lancet*.

In the latter part of the nineteenth century, Renon (1897) described aspergillosis in six patients. Five of these patients had an occupational history either as wig cleaners or pigeon feeders. Darling, in 1906, characterised histoplasmosis and, in 1910, Sabouraud published his classic on dermatophytes, *Les Teignes*. Coccidioidomycosis was described in an Argentinian soldier in 1892 and later the importance of this infection in Southern California, particularly in the San Joaquin valley, was described by Dickson & Gifford (1938).

Cryptococcal infection in a sarcoma-like lesion in a tibia was described during the 1890s, but it was not until 1914 that Verse described the classical form of meningitis. Blastomycosis, a disease known as the 'Chicago disease' on account of its geographical distribution, was described by Gilchrist in 1896. Other fungal diseases were recognised in the earlier part of this century, but for many years it was only possible to describe the various forms of disease related to fungi, to recognise predisposing factors and to study their geographical distribution.

This situation altered with the introduction of the polyene antifungals followed by flucytosine and then the imidazoles. It is now possible for the clinician to alter the course of fungal infections in man by specific antifungal therapy. Fungal disease in man is not, however, synonymous with fungal infection although this paper will be primarily concerned with infections, their pathogenesis and their control.

Fungi and fungal spores can be powerful allergens, especially in the respiratory tract. Extrinsic allergic alveolitis results from exposure to various fungal spores, usually in association with the patient's occupation (Table 2.1); the disease is an Arthus Type III reaction in alveolar capillaries. Continued exposure to the allergen leads to pulmonary fibrosis. *Aspergillus fumigatus* and less frequently other species of *Aspergillus* can cause severe allergic diseases in sensitised patients. The exposure to environmental spores may be promptly followed by bronchospasm, fever and malaise. The sputum will frequently contain large numbers of eosinophils. In patients with an aspergilloma the disease is often due to both the fungal growth and an allergic element. Allergic lesions of erythema nodosum and erythema multiformae are found in association with infections due to *Coccidioides immitis*.

Many fungal species are producers of potent toxins. Mushroom poisoning is well recognised by mycophagists and the ingestion of certain species can have fatal consequences. *Amanita phalloides* produces a protoplasmic poison. *Amanita muscaria* contains muscarine which acts on the parasympathetic nervous system as well as pilzatropine which has an atropine-like effect. There are many other toxins and more than one type may be present in a fungus (Tyler, 1963; Wieland, 1968).

Moulds may grow in food, producing a mycotoxin which is later ingested. The food (often flour, grain or seeds) has usually been stored in unsatisfactory conditions. Mycotoxins can cause a variety of diseases in both man and animals. The literature on mycotoxins is reviewed by Purchase (1971), Chick, Balows & Furcolow (1975) and Ciegler (1975).

The study of all infections and infectious diseases must be concerned not only with the individual patient and the disease process, but also with the spread of the infection from one host to another. Management of individual patients requires an understanding of the pathogenesis of the infection, but control of the disease requires knowledge of the links in the chain of infection. Fungi are widespread throughout the environment, they occur in many different forms and cause disease in plants and animals. Only a comparatively few species, however, are recognised as

Table 2.1. *Extrinsic allergic alveolitis due to fungal spores*

	Clinical entity	Antigen source
Thermophilic actinomycetes		
Micropolyspora faeni	Farmer's lung; ventilation pneumonia;	Mouldy hay or vegetable produce;
Thermoactinomyces vulgaris	?Mushroom worker's lung	Growth in humidified hot-air ventilation systems
T. sacchari	Bagossosis	Mouldy sugar cane bagasse
Fungal spores		
Cryptostroma corticale	Maple bark pneumonitis	Mouldy maple bark
Aspergillus clavatus	Malt worker's lung	Mouldy barley or malt dust
A. fumigatus		
Graphium	Sequoiosis	Mouldy redwood sawdust
Aureobasidium pullulans		
Penicillium casei	Cheese worker's lung	Mould on cheese
P. frequentans	Suberosis	Mouldy cork bark

pathogenic to man, and the number of species which cause a primary disease are even fewer. Epidemiological studies of fungal infections indicate that they can be divided into broad groups:

(i) those which are primary infections and have a geographically defined distribution,

(ii) those which may be a primary infection but are frequently associated with a disturbance in the host defence mechanisms,

(iii) those which invariably require a predisposing factor in the host.

These groupings are not absolute but are useful when considering the control and management of fungal diseases. *Histoplasma capsulatum* may, for example, cause a mild or asymptomatic primary disease but the infection may be very severe or overwhelming if the exposure is by the inhalation of a large number of spores or if a patient's cellular immunity is diminished.

The control of fungal diseases

The theory of the control of communicable diseases, including those due to fungi, is complex and multifactorial. The essentials can be stated simply as the breaking of one or more of the four links in the infection chain. The links represent:

(i) the source of infection,

(ii) the depot or reservoir,

(iii) the route of spread and

(iv) the susceptible host.

A successful parasite must travel along this chain.

Eradication or isolation of an infection source will break the first link. *Coccidioides immitis* grows in the desert soil of the lower sonoran life zone of the southern United States of America, Central America and South America. Apart from a few secondary cases in medical staff treating patients with coccidioidomycosis, all infections have been associated with persons visiting the endemic area. The source can be defined in geographical terms. *Cryptococcus neoformans* is present worldwide and pigeons are a recognised source of infection. The pigeons themselves are not the hazard but the reservoir of infectious spores is in their droppings and nest material. Cryptococcosis is not a disease of pigeons, probably because their body temperature is too high for growth of the fungus. Dermatophytes may have their source in soil or on animals. Ringworm of domestic animals is a common source of human infection. Liaison with veterinary surgeons to ensure treatment of a family pet may prevent human ringworm infections in the family.

function described by Lehrer & Cline (1969) used the ability of the cells to kill ingested *Candida albicans*.

Patients with a circulating PMN cell count of 1×10^9 cells l^{-1} or less are particularly prone to infections. These infections are primarily bacterial, and an increase in fungal infections is not well defined. Bodey (1966) reported that a group of leukaemic patients with aspergillosis, candidosis and mucormycosis had PMN cell counts of $1.5 \times 10^9 l^{-1}$ or less for 92% of the month prior to infection. A control group had similarly reduced counts for only 68% of the time. It may be that modern therapy usually ensures that the PMN cell counts remain very low for a minimal length of time and most fungal infections occur in patients with chronic protective defects.

Deficiencies in neutrophil function are primarily related to a disease although studies *in vitro* have shown an inhibitory action by various antimicrobial agents. Amphotericin B may inhibit neutrophil migration (Björksten, Ray & Quie, 1976), but ketoconazole in mixed cultures of PMN and *Candida albicans* enhances the elimination of the yeast (De Brabrander *et al.*, 1980). Quantitative deficiencies are invariably related to the treatment of leukaemias and other malignancies. Cytotoxic therapy is titrated to provide the maximum antineoplastic activity without over-extending the body's defence mechanisms.

Lymphocyte deficiencies are more closely linked with fungal infection. Defects in both T and B cells occur in a number of malignancies but the majority of fungal infections are seen in association with drug therapy affecting lymphocytes. Anticancer drugs and corticosteroids have profound effects on T and B cells. The effects can be diminished if the therapy is intermittent rather than continuous. In many patients, the defects are due to both disease and the drugs used in its treatment.

The most striking fungal infection associated with T lymphocyte dysfunction is chronic mucocutaneous candidosis in children. These children have severe infection of mucous membranes and the skin of the face, scalp, hands and feet. A surprising feature is the relative lack of systemic infection.

The association of malignant haematological conditions with fungal infections is probably due to the combined effect of disease and therapy on T cell proliferation and function. Fungi such as aspergilli and candida are opportunistic pathogens in these patients and may spread throughout the patient causing a severe systemic infection. Histoplasmosis and coccidioidomycosis, which are normally primary fungal infections causing localised disease, have a tendency to spread and are frequently fatal.

The advances in organ transplantation and especially the development of effective immunosuppressive drugs have highlighted the importance of cellular immunity in the prevention of fungal infection. Infection is usually localised and low grade at first but may not be controlled by the host defences and frequently becomes well established or spreads systemically. Successful treatment may only be achieved by reducing or stopping the immunosuppressive drugs, with the consequent rejection of the transplanted organ. In this group of patients, even apparently localised lesions may cause a severe systemic upset.

Although great interest has been shown in defects of immunological systems and fungal infections, many of the patients in British hospitals with fungal infections have other predisposing causes. Anatomical damage, especially in the lung, is the commonest lesion predisposing to pulmonary aspergillosis. The formation of cavities following tuberculosis or sarcoidosis provides a ready site for an aspergilloma to develop. It is cut off not only from the humoral and cellular defences but also from systemically administered antifungal drugs. Similarly, abnormalities of the urinary tract (especially bladder diverticulae) may become infected by *Candida albicans* and be very difficult to eradicate. For urinary candidosis to become established, there is invariably a history of broad spectrum antibacterial drug usage.

The relationship between broad spectrum antibacterial drugs and candidosis illustrates the role of the normal bacterial flora in maintaining the level of candida in the gut, vagina, mouth and body surfaces to below the infective level. The exact course of events between antibacterial therapy and candida infection is not precisely known but the hypothesis is that suppression of bacteria enables the yeasts to multiply to a level where they can cause infection. Mixed infections by fungi and bacteria are extremely uncommon although they may occur in the interdigital spaces of the feet (Leyden, 1982).

Systemic candidosis in surgical patients is invariably associated with the use of broad spectrum antibacterial drugs. These patients frequently have had complicated surgical procedures, have multiple catheters, drains and intravascular lines and have received many different antibacterial drugs. Management includes stopping the antibacterials and removing as many tubes as possible. Surprisingly, this is often all that is necessary, specific antifungal drugs not being required. This would suggest that antibacterial drugs play a greater part than just enabling the candida load to be sufficient to start the infection. The normal bacterial flora must be recognised as an important part of the host defence system against fungal infection.

Specific fungal infections

Detailed descriptions of fungal infections can be obtained from medical mycology textbooks. In this section some of the more common infections, or those causing therapeutic problems, will be considered. The fungi and antifungal drugs will be considered in a clinical context as it is important to remember that the drugs are used to treat infections in patients.

Infections of the skin and mucous membranes

Fungal infections of the skin and mucous membranes are the most common fungal infections. They are generally of a mild nature, causing discomfort but not debilitating disease or death. Some may spread to cause systemic disease but this is very uncommon. Immunological responses to these infecting fungi are difficult to detect. The majority of antifungal agents available commercially are marketed for the management of superficial infections.

Dermatophytosis. The largest group of these infections is the dermatophytoses in which the fungi are localised in keratinised tissues. The species of fungi causing dermatophytosis were listed by Ajello (1968) and contained one species of *Epidermophyton*, 14 of *Microsporum* and 20 of *Trichophyton*. The route of infection is either by direct contact with other infected persons and animals, by indirect contact (via fallen infected hairs and desquamated epithelium) or by infected soil. Soil is probably the natural habitat of *Microsporum gypseum* and *Microsporum fulvum*. The *Trichophyton* spp., *Microsporum audouinii* and *Epidermophyton floccosum*, are primarily pathogens of man or animals. Infected patients frequently give a history of keeping domestic animals which, on examination, also have a dermatophyte infection. Athlete's foot is primarily acquired as a cross infection and is associated with foot baths, shower stalls and changing rooms (Gentles & Evans, 1973); this route of spread is particularly recognised in pithead baths.

The fungi can infect any part of the keratinised layers of skin, nails and hair causing discrete or widespread lesions. There may be a localised host reaction in the skin but the fungus does not spread into deeper tissues except in rare instances.

Various clinical forms of dermatophyte infections are recognised. The classical ringworm is found in most body sites. The lesion is circular, with varying degrees of inflammation, and expands in a radial manner, the periphery containing the fungal elements. The infections are usually minimal, responding rapidly to treatment. Domestic animals are a common source of infection. The aetiological agent in the cat and dog is usually *Microsporum canis*. *Trichophyton mentagrophytes* is also found in dogs, horses, guinea pigs and mice.

Trichophyton rubrum may cause a more widespread lesion and is associated with groin infections. This fungus is also commonly isolated from patients with athlete's foot. The acute lesions on the feet are frequently vesicular and usually start between the fourth and fifth toes. Spread between the other toes is common and may eventually involve the whole foot. The lesions between the toes are usually worse in hot, sweaty feet and secondary bacterial infection is common. This is one situation where a primary fungal infection may become a mixed infection with bacteria, requiring both antibacterial and antifungal therapy (Leyden, 1982). Athlete's foot is readily spread in communal showers and changing rooms.

The involvement of nails by dermatophytes, usually *Trichophyton rubrum* or *Trichophyton mentagrophytes*, produces onychomycosis. Hypertrophy of the nail bed may result and the infection can be difficult to cure. Tinea cruris is a dermatophytosis of the groin, perineal and perianal area. The lesions frequently itch and can be very uncomfortable.

An uncommon chronic dermatophytosis is favus caused by *Trichophyton schoenleinii*, *Trichophyton violaceum* and *Microsporum gypseum*. The fungus forms a dense mass of mycelium and arthrospores originating in a hair follicle. Removal of this mass leaves a red, moist and oozing base; extensive crusting and scarring follows.

Ringworm of the scalp, tinea capitis, is caused by a variety of fungi. Bald patches are common, the fungi infecting the hairs. This condition was well recognised by Sabouraud (1897) when the principal fungus was *Microsporum audouinii*.

As a cardinal feature of all the dermatophytes is the localisation of the infection to the keratin, and as all keratinised structures are continually growing and being shed, these properties are used in treatment. The hypothesis is that if fungal growth can be arrested, the body will eventually shed the infected keratin. Antifungals are used to arrest fungal growth, and the shedding may be aided by the application of keratolytic agents.

Sabouraud (1897) described the use of tincture of iodine which had a local antiseptic action and caused exfoliation. In the 1920s, benzoic acid compound ointment (Whitfield's ointment) was introduced and still remains as effective as topical imidazoles (Clayton & Connor, 1973). It is however, less acceptable to the patient on account of local irritation.

The introduction of oral griseofulvin (Williams, Marten & Sarkany, 1958) greatly improved the prognosis in dermatophytosis. Griseofulvin is selectively concentrated in the keratin where it exerts a fungistatic effect. The length of treatment will depend on how rapidly the infected keratin is shed; it may extend over many months. Patients with onychomycosis, especially of toenails, may fail to respond. Infections with *Trichophyton rubrum* may also be very slow to respond or relapse as soon as therapy is stopped.

Topical imidazoles are very acceptable to patients, producing good results. Nail infections still remain difficult to treat. Ketoconazole, an oral imidazole, has been a useful addition to griseofulvin for oral therapy and is particularly of use in 'griseofulvin resistant' infections (Robertson, Hanifin & Parker, 1980).

The majority of dermatophyte infections can be controlled with modern antifungals, although some patients with nail infections or severe *Trichophyton rubrum* infections remain difficult to treat. The disease has not been controlled as it is difficult to eradicate the non-human reservoir of infection and reinfection can easily occur. In mining communities the prevention of cross infection in pit-head baths would do more to control the disease than the treatment of the individual patient.

Candidosis. The mucous membranes provide a favoured environment for candida infections. Numerically, vaginal candidosis is by far the most important candida infection with which patients present to their general practitioner. Infections of the mouth, oropharynx and oesophagus are mainly seen in hospital patients.

Vaginal candidosis has two distinct clinical forms: the acute and the chronic or recurring form. Both forms of the disease are commonest in women of child-bearing age. The pathognemonic symptom is itching, which can be of such intensity as to cause severe discomfort. A vaginal discharge may also be present. The infection spreads in some patients to the perianal region. Acute infections are usually associated with sexual intercourse, but may also follow treatment with antibacterial drugs (especially those with a broad spectrum such as amoxycillin and cotrimoxazole). Vaginal candidosis is not uncommonly associated with pregnancy, especially during the last trimester.

The majority of patients respond rapidly to a short course of a locally applied antifungal drug. Nystatin (Barr, 1957) was the first specific drug used, and has become the standard against which effectiveness of other drugs has been measured. Clotrimazole was the first imidazole to be marketed, followed by miconazole, econazole and isoconazole. The imidazoles generally are an improvement over nystatin and other polyenes in that the resolution rate is more rapid and the relapse rate less (Weuta, 1972; Tan *et al.*, 1974). An important advance in treating patients with vaginal candidosis was the observation by Masterton *et al.* (1977) that patients frequently discontinued treatment within the first week. Although the accepted length of treatment had been 10–14 d, the use of clotrimazole pessaries for 3 d gave satisfactory results. Trials have now been reported indicating that the insertion of one or two pessaries is all that is necessary in most patients. Studies which I have undertaken show that, in most of the volunteers studied, antifungal activity can be detected in the vagina for 3–5 d following the insertion of a single imidazole pessary.

The management of patients with acute vaginal candidosis is rarely a problem as they can be effectively treated. The disease cannot, however, be considered to be under control as the incidence continues to increase.

The situation with patients suffering from chronic, recurring or 'treatment-resistant' vaginal candidosis is, however, different. These women have regular attacks of infection with a history which may go back for a number of years. They have a lot of discomfort and pain from the inflamed vagina which may prevent intercourse, with consequent marital problems. There is only very rarely any factor which can be demonstrated to account for this form of continuing infection although the initial infection frequently follows antibacterial treatment. The pathogenesis is as yet unknown. The infection usually relapses at a specific time in the menstrual cycle – frequently just after or just before menstruation; the relapse may not occur every month. Normal courses of antifungal treatment may give temporary relief and there is no evidence of candida resistant to antifungal drugs.

Various treatment regimes have been suggested – including topical treatment of the sexual partner and oral nystatin or miconazole – to reduce the gut carriage of candida in the patient. In my own practice, I have had few failures (as judged by freedom from infection for 6–9 months) by asking the patient to insert a clotrimazole pessary high into the vagina nightly throughout a complete menstrual cycle. Occasionally a second course is needed. I would expect similar results with other imidazoles.

Ketoconazole has been used successfully in both acute and chronic vaginal candidosis (Scudamore & Blatchford, 1981). The use of oral medication is preferred by many women although there is a small risk of hepatotoxicity.

The therapy of vaginal candidosis is effective, although failure to recognise and understand the different clinical forms has led to problems in the management of individual patients. The acute disease requires only a short exposure to an antifungal. The need for prolonged treatment in chronic cases is not understood but is effective if continued throughout a menstrual cycle. Further knowledge of the pathogenesis may lead to more rational therapy.

Candida infections of the mouth and oropharynx differ in one important clinical aspect from vaginal infections – they do not itch. Oral infections are not uncommon in debilitated patients, especially during or following the use of antibacterial drugs; the mouth may be covered with white plaques and be extremely sore. The disease is also found in the newborn associated with maternal vaginal candidosis. Response to local antifungals, nystatin, amphotericin B or miconazole is rapid, providing the antifungal is in contact with the candida for a sufficient time to be effective. Failures of treatment are often due to the drug being administered only 3 or 4 times daily. To be effective it needs to be placed in the mouth every 2 or 3 h. The establishment of a normal flora is important in the prevention of further infection.

If the infection does not clear or recurs the underlying factor may be ill-fitting dentures. Budtz-Jorgensen (1974) estimated that several thousand out of a million denture wearers in Denmark were affected with denture stomatitis. Effective treatment requires attention to the fit of the denture and regular thorough cleaning.

Candidosis of the oropharynx and oesophagus is commonly related to defects in immunity (either local or generalised). The use of corticosteroid aerosols in the management of asthma has been demonstrated by Milne & Crompton (1974) to be associated with oropharyngeal candidosis. Oesophageal infections are often recognised in patients with diseases of the reticulo-endothelial system who are receiving chemotherapy.

The infections may not require any specific treatment, improving as immunosuppressive therapy is reduced. In severe cases, local antifungals or even intravenous amphotericin B or miconazole have been successfully used. Flucytosine, administered either orally or intra-venously, is valuable when the yeast is sensitive. The route of adminis-tration must be chosen to ensure that the active antifungal drug reaches the infecting fungus. Non-absorbable drugs given by mouth spend very little time in the oesophagus.

Candida infections of skin occur primarily in moist areas and are usually secondary to candida infection elsewhere or to the use of antibacterial drugs. Infections in the nappy area of babies results from maternal vaginal infection or occasionally cross infection in a nursery; the babies frequently have concomittant oral thrush. Topical antifungals may be necessary but many babies respond to keeping the area as dry as possible. In older patients skin candidosis is usually found in severely ill patients and especially those who are obese or unconscious for long periods. Topical antifungals may be necessary but, again, good cleans-ing and keeping the skin as dry as possible is of equal or greater importance.

Candida onychomycosis and paronychia can be difficult to treat, as antifungal drugs do not penetrate nails. Topical antifungals can be effective but, as in oral candidosis, need to be applied frequently. The polyenes and topical imidazoles have been used successfully. Success has also been reported with oral ketoconazole (Botter & Nuijten, 1980).

The uncommon condition of chronic mucocutaneous candidosis is an absolute indication for the oral imidazole, ketoconazole. The principal underlying defect relates to deficient T cell function. The candida infects the skin and mucous membranes, often producing encrusted tumour-like lesions. The only therapy of lasting value has been ketoconazole. Oral lesions respond first, then skin lesions, followed by nail infections. Recurrences frequently follow discontinuation of therapy which may need to be continuous (Horsburgh & Kirkpatrick, 1983; Rosenblatt & Stiehm, 1983).

The eyes can be infected by a wide range of fungi, the route of infection being either direct, following injury, or via the blood stream. Direct infections have been caused by over 100 different fungal species, many normally regarded as saprophytes. *Aspergillus fumigatus* and *Candida* spp. account for the majority of cases. The infection initially involves the conjunctiva and cornea, but untreated or incorrectly treated may spread throughout the eye. Infections due to *Fusarium solani* and *Acremonium Cephalosporium* spp. may progress within a few weeks, whereas those due to *Phialophora* spp. take months. Treatment is with topical antifungals, using either an imidazole or natamycin.

Penetration into the eye is poor and deep infections may require enucleation of the eye (Jones, 1975). Systemic candidosis may spread through the blood and cause retinal lesions. Patients with candida oculomycosis should be treated with flucytosine orally as the drug is well absorbed and will penetrate into the eye. Its use is, however, limited if the infecting candida is resistant.

Otitis externa. Otitis externa is associated with *Aspergillus niger* and *Candida* spp. and responds well to topical antifungals.

Mycetoma. Mycetoma are chronic infections originating in the skin or subcutaneous tissues, usually of a limb, and may progress to involve and destroy the bone. Two groups of microorganisms produce a similar clinical picture: true fungi and actinomycetes. The true fungi include *Madurella mycetomates, M. grisea, Phialophora jeanselmei, Petriellidium boydii* and *Acremonium* spp. These fungi may show sensitivity to antifungals *in vitro*, but the essential of treatment remains amputation. This is one fungal infection in which chemotherapy has not controlled the disease. Mycetoma due to actinomycetes respond to appropriate antibacterial drugs.

Respiratory tract fungal infections

Fungal spores are present in the air and are regularly inhaled by everyone, yet fungal infection of the lungs is uncommon. A few fungal species can cause primary lung infection, whereas others require a damaged pulmonary tree or an immunocompromised host.

The fungal pathogens causing primary pulmonary infections are *Histoplasma capsulatum, Coccidioides immitis, Paracoccidioides brasiliensis, Blastomyces dermatitidis, Sporothrix schenkii* and *Cryptococcus neoformans*. The infections may produce a broad spectrum of disease ranging from an asymptomatic state to destructive lung lesions with systemic spread. There may be a varying degree of hypersensitivity reaction with corresponding symptoms. The final disease caused depends on the state of the host responses and the infecting dose of fungal spores.

Many primary pulmonary mycoses are unrecognised as such, the patient having an influenza-like illness with a cough and fever. The appearance of erythema nodosum or erythema multiforme and joint pains may suggest the fungal aetiology. The history will usually reveal exposure to a recognised source of infection. This acute form is self limiting and requires no specific treatment; indeed, the diagnosis is frequently made retrospectively as a result of antibody studies. If the fungus becomes established in the lung there will be both local damage and a risk of systemic spread. Amphotericin B is the drug of choice in these patients but it penetrates poorly into bronchial secretions. It is also toxic to the kidneys, especially in high concentrations.

Intravenous miconazole has been used in patients with disseminated coccidioidomycosis (Stevens, 1977) and pulmonary sporotrichosis (Rohwedder & Archer, 1976). Improvement has been reported but the relapse rates after stopping the drug are high. Miconazole has also been used with some success in amphotericin B failures, for primary pulmonary mycoses.

Ketoconazole has given encouraging results in patients with coccidioidomycosis, paracoccidioidomycosis, histoplasmosis and blastomycosis. Clinical improvement occurred in all groups of patients. In many patients, however, the mycological results were disappointing as they remained culture positive, explaining the relapse rate when the drug was discontinued. The low toxicity and lack of resistant fungal strains has enabled investigators to give ketoconazole for periods in excess of two years (Restrepo et al., 1983; Slama, 1983; Stevens et al., 1983). Ketoconazole is a useful addition to the few drugs available for treatment of the pulmonary mycoses and may enable patients to live until an effective fungicidal drug is available. Patients who have chronic lung disease or destructive lung lesions may require surgical resection of the affected part of the lung.

The outcome of a primary pulmonary mycosis is determined by the host response to the infection. If this response is impaired or overwhelmed by a large infecting dose of fungal spores, the disease may become progressive or chronic. Progressive disease may remain within the lungs but is more likely to become disseminated throughout the body. The lung lesions may be dense infiltrates, often in a segmental or lobar pattern, or they may produce widespread miliary infiltration. Chronic infections which are localised in the lung may cause local breakdown of tissue and lead to cavity formation.

The management of patients with disseminated infections is by restoring or improving the immunological response, if possible, and by the use of specific antifungal drugs. The drug of choice in progressive or severe infections is amphotericin B. Amphotericin B is administered intravenously on a daily or alternate-day basis. There are numerous dosage schedules but the limiting factors are a combination of side effects while the drug is being administered and renal toxicity. It is necessary to continue with treatment for weeks or months, and even then there is a risk of relapse. If the infection is due to a strain of *Cryptococcus neoformans* sensitive to flucytosine, this drug is used in addition to the amphotericin B. Flucytosine is not used alone as resistance may develop during treatment.

Aspergillus spores and especially those of *Aspergillus fumigatus* may produce an allergic pulmonary disease. Most of the patients have had atopic asthma since childhood. The spores are a contributory factor in the asthma, and extrinsic allergic alveolitis or allergic bronchopulmonary aspergillosis may develop. Widespread pulmonary infiltrates which are transient accompany febrile episodes with bronchospasm and purulent sputum. The disease may progress to irreversible lung damage and bronchietasis. Therapy is with corticosteroids, antifungal drugs having no place in this allergic disease.

In lungs already damaged by tuberculosis, sarcoidosis, lung abscesses, pneumoconiosis and other diseases causing irreversible lung damage, aspergillus colonisation may occur. The presence of the fungus is usually not recognised until it has formed an aspergilloma. The aspergilloma may remain in a cavity without causing symptoms or may erode a blood vessel producing a haemoptysis. The pleura may also be breached, with a resultant pneumothorax.

The use of antifungal drugs is hampered by the difficulty in getting the drug to the infecting fungus. Aspergillomata are usually in a cavity with little or no connecting blood supply. Intravenous amphotericin B has been used in patients with evidence of tissue invasion. Amphotericin B has also been instilled directly into a pulmonary cavity containing an aspergilloma. The most successful therapy is the surgical removal of the damaged lung. This not only removes the infection but also the potential site of a further infection.

Patients who are difficult to treat are those with widespread lung damage, which precludes surgery, and widespread colonisation by *Aspergillus fumigatus*. I have given long-term amphotericin B intravenously to a few such patients with temporary relief. Ketoconazole, although showing activity *in vitro* against infecting *A. fumigatus*, failed in three patients with widespread pulmonary aspergillosis to produce any clinical improvement. Antifungal activity could be detected in the sputum using a *Candida pseudotropicalis* bioassay.

Patients who are immunosuppressed either by disease or immunotherapy are susceptible to invasive aspergillosis and mucormycosis. The infection not only spreads in the lung but, as a primary pulmonary mycosis, may be disseminated throughout the body. Amphotericin B is the only antifungal agent which is effective in these infections. It must be given intravenously in the maximum tolerated dose. Total doses of 1.6–3.1 g of amphotericin B have been suggested (Sinclair, Rossef & Coltman, 1978). Unless the underlying immunological defect can be improved, the amphotericin B is rarely effective.

The upper respiratory tract is not a common site for fungal infection but, when it occurs, the infection may spread into the cranium. Primary infections with *Aspergillus flavus* occur in subtropical areas. The site of infection is commonly the sinuses. I have been involved in the unsuccessful treatment of three such patients in whom surgical excision had failed to remove all the fungus. In spite of parenteral treatment with amphotericin B and econazole, the infection extended into the cranium causing cerebral aspergillosis. Similar infection due to *Aspergillus fumigatus* is commoner in immunocompromised patients.

Rhinocerebral mucormycosis is the commonest major clinical mucor infection. The infection involves the nose, orbit and brain. It is especially associated with patients who have uncontrolled diabetes mellitus. This infection is also recognised in patients with malignant haematological disease. An outbreak of rhino-orbital phycomycosis, involving three leukaemic patients infected by *Rhizomucor pusillus*, occurred in a haematological unit. The fungal spores were isolated from the air conditioning. A combination of surgery and amphotericin B has improved the prognosis in these patients but the best results are obtained when the underlying condition can be controlled.

Systemic mycoses

In compromised hosts, fungi can spread and infect any organ of the body. The clinical manifestation will largely depend on the organ which is involved. The ability of the primary pulmonary mycosis to spread in compromised patients has already been mentioned. These infections, however, are uncommon in Europe. The fungi which cause the majority of systemic and deep-seated infections are the *Candida* spp., *Aspergillus fumigatus* and *Cryptococcus neoformans*.

Although the severest forms of disseminated candidosis occur in immunosuppressed patients, the greatest number of infections occur in patients who receive broad spectrum antibacterial drugs and have indwelling peripheral and central venous catheters. The longer the course of antibacterial drugs and the longer the time the catheters are *in situ*, the greater the incidence of candidosis. Antibacterials selectively increase colonisation of the gastrointestinal tract by candida. Normal persons can have a transient candidaemia from the persorption of yeasts through the mucosa but infections do not develop. Indwelling catheters are foreign bodies which may be colonised by the candida. Central venous lines can also traumatise the endocardium and predispose to candida endocarditis. As well as an increase in gut candida, yeasts can be readily isolated from the skin and may cause local infections especially in the groins, in the axillae, and under the breasts in women.

Major surgery of the gastrointestinal tract or the heart has been associated with systemic candidosis, but the predisposing factor is probably not the surgical procedure but the associated use of antibiotics, often for prophylactic reasons.

A distinct group of patients who are at risk from candida endocarditis are drug addicts who inject intravenously. The source of infection is contaminated needles and syringes, and the use of saliva to ease the injection.

The most important factor in making the diagnosis and hence commencing adequate therapy of disseminated candidosis is the recognition that the patient may have a fungal infection. It is not uncommon for the possibility of a fungal infection to be considered only when the infection is well established. The clinical signs of fever, general malaise and a raised peripheral blood leukocyte count are usually regarded as indicative of a bacterial infection. The use of antibacterial drugs, usually with a broad spectrum, does not alter the course of disease. The diagnosis is confirmed by isolating candida from blood cultures, biopsies or aspirations. It can also be made by demonstrating a rise in candida antibodies. Antibody estimation can be particularly useful in candida endocarditis, both in diagnosis and in monitoring the effectiveness of therapy.

The majority of patients with systemic candidosis respond well to the cessation of antibacterial drugs and the removal of intravenous lines, provided there is no underlying immunological defect. The use of specific antifungal drugs is indicated if there is an immunological defect, if the patient is seriously ill or fails to respond to the simpler measures.

There are three effective drugs in systemic candidosis but there are problems associated with all of them. All candida strains are sensitive to amphotericin B and it is a fungicidal drug. It does, however, have undesirable side effects and toxicity. The parenteral preparation is a complex of amphotericin B and sodium desoxycholate which is usually administered intravenously; it has, however, been given into joints, the pleural cavity, the peritoneum and intrathecally. Local inflammation is common and generalised reactions of fever, nausea and vomiting, will curtail therapy in some patients. Local inflammation in the veins causes a phlebitis which can be reduced by the incorporation of hydrocortisone in the infusion fluid. Arachnoiditis and nerve damage may follow intrathecal or intraventricular administration.

Amphotericin B has a molecular weight of 924 and is highly protein bound. Penetration into cerebrospinal fluid, parotid gland fluid, aqueous humour and haemodialysis solutions is poor. Its value is therefore limited to infections where the fungus is exposed to the blood stream. In candida endocarditis, amphotericin B does not penetrate vegetations even when they are immersed in a high concentration of drug (Rubinstein *et al.*, 1974).

The renal toxicity of amphotericin B prevents high serum levels being attained. It is my practice to increase amphotericin B dosage until renal insufficiency is detected by a rising plasma urea or creatinine. The dosage is then reduced until a steady renal function is maintained.

Flucytosine, with a molecular weight of 129 and protein binding of less than 5%, is well distributed throughout all body cavities. The majority of candida strains are sensitive to flucytosine, but up to 15% of fresh isolates in the United Kingdom are resistant; resistance may develop while the patient is on treatment. Flucytosine can be used in combination with amphotericin B having a synergistic activity against some yeasts. Excretion is through the kidneys, renal insufficiency causing a build-up of serum levels. High serum levels are associated with bone-marrow toxicity.

Miconazole is the only imidazole available for parenteral administration. It is not soluble in aqueous solutions, Cremophor EL being used as a carrier. It is not well distributed throughout body cavities. Resistance of yeasts is not a problem but its clinical efficacy is mediocre. It has, however, been used successfully in patients with systemic candidosis although it is not to be recommended in life-threatening infections.

The use of ketoconazole, highly effective in chronic mucocutaneous candidosis, has not been extensively reported for systemic mycoses. It should not be used in severe infections as gut absorption can be erratic. The use of oral ketoconazole to prevent candida infections in compromised hosts has been advocated but is not fully substantiated.

Aspergillus fumigatus behaves differently to *Candida albicans*. Systemic infections usually being associated with neutropaenia or immunosuppression. The infection may involve any organ, causing a wide spectrum of disease. Clinical features are due to the tumour effects of local aspergillomata causing pain and pressure effects. Ulceration into blood vessels may exacerbate local symptoms or cause bleeding into body cavities or passages. General symptoms may be those of an abscess which fails to respond to antibacterial drugs. In many patients the disease may produce non-specific symptoms and the diagnosis is not made until the fungus is firmly established.

Systemic aspergillosis is a serious infection with a poor prognosis. Treatment is by surgical excision if possible and intravenous amphotericin B. Any immunosuppressive therapy must also be reduced or stopped. In post-transplant patients, this may mean a choice between trying to control the infection with amphotericin B or risking rejection of the transplant. Amphotericin B has the disadvantages described earlier. Patients receiving amphotericin B should be followed carefully after the drug is stopped as relapses of infection occur. At present there is no effective antifungal drug for the management of systemic aspergillosis.

The central nervous system is the most common site of infection for disseminated cryptococcosis although cutaneous, ocular, articular and genitourinary infections have been described. Cryptococcal meningitis, the most frequent clinical infection, is usually associated with an underlying disease, including Hodgkins disease, leukaemia, diabetes mellitus, cancer, sarcoidosis and silicosis. The infection is most prominent over the base of the brain and involves the meninges. A thick mucoid material may accumulate in the subarachnoid space. Invasion of the brain substance may occur. The presentation is usually that of a slowly developing meningitis with headache and meningeal irritation. Diagnosis is made by demonstrating either cryptococci or cryptococcal antigen in the cerebrospinal fluid.

Regardless of whether the underlying cause can be improved, patients with cryptococcal meningitis require antifungal chemotherapy. The only drugs which are of proven value are amphotericin B and flucytosine. They should be used in combination unless the yeast is flucytosine-resistant when amphotericin B only is given. Even with combined therapy the response in patients treated in the United Kingdom is not satisfactory. The reasons for the poor response are partly related to the underlying diseases but also to the poor penetration of amphotericin B to the site of infection (Hay, 1982).

Other systemic mycoses occur and specific deep-seated infections due to a wide range of fungi have been reported. The majority of these infections have in common predisposing factors and a limited number of antifungal drugs for their management.

Conclusion

Fungi of many species can cause disease in man. The disease may be due not only to infection, but also to fungal toxins or to an allergic host response to the presence of fungal antigen. Infections due to fungi cover a wide clinical spectrum, their diversity depending largely on the effectiveness of the host mechanisms against infection.

The advent of specific antifungal drugs has been a step forward in the management of patients with fungal infections. There remains, however, considerable room for improvement. What are the requirements for new antifungals? Their ability to stop or restrict fungal growth while not harming the host is apparent, but further properties must take into account the whole of the disease processes for which the drug is being developed. As dermatophyte infections are restricted to the keratinised tissue, keratinophilic compounds have an advantage. Eradication of the dermatophyte using present therapy depends on the shedding of infected skin, hairs or nails. An agent which was fungicidal might make shorter courses of treatment possible.

Deep-seated mycoses that cause clinical problems are difficult to treat with available antifungals. Amphotericin B, the most useful drug in these infections, is not well distributed on account of its large molecular weight and high protein binding. It has a degree of toxicity which would make it unlikely to be passed for general use by present-day licensing authorities. Miconazole has high protein binding and is not well distributed. Flucytosine passes readily into all body compartments, but has a narrow spectrum of activity, and resistant strains may readily emerge.

A clinically valuable drug will have a broad spectrum of activity, resistance will not easily develop and it will pass readily into all body compartments in an active form. It should also be fungicidal, as the host defences in infected patients are invariably suppressed.

Fungal diseases are steadily becoming more important. Acute infections either do not require treatment or respond well to available drugs. Chronic and progressive infections are not well controlled at present. As the population of compromised patients increases, so will problems of fungal infection. Present antifungal drugs are inadequate in the management of these patients and, although management will improve as pathogenesis of the infections is better understood, there will remain a need for effective, non-toxic antifungal drugs.

References

Ajello, L. (1968). A taxonomic review of dermatophytes and related species. *Sabouraudia*, **6**, 147–59.

Barr, W. (1957). Nystatin. *Practitioner*, **178**, 616–21.

Björksten, R., Ray, C. & Quie, P. G. (1976). Inhibition of human neutrophil chemotaxis and chemiluminescence by amphotericin B. *Infection and Immunity*, **14**, 315–17.

Bodey, G. P. (1966). Fungal infections complicating acute leukemia. *Journal of Chronic Diseases*, **19**, 667–87.

Botter, A. A. & Nuijten, S. T. (1980). Further experiences with ketoconazole in the treatment of onychomycosis. *Mykosen*, **24**, 156–66.

Budtz-Jorgensen, E. (1974). The significance of *Candida albicans* in denture stomatis. *Scandinavian Journal of Dental Research*, **82**, 151–90.

Chick, E. W., Balows, A. & Furcolow, M. L. (1975). *Opportunistic Fungal Infections*. Springfield. Charles C. Thomas.

Ciegler, A. (1975). Mycotoxins: occurrence, chemistry, biological activity. *Lloydia*, **38**, 21–35.

Clayton, Y. M. & Connor, B. L. (1973). Comparison of clotrimazole cream, Whitfield's ointment and nystatin ointment for the topical treatment of ringworm infections, pityriasis versicolor, erythrasma and candidiasis. *British Journal of Dermatology*, **89**, 297–303.

Darling, S. T. A. (1906). A protozoon general infection producing pseudotubercles in the lungs and focal necroses in the liver, spleen and lymph nodes. *Journal of the American Medical Association*, **46**, 1283–5.

De Brabander, M., Aerts, F., Van Cutsem, J., Van den Bossche, H. & Borgers, M. (1980). The activity of ketoconazole in mixed cultures of leukocytes and *Candida albicans*. *Sabouraudia*, **18**, 197–210.

Dickson, E. C. & Gifford, M. A. (1938). Coccidioides infection. *Archives of Internal Medicine*, **62**, 853–71.

Gentles, J. C. & Evans, E. G. V. (1973). Foot infections in swimming baths. *British Medical Journal*, **3**, 260–2.

Gilchrist, T. C. (1896). A case of blastomycetic dermatitis in man. *Johns Hopkins Hospital Review*, **1**, 269–98.

Hay, R. J. (1982). Clinical manifestations and management of cryptococcosis in the compromised host. *Fungal Infection in the Compromised Patient 1982*, ed. D. W. Warnock & M. D. Richardson, pp. 93–117. Chichester: John Wiley & Sons.

Horsburgh, C. R. & Kirkpatrick, C. H. (1983). Long-term therapy of chronic mucocutaneous candidiasis with ketoconazole: experience with twenty-one patients. *American Journal of Medicine*, **74**, 23–9.

Jones, B. R. (1975). Principles in the management of oculomycosis. *American Journal of Ophthalmology*, **79**, 719–51.

Langenbeck, B. (1839). Auffingung von Pilzen aus der Schleimhaut der Speiseröhre einer Typhus-Leiche. *Neue Notizen auf dem Gebiet der Natur und Heilkunde (Froriep)*, **12**, 145–7.

Lehrer, R. I. & Cline, M. J. (1969). Interaction of *Candida albicans* with human leukocytes and serum. *Journal of Bacteriology*, **98**, 996–1004.

Leyden, J. J. (1982). The interaction of dermatophytic fungi and bacteria in the pathogenesis of interdigital athlete's foot. *Bacterial Interference*, ed. R. Aly & H. R. Shinefield, pp. 99–109. Boca Raton, Florida: CRC Press.

Masterton, G., Napier, I. R., Henderson, J. N. & Roberts, J. E. (1977). Three-day clotrimazole therapy in candidal vulvovaginitis. *British Journal of Venereal Diseases*, **53**, 126–8.

Milne, L. J. R. & Crompton, G. K. (1974). Beclomethasone dipropionate and oropharyngeal candidiasis. *British Medical Journal*, **3**, 797–8.

Purchase, I. F. H. (1971). *Symposium on Mycotoxins in Human Health*. London: MacMillan.

Renon, L. (1897). *Étude sur l'Aspergillose chez les Animaux et chez l'Homme*. Paris: Masson & Cie.

Restrepo, A., Gomez, I., Cano, L. E., Arango, M. D., Gutierrez, F., Sanin, A. & Robledo, M. A. (1983). Post therapy status of paracoccidioidomycosis treated with ketoconazole. *American Journal of Medicine*, **74**, 53–7.

Robertson, M. H., Hanifin, J. M. & Parker, F. (1980). Oral therapy with ketoconazole for dermatophyte infections unresponsive to griseofulvin. *Reviews of Infectious Diseases*, **2**, 578–81.

Rohwedder, J. J. & Archer, G. (1976). Pulmonary sporotrichosis: treatment with miconazole. *American Review of Respiratory Disease*, **114**, 403–6.

Rosenblatt, H. M. & Stiehm, E. R. (1983). Therapy of chronic mucocutaneous candidiasis. *American Journal of Medicine*, **74**, 20–2.

Rubinstein, E., Noriega, E. R., Simberkoff, M. S. & Rahal, J. J. (1974). Tissue penetration of amphotericin B in candida endocarditis. *Chest*, **66**, 376–7.

Sabouraud, R. (1897). *A Pictorial Atlas of Skin Diseases*, ed. J. J. Pringle, p. 117. London: Rebman Publishing Co. Ltd.

Sabouraud, R. (1910). *Les Teignes*. Paris: Masson & Cie.

Scudmore, J. & Blatchford, N. (1981). An open assessment of ketoconazole in the treatment of recurrent vaginal candidosis in general practice. *Clinical Research Reviews*, **1**, 203–6.

Sinclair, A. J., Rossef, A. H. & Coltman, C. A. (1978). Recognition and successful management of pulmonary aspergillosis in leukaemia. *Cancer*, **42**, 2019–24.

Slama, T. G. (1983). Treatment of disseminated and progressive cavitary histoplasmosis with ketoconazole. *American Journal of Medicine*, **74**, 70–3.

Stevens, D. A. (1977). Miconazole in the treatment of systemic fungal infections. *American Review of Respiratory Disease*, **116**, 801–6.

Stevens, D. A., Stiller, R. L., Williams, P. L. & Sugar, A. M. (1983). Experience with ketoconazole in three major manifestations of progressive coccidioidomycosis. *American Journal of Medicine*, **74**, 58–63.

Tan, C. G., Milne, L. J. R., Good, C. S. & Loudon, J. D. O. (1974). A comparative trial of six day therapy with clotrimazole and nystatin in pregnant patients with vaginal candidiasis. *Postgraduate Medical Journal*, **50**, 102–5.

Tyler, V. E. (1963). Poisonous mushrooms. *Progress in Chemical Toxicology*, **1**, 339–84.

Verse, M. (1914). Uber einen Fall von generalisierter Blastomykose beim Menschen. *Verhandlungen der Deutschen Gesellschaft für Pathologie*, **17**, 275–8.

Weuta, H. (1972). Clotrimazol-Vaginaltabletten – klinische Prüfung im offenen Versuch. *Arzneimittel-Forschung*, **22**, 1291–4.

Wieland, T. (1968). Poisonous principles of mushrooms of the genus *Amanita*. *Science*, **159**, 946–52.

Wilkinson, J. S. (1849). Some remarks upon the development of epiphytes with the description of new vegetable formation found in connection with the human uterus. *Lancet*, **2**, 448–51.

Williams, D. I., Marten, R. H. & Sarkany, I. (1958). Oral treatment of ringworm with griseofulvin. *Lancet*, **ii**, 1212–13.

3

Potential targets for the selective inhibition of fungal growth

B. C. BALDWIN

ICI Plant Protection Division, Jealott's Hill Research Station, Bracknell, Berkshire RG12 6EY, UK

Introduction

One of the great advances of mankind in the twentieth century has been the discovery of selective chemicals for use against the pests and diseases of our bodies and food. This selectivity has not been designed on biochemical or physiological principles by and large, but has been discovered by chance. An understanding of the means whereby selectivity has been achieved has often taken much patient and painstaking research. As we have become more knowledgeable and sophisticated, there have been examples of modifications to toxic moieties to improve their properties to our advantage, but no one yet can truly claim to have designed a pesticide or chemotherapeutic agent (the classical text is Albert, 1979).

Fungal diseases of animals have traditionally been studied by a different group of scientists from those concerned with fungal pathogens of plants. Add to this the biochemistry and physiology of fungi studied in culture in the laboratory without any reference to the ecological niche of the organism, and it becomes clear that background knowledge must be drawn from a wide range of sources if we are to pin-point the places where selective inhibition of fungal growth may be achieved. Some of the information obtained *in vitro* may not be relevant to choosing targets for a biochemical or physiological approach to therapy *in vivo*, and will have to be disregarded if it is not to obscure our perception. The rules governing this selection are, of course, not defined!

Despite this uncertain outlook, discoveries made recently have re-volutionised treatment of both human and plant pathogens, and we now have some very effective selective and systemic chemicals at our disposal (Siegal & Sisler, 1977; Ryley *et al.*, 1981). These have been discovered by the traditional means of routinely examining chemicals in screening tests *in vitro* or *in vivo* and then chemically modifying the 'lead' to give a product with the desired effect in the real-life situation. The hope is that with more background understanding it will be possible to improve the selection and guide the synthesis of chemicals for such tests. When fungicides with high levels of activity are discovered, we hope that we have the ability to improve their properties by chemical modification to match the required profile of activity and safety.

In many cases of selective action, delivery or degradation are just as important as differences at the active site in a biochemical reaction (Corbett, 1979), so to limit the discovery process at the outset to one type of selectivity would mean that potentially useful compounds would be missed. When discovery is such a slow and costly process (Braunholtz, 1977; Ryley *et al.*, 1981) we cannot afford to neglect any opportunities. The emphasis of this chapter is, therefore, that the traditional approach of screening will not be displaced, but that biochemical and physiological information will be used in ensuring that screening tests cover novel ways of affecting fungi, and will be incorporated into the thinking of synthetic chemists.

The search for selectivity
The main body of this chapter will deal with the search for new targets for fungicidal action; the emphasis will be on plant pathogenic fungi – due to the background of the author – but much will be relevant to human mycoses. There are four starting points to be considered.

Starting from plant pathology
The study of the interaction between the host plant and the fungal pathogen has blossomed during the last 20 years, and many books and reviews cover the genetics, physiology and biochemistry of the complex interplay between the two organisms (for example Wood, 1967; Heitefuss & Williams, 1976; Friend & Threlfall, 1976; Vanderplank, 1978). Leading from this background information, several reviewers have discussed the ways in which we can disrupt vital parts of the integrated process to give selective control of plant diseases without needing to kill the fungus directly (van der Kerk, 1963; Horsfall, 1972; Dimond & Rich, 1977; Sisler, 1977; Hancock & Huisman, 1981; Geissbühler, Brenneisen & Fischer, 1982; Hedin, 1982; Fuchs *et al.*, 1983). Most of these writers have divided the topic into:

(i) changing the metabolism of the host;

(ii) interfering with the relationship of the plant with the parasite;
(iii) inactivating the toxins and enzymes produced by the fungus in its attack upon the plant.

The basis for such an approach is the specificity of the interaction between host and pathogen; for, although many fungi may be present in the immediate environment of the plant, the ability to cause disease will be present in only a few. Furthermore, only specific races of pathogen will be able to attack successfully specific varieties of a crop. If the genetical and biochemical bases for the specificity are known, it may then become possible to design a control measure.

One immediate problem, because of the specificity of the interaction, is that economically important diseases have to be chosen for the studies if they are aimed at devising control measures; this has not always been the case. Rathmell (1983) lists vine diseases (mostly downy mildew, *Plasmopora viticola*, and grey mould, *Botrytis cinerea*), rice blast (*Pyricularia oryzae*), cereal diseases (mildew, *Erysiphe graminis*; rusts, *Puccinia* spp.; and others such as *Septoria* spp. and *Pyrenophora* spp.) and the diseases caused by *Cercospora* spp. as those that growers spend most money controlling. Rice, wheat and other grains, coffee and peanuts were the most important single crops in terms of fungicide usage in 1980. However, more money was spent on controlling diseases of fruit and nuts, followed by potatoes and vegetables, than on any of the single crops (Anon., 1981). Pathologists should therefore choose wisely the diseases and crops they are to study.

However, even if an economically important crop–disease combination is studied, and the details of the biochemical events during the attack of the pathogen are known, we still have the problem of discovering a treatment, which may be chemical, to disrupt the process. It may be more productive to conduct a routine screening test *in vitro*, as used to detect directly acting fungicides, but ensuring that indirect action is also examined. Thus the immediate development from fundamental studies should be that screening tests are checked to ensure that they are capable of detecting chemicals which alter the disease process as well as those that are directly toxic. Rathmell (1983) has indicated that many chemicals which have an indirect action were detected in the course of a programme of screening *in vivo*. Thus tests *in vitro* are now not used to any significant extent by the agrochemical industry. Less is known about the interaction between fungus and animal host, and economic considerations often cause pharmaceutical companies to rely on testing *in vitro* (Ryley *et al.*, 1981; Ryley &

Table 3.1 *The process of fungal attack*

Activity	Processes involved
Finding host	Tropic and taxic responses. Chemical stimuli. Landing of aerial spores.
Adhesion	Chemical and physical forces. Leaf-surface chemistry. Moisture.
Germination	Interactions with other spores and pollen. Germination inhibitors. Dormancy. Switching on germination. Chemical recognition.
Penetration	Finding stomata or sites for penetration. Enzymic or mechanical forces? Switching on of enzyme synthesis. Inhibitory substances from plant.
Internal growth	Specific nutrients required. Formation of haustoria. Uptake mechanisms. Immune response of plant. Can fungus suppress it? Maintaining biotrophic state. Can fungus suppress it? Manipulation of host metabolism. Recognition of fungus. Elicitation of synthesis of inhibitory factors and phytoalexins. Suppression by fungus. Can they be enhanced? Oxidase and peroxidase action. Lignification. Must necrotrophs kill plant to survive? Toxin production. Growth of fungus to specific organs. Age, hormone and nutritional status of plant.
Reproduction	Stimulus to form spores. Sexual or asexual. Recognition of opposite sex. Fungal hormones. Can spore formation be prevented? Alternate hosts and reproduction.

Rathmell, this volume). Screens against a biochemical target such as an isolated enzyme or receptor preparation *in vitro* have not been used to any great extent, and are further removed from the physiological state than is a whole-organism test *in vitro*.

Table 3.1 details some of the events which take place during the course of a fungal infection. Primarily based upon the plant–fungus interaction, it reveals the many steps that have to succeed before an infection is established and which, if blocked, will prevent pathogenesis. A similar list could be drawn up for human mycoses, and many of the steps would in fact be the same. It is obviously intellectually stimulating to plan new approaches to disease control from such a list. At the same time, we have to be practical and we will only be successful in our search for new targets if we can answer some practical questions.

(i) Will the proposed target lead to a marketable commodity?

(ii) Will the proposed fungicide offer a technical advantage over existing toxophores?

(iii) Is it possible to make worthwhile progress in the search?

Only when the answers are satisfactory will the target be worth consideration.

Thus, there has been a great deal of research into pre-existing, naturally occurring antifungal chemicals and into chemicals induced post-infection (phytoalexins). Many phytoalexins are found in high concentrations in very localised tissues of the plant, and are phytotoxic. This target has been chosen by researchers with the idea of switching on the mechanism prior to infection, or of stimulating the production of phytoalexins subsequent to infection, so that the balance may be tipped in favour of the host and against the pathogen (see for example Geissbühler *et al.*, 1982). If, however, the metabolic-energy demand for the extra biosynthesis diverts photosynthate and reduces crop yield (Smedegaard-Petersen & Stølen, 1981; Tuleen & Fredericksen, 1982), or if it causes phytotoxicity to the crop, no farmer is likely to buy the treatment. Antisporulant chemicals (Horsfall, 1977) will only be in demand if they prevent epidemics occurring; farmers will otherwise resort to a conventional fungicide.

Chemicals of commercial importance have been discovered which have an indirect action on the infection process, but this understanding has only come after their discovery (Langcake, 1981; Wade, this volume). Chemicals falling into this category are the blockers of melanin biosynthesis, which control rice blast (*Pyricularia oryzae*) by preventing penetration of the epidermal wall of the plant by infection hyphae (Woloshuk & Sisler, 1982); and aluminium tris-*O*-ethyl phosphonate, which increases defence reactions in vines infected with *Plasmopara viticola*, also decreasing the liberation of zoospores (Raynal, Ravise & Bompeix, 1980), but which does not have any direct toxicity *in vitro* to Oomycete species (Farih, Tsao & Menge, 1981*a*). These last workers suggest it may have antisporulant activity.

Searching for agents with an indirect action on host-defence reactions may be self-defeating if they have high specificity, as there are few single diseases sufficiently important to bear the costs of the research and development (Sisler, 1977; Langcake, 1981). Fortunately the chemicals mentioned in these two examples both act upon important target diseases. 'There are indeed many possibilities but a case of practical success would greatly stimulate investigation of this approach to chemical control' (Sisler, 1977). I believe, with our present state of knowledge, we do not have the ability to choose a target to block from any of these areas; we shall only find this type of chemical as a result of general screening tests *in vivo*.

Nevertheless, starting from an observation of interference with the disease process may lead to a new test or into a new area of chemistry. In human and animal mycoses, the transformation of *Candida albicans* from yeast to mycelial form is associated with pathogenicity. Imidazoles, which have a direct action as inhibitors of ergosterol biosynthesis, are also potent inhibitors of this transformation, and may thereby lead to an understanding of how it is controlled (Davies & Marriott, 1981). Ketoconazole, an azole of this type, is also reported to facilitate the ability of host-defence cells to eradicate *C. albicans* (Borgers & Waldron, 1981). Structural cell-wall carbohydrates are, or give rise to, important recognition molecules which may elicit the host response in plant pathogens (Rathmell, 1983). Chemical control agents based on sugar chemistry are not the types of molecule which fungicide chemists usually synthesise; a side effect of responding to background knowledge may be to venture into new areas of chemistry!

Starting from fungal biochemistry

Saccharomyces cerevisiae, Neurospora crassa and *Aspergillus nidulans* have been standard subjects for biochemical research for many years; much of our knowledge of intermediary metabolism has been elucidated using wild and mutant strains of such fungi. The usefulness of this research for picking out specific biochemical differences between fungal and plant, or fungal and animal, metabolism that may then be exploited is questioned. Fungal cultures in nutrient-rich media, maintained at steady temperatures, are not satisfactory models for organisms fighting for survival on exposed leaf surfaces under varied environmental pressures. In particular, their nutrient status is likely to be very different in the two environments.

Many reviews have attempted to list biochemical differences between fungi and their hosts (with the emphasis on plant hosts) which are potential sites for selective inhibition, and a compilation is given in Table 3.2 (Cowling, 1963; Dimond, 1963; van der Kerk, 1963; Horsfall & Lukens, 1969; Somers, 1969; Sbragia, 1975; Horsfall, 1977; Hedin, 1982). There is little evidence that any of these targets have inspired any of the newer fungicides, although the mode of action of at least two major classes (inhibition of the biosynthesis of ergosterol and of chitin) were predicted. The deficiency in our knowledge is that, given a genuine biochemical difference, we do not have sufficient information about the levels, turnover and activity of the enzymes in the pathway to choose the correct target. Some of the differences may not be in critical areas of

Table 3.2 *Biochemical attributes of fungi which may provide a key to selective chemotherapeutic agents*

Biochemical or physiological step	Reasons or observation
Chitin synthesis and cell wall protein	Cell wall synthesis and organisation[a]
Hydrolytic enzyme synthesis	To aid penetration and to release substrates for growth[a]
Rate of cell division	Faster than host, try antimitotic agents[a]
Ergosterol is fungal sterol	Interfere with sterol metabolism[a]
Fungal cell membranes	Many antibiotics affect bacterial cell permeability[a]
Exploit host defence mechanisms	Phytoalexin production, immune response[a]
Vitamin requirements	Plants make their own[a]
Specific nutrient requirements	Especially for obligate pathogens[b]
Hormones	Upset fungal reproduction[a]
Self-inhibitors	Prevent spore germination[c]
Sporulation and conidiophore formation	Inhibitors were found[d]
Nitrogen metabolism	Toxicity of D-amino acids,[e] lysine biosynthesis[f]
Accumulation mechanisms	Specificity of uptake processes[b]
Water relations	Especially in powdery mildews[b]
Glycolate oxidation	Sporulation in *Alternaria*,[g]
Sugar levels	low and high sugar diseases[b]
Polyphosphate	Phosphagen or phosphate store in fungi[h]
Thymidine kinase	Absent in some fungi[i]

[a]Cowling, 1963 [d]Horsfall, 1977 [g]Horsfall & Lukens, 1969
[b]Dimond, 1963 [e]van Andel, 1958 [h]Harold, 1966
[c]Allen, 1976 [f]LéJohn, 1971 [i]Grivell & Jackson, 1968

biochemistry for the fungus. Even if we were certain that we had chosen a critical enzyme, we do not have the ability to design inhibitors without some prior knowledge of an existing inhibitor. If that were possible, the inhibitor we designed has still to have the correct properties to move in the host to the site of fungal attack, to be sufficiently metabolically stable, and then to penetrate into the pathogen and inhibit the target enzyme.

Considering first the nutrition of the pathogen, we find that information about nutrient transfer is scarce. There is evidence that biotrophic interactions lead to increased respiration, and to a change in the balance between normal Embden–Meyerhof–Parnas glycolysis and the hexose monophosphate shunt in the host, but the interpretation is complicated by the presence of both host and pathogen in the experimental material (Daly, 1976; Manners, 1982). Using techniques with radiolabelled substrates and following transfer to the mycelium (for example, studies by Andrews, 1975; Martin & Ellingboe, 1978), or by isolating the absorptive haustorial complexes and studying their permeability properties (Manners & Gay, 1982), it may be possible to determine how molecules from the host are transferred to the pathogen. However, such information will not lead directly to knowledge of which particular nutrients are critical for fungal development. Woods & Gay (1983) present evidence that ATPase activity is present at the host plasmalemma of the haustorial complex of *Albugo candida* and so uptake is dependent on active processes of the host. Plant pathogenic fungi of all classes have specialised absorptive organs – but our knowledge of their function is rudimentary. Hohl (1983), in discussing the nutrition of *Phytophthora*, concludes that 'the relation between nutrition and pathogenesis is still unclear, and the possible role of nutrients in horizontal resistance needs further exploration'. The same conclusion appears to be true for other pathogens. Thiamine and biotin are vitamins commonly required by fungi in culture (Koser, 1968; Ridings, Gallegly & Lilly, 1969) but, as they are synthesised by plants, they may be freely available. The vitamin status of human mycoses may therefore be worth investigation, for the animal diet will have to supply both these vitamins.

Lysine biosynthesis is one example of a biochemical difference between the higher fungi which use the aminoadipic acid route, and the host plants which use the diaminopimelic acid pathway (LéJohn, 1971). To exploit this difference would suggest that the final enzyme in the pathway, saccharopine dehydrogenase, should be a fungicidal target. However, it has been shown to be a reversible enzyme, probably more involved in lysine degradation than synthesis! Saccharopine is present in substantial amounts in the mycelia of the species studied by Wade, Thomson & Miflin (1980). There is no evidence that if the enzyme were inhibited lysine would become limiting for growth – as it may be available from the host – and there is no evidence that the dehydrogenase is an enzyme present in critical amounts, or that it is a key regulatory enzyme.

Corbett (1975) has suggested four criteria that must be met for a biochemical target to be chosen:

  (i) the target site should be present at low concentration in the organism;

(ii) the site should be of short-term importance to the life of the pest;

(iii) a lead for chemical synthesis should be available;

(iv) an *in vitro* test should be available.

The first three of these criteria have not been met, and so much more information would have to be available before this pathway could be chosen as a basis for fungicide design. There is even less information available for some of the other differences listed in Table 3.2. Thus a knowledge of different biochemical pathways is only the starting point for a detailed study that will have to be made, and so is considered of minor importance in selecting new targets.

Starting from natural products

Natural products with fungicidal activity arise in plants either as preformed secondary metabolites, or from pathogen-induced metabolism (phytoalexins). The phytoalexins which have been tested are not as active on plant or human pathogens as some modern synthetic chemicals, and so are unlikely to be chosen as conventional fungicides (Rathmell & Smith, 1980; Ryley *et al.*, 1981). They could be suitable starting points for chemical synthesis, but little is known about their mode of action other than that they damage membranes (VanEtten & Puekke, 1976). If the biochemical basis of the effect were understood it would give a stimulus to synthetic work. What is known suggests that phytoalexins are not selectively toxic to the fungus, but that locally they destroy both the host tissue and the pathogen. In the plant they are effective because they are present in the cells immediately involved in the lesion. Another factor to be considered is that some fungi have developed the ability to degrade phytoalexins, thereby overcoming their toxicity (Zhang & Smith, 1983). The presence of degradative mechanisms already in existence within a fungal population will decrease the attraction of using a fungicide based upon a similar chemical structure. Taken together, these factors go against this as an exciting target area.

Antifungal antibiotics are included in the review by Ryley *et al.* (1981). Only the polyenes and griseofulvin of the major classes have achieved medical importance; blasticidin, the polyoxin group and kasugamycin are used in agriculture. The testing of microbial cultures for antibiotics continues to reveal antifungal activity in many chemicals (see, for example, the *Index of Antibiotics and other Microbial Products* in issues of the *Journal of Antibiotics*). Many of these initial tests look for antibacterial activity as well as for inhibition of the growth of

Table 3.3 *Some simple antifungal antibiotics*

OH
|
HCO—N—CH₂—COOH

Hadacidin
Aspartate analogue in
adenylosuccinate synthesis
(Shigeura & Gordon, 1962*a*, *b*)

Pyrrolnitrin
Respiratory inhibitor

(Gottlieb & Shaw, 1970)

Pyoluteorin

(Howell & Stipanovic, 1980)

Thiolutin
Inhibitor of RNA synthesis
(Jimenez, Tipper & Davies, 1973)

selected fungi and, as they are conducted *in vitro*, activity will have to be confirmed in subsequent tests *in vivo*. Antibiotics have been selected in nature because they are able to help one organism to compete with another, and many have extremely high activity. Resistance to antibiotics can be a problem, just as is resistance to synthetic chemicals, and it is a problem in the use of polyoxin and kasugamycin (Misato, 1982).

The attraction of looking for antibiotics is that a wide range of chemical structures are discovered, some of which are very simple in chemical terms. Because of their potency they represent attractive starting points for chemical synthesis, and many are new types of toxophore. Thus hadacidin, pyrrolnitrin and pyoluteorin, and thiolutin are shown as examples well documented in the literature (Table 3.3). The second advantage of antibiotics is that studies on their mode of action have identified target enzymes which are of critical importance to the fungus, and where inhibition will prevent the pathogen from attacking the host. In addition, they provide chemists with the structure of a known inhibitor and biochemists with a tool for enzyme studies on the mechanism of the inhibition.

The difference between fungal and plant cell walls was mentioned as a possible target site (see Table 3.2). That this is indeed a critical site was demonstrated by the action of the polyoxin antibiotics, shown to be potent inhibitors of chitin synthase (Hori, Kakiki & Misato, 1974). Subsequently the related neopolyoxins (Uramoto *et al.*, 1980) and nikkomycins (Brillinger, 1979) have been discovered, and have the same mode of action – competitive inhibition of chitin synthase, due to the similarity of the inhibitor to the substrate UDP-*N*-acetylglucosamine. Insects share with the higher fungi the ability to use chitin as a structural polysaccharide, and the benzoylurea insecticides block insect chitin synthase (Sowa & Marks, 1975). Unfortunately there is conflicting evidence about their ability to block fungal chitin synthase but, in any event, the compounds are not markedly fungicidal (Leighton, Marks & Leighton, 1981).

Cell-wall formation in fungi, with an interplay between synthesising and hydrolysing enzymes at the growing mycelial tip, is a delicately balanced system which has attracted much study (see Burnett & Trinci, 1979). Cell walls of fungi contain not only chitin but also cellulose and glucans (LéJohn, 1971). Papulacandin is an antibiotic which inhibits glucan synthesis, and is particularly active against yeasts (Baguley *et al.*, 1979). Tunicamycin has been shown to inhibit protein glycosylation by interfering with the transfer of *N*-acetylglucosamine to dolichol phosphate (Elbein, 1981). There is thus potential for discovering new fungicides active as inhibitors of cell-wall synthesis and organisation, and the challenge will be to simplify the complex structures of these antibiotics to something more easily synthesised and yet cost-effective as a commercial fungicide, or to enhance the production of the antibiotics by fermentation.

Starting from known fungicides

In the last 15–20 years, several new and important groups of fungicides with novel modes of action have been discovered (Bent, 1979; Edgington, 1981). These also indicate to us areas of biochemical susceptibility which are worth careful examination, and where further fungicidal discoveries may be expected. When a new type of fungicide is discovered and announced to the world, there is a great temptation for synthetic chemists in competing companies to attempt to modify the structure, whilst still retaining the toxophore and associated recognition groups.

An up-to-date example is in the azole inhibitors of the 14-demethylation reaction of ergosterol biosynthesis. Here we have a great proliferation of very active chemicals, effective for both plant and animal mycoses (Baldwin, 1983). Cowling (1963) was correct in identifying ergosterol as a target in fungi, though for the wrong reasons. The 14-demethylation reaction which is blocked is present in the biosynthesis of both plant and animal sterols. The selectivity which is achieved is due not so much to the presence or absence of a specific enzyme, but to the ability to inhibit the pathway only to the disadvantage of the fungus. Van den Bossche *et al.* (1980) found that a six-times higher dose of ketoconazole was required to affect cholesterol biosynthesis by rat liver than erogsterol biosynthesis in the infecting *Candida albicans*, and Buchenauer (1977) found that plant sterols were changed only slightly by doses of related azoles which were fungicidal.

Investigation of the mechanism of the reaction inhibited by these fungicides shows that it is a cytochrome P-450 dependent oxidation. There is evidence that the interaction of the N4 heteroatom of the triazole ring system (and the analogous nitrogen in the related imidazole, pyrimidine and pyridine fungicides) with the iron atom to the haem prosthetic group of the cytochrome is responsible for the inhibition (Gadher *et al.*, 1983; Wiggins & Baldwin, 1984). As modern computer techniques are used to model the interaction (Marchington, 1984), and with knowledge of the active site of the enzyme (an X-ray study is not yet available), it should be possible to design better inhibitors of the enzyme. If these retain the same physical properties as the existing fungicides, they may well be active *in vivo*. There are two arguments against this approach. Modelling from known 14-demethylation inhibitors is only likely to give rise to another chemical with similar fungicidal spectrum and type of activity. Secondly, tolerance to azole types has been observed in the laboratory already, with cross-tolerance to other chemicals having the same mode of action (Fuchs *et al.*, 1983), so the presence in nature of mechanisms diminishing the efficacy of the new chemical may detract from the introduction of another azole.

However, another approach is to consider ways of blocking the biosynthesis of ergosterol at a different stage in the sequence. Based upon Corbett's criteria (1975), we have established the importance of the pathway, and other inhibitors are known (Baldwin, 1983), so it should be possible to find new fungicides starting from this base. It is encouraging that the morpholine group of fungicides, blocking other stages in the pathway, are not cross-tolerant to the resistant strains (de Waard & van Nistelrooy, 1982).

Ergosterol-biosynthesis inhibitors are not active on the diseases caused by the Peronosporales, which do not contain sterols as normal membrane constituents (McCorkindale *et al.*, 1969). Hendrix (1974) and Nes & Stafford (1983) discuss the role of exogenous sterols in stimulating growth and reproduction in *Pythium* and *Phytophthora*. There now is available a group of systemic fungicides highly active on Peronosporales, including prothiocarb, hymexazol, cymoxanil and metalaxyl (Schwinn, 1983). Edgington (1981) has drawn attention to the fact that this group of chemicals has high water solubility in comparison with fungicides of other classes, and has related this to the aqueous habitat of the Oomycetes. The chemicals active on Oomycetes have significant phloem mobility (Rathmell, 1983).

Both cymoxanil (DPX 3217) (Despreaux, Fritz & Leroux, 1981) and metalaxyl (Davidse, Gerritsma & Hofman, 1981) are reported to block RNA synthesis. Davidse (1983) has subsequently reported that the site of inhibition of RNA synthesis caused by metalaxyl is at the nuclear α-amanitin insensitive RNA-polymerase–template complex in *Phytophthora megasperma*. Chloroneb (Hock & Sisler, 1969) and pyroxyclor (Tillman & Ferguson, 1980) are also reported to interfere with DNA synthesis. It thus appears that synthesis of nucleic acids is a biochemical soft spot in the Oomycetes, and it therefore becomes a target of high priority, especially as high levels of field resistance to metalaxyl and related acylanilides has developed, which is thought to be due to an insensitive site of action (Davidse, 1983).

RNA synthesis is essential for the germination of spores in *Neurospora crassa* and *Aspergillus nidulans* but not in the Oomycete *Peronospora tabacina* (Hollomon, 1970). Inhibition of the synthesis of RNA in the actively growing mycelial tips was reported by Grohmann & Hoffmann (1982) when *Pythium* and *Phytophthora* spp. were treated by metalaxyl. Metalaxyl was reported to be highly inhibitory to germ-tube growth by Farih, Tsao & Menge (1981*b*). This suggests that nucleic acid synthesis at an early stage of the active growth needs further examination to characterise the mechanism of the inhibition and to assist in the development of other toxicants. It is noteworthy that benomyl and carbendazim, which inhibit DNA synthesis in higher fungi (Clemons & Sisler, 1971) by binding to tubulin and preventing mitosis (Davidse & Flach, 1977), have no activity on Oomycetes (Bollen & Fuchs, 1970). There may thus be significant differences in the processes of mitosis and the biochemistry of replication between the higher and the lower fungi, both of which are susceptible to inhibition by fungicides.

The final example chosen to illustrate this area is the action of a series of chemicals effective against rice blast, *Pyricularia oryzae*, which have already been referred to as indirectly acting compounds. Sisler and his co-workers (Woloshuk *et al.*, 1980; Woloshuk & Sisler, 1982; Wolkow, Sisler & Vigil, 1983; Woloshuk, Vigil & Sisler, 1983) have demonstrated by biosynthetic studies in *P. oryzae* and *Verticillium dahliae*, and by electron microscopy of the penetration process by *P. oryzae* and *Colletotrichum lindemuthianum*, that melanin biosynthesis is blocked between 1,3,8-trihydroxynaphthalene and vermelone after treatment with tricyclazole and related chemicals. The lack of melanin leads to the inability of the fungus to penetrate through the epidermal wall of the plant; mutant strains lacking melanin were also unable to penetrate. The target site exposed by these studies could provide the lead for more active fungicides based upon this idea. There is no evidence to date that resistance has developed to this indirect mode of action. The indirect action of probenazole, another highly active rice-blast fungicide, has another basis which would repay investigation (Yamaguchi, 1982).

Biological control

Whilst outside the experience of the author and so not discussed in detail in this review, one area of plant pathology which is receiving research attention is the possibility of biological control. In nature, not only does the fungus interact with the host, but many other micro-organisms are present in the infection zone. To be successful, the fungus must attack the host and fight off or compete with other organisms at the same time. If we can introduce a biological antagonist, either on its own or in conjunction with a chemical treatment, we may achieve good control. Here selectivity of action of the chemical will have to be such that the introduced biological agent is not damaged by the chemical. This principle has been established against insect pests but has only recently emerged in fungal control; it is likely to be used first against pathogens that are not adequately controlled by existing measures. Thus Pseudomonads were found to protect elm trees against *Ophiostoma* (= *Ceratocystis*) *ulmi* when injected before fungal attack (Scheffer, 1983). Recent reviews covering leaf and soil interactions of microbes with biological control in mind are Blakeman (1981), Schroth & Hancock (1981) Ponchet (1982) and Cook & Baker (1983).

Conclusions

Much biochemical and physiological research is carried out in the area of plant pathology, though there are fewer investigations into the relationship of the animal host and pathogen. To use much of this information directly to choose potential targets for the selective inhibition of fungal growth is difficult, as any new control measure will have to demonstrate technical advance over existing control measures and prove to be marketable. Advances in the next decade are most likely to be made in the areas of chemistry based upon natural products and antibiotics, and by following in detail the 'Achilles heels' revealed by the mode of action of existing fungicides. New toxophores are more likely to be discovered by random screening, but knowledge of mode of action should enable them to be exploited more effectively than in the past.

Acknowledgements. The author is indebted for many discussions with colleagues on these topics. Nevertheless, the points of view put forward are personal.

References

The references given are to review articles wherever possible.

Albert, A. (1979). *Selective Toxicity: the Physio-chemical Basis of Therapy*, 6th edn, London: Chapman & Hall.

Allen, P. J. (1976). Control of spore germination and infection structure formation in the fungi. *Encyclopedia of Plant Physiology, New Series* 4, ed. R. Heitefuss & P. H. Williams, pp. 51–78. Berlin: Springer Verlag.

Andrews, J. H. (1975). Distribution of label from ^{3}H-glucose and ^{3}H-leucine in lettuce cotyledons during the early stage of infection with *Bremia lactucae*. *Canadian Journal of Botany*, **53**, 1103–15.

Anon. (1981). A look at world pesticide markets. *Farm Chemicals*, **144** (9), 55–60.

Baguley, B. C., Römmele, G., Gruner, J. & Wehrli, W. (1979). Papulacandin B: an inhibitor of glucan synthesis in yeast spheroplasts. *European Journal of Biochemistry*, **97**, 345–51.

Baldwin, B. C. (1983). Fungicidal ergosterol biosynthesis inhibitors. *Biochemical Society Transactions*, **11**, 659–63.

Bent, K. J. (1979). Fungicides in perspective: 1979. *Endeavour, New Series*, **3**, 7–14.

Blakeman, J. P. (1981). *Microbial Ecology of the Phylloplane*. London: Academic Press.

Bollen, G. J. & Fuchs, A. (1970). On the specificity of the *in vitro* and *in vivo* antifungal activity of benomyl. *Netherlands Journal of Plant Pathology*, **76**, 299–312.

Borgers, M. & Waldron, H. A. (1981). The action of ketoconazole on fungi. *Clinical Research Reviews*, **1**, 165–71.

Braunholtz, J. T. (1977). The crop protection industry; products in prospect. *Proceedings of the 9th British Crop Protection Conference; Pests and Diseases*, pp. 659–70. London: British Crop Protection Council.

Brillinger, G. U. (1979). Metabolic products of microorganisms 181: chitin synthase from fungi, a test model for substances with insecticidal properties. *Archiv für Microbiologie*, **121**, 71–4.

Buchenauer, H. (1977). Mode of action and selectivity of fungicides which interfere with ergosterol biosynthesis. *Proceedings of the 9th British Crop Protection*

Conference: Pests and Diseases, pp. 699–711. London: British Crop Protection Council.

Burnett, J. H. & Trinci, A. P. J. (eds) (1979). *Fungal Walls and Hyphal Growth*. Symposium of the British Mycological Society, April 1978. Cambridge University Press.

Clemons, G. P. & Sisler, H. D. (1971). Localisation of the site of action of a fungitoxic benomyl derivative. *Pesticide Biochemistry and Physiology*, **1**, 32–43.

Cook, R. J. & Baker, K. F. (1983). *Nature and Practice of Biological Control of Plant Pathogens*. St Paul: American Phytopathological Society.

Corbett, J. R. (1975). Biochemical design of pesticides. *Proceedings of the 8th British Insecticide and Fungicide Conference*, pp. 981–93. London: British Crop Protection Council.

Corbett, J. R. (1979). Resistance and selectivity: biochemical basis and practical problems. *Proceedings of the 1979 British Crop Protection Conference: Pests and Diseases*, pp. 717–30. London: British Crop Protection Council.

Cowling, E. B. (1963). Discussion on paper by A. E. Dimond on the modes of action of chemotherapeutic agents in plants. *Bulletin of the Connecticut Agricultural Experimental Station*, **663**, 72–5.

Daly, J. M. (1976). The carbon balance of diseased plants. *Encyclopedia of Plant Physiology, New Series*, **4**, ed. R. Heitefuss & P. H. Williams, pp. 450–79. Berlin: Springer Verlag.

Davidse, L. C. (1983). Mechanism of action and resistance to metalaxyl. *Proceedings of the Rheinhardsbrunn Conference on systemic fungicides, 1983* (in press).

Davidse, L. C. & Flach, W. (1977). Differential binding of methyl benzimidazol-2-yl carbamate to fungal tubulin as a mechanism of resistance to the antimitotic agent in mutant strains of *Aspergillus nidulans*. *Journal of Cell Biology*, **72**, 174–93.

Davidse, L. C., Gerritsma, O. C. M. & Hofman, A. E. (1981). Mode d'action biochimique du métalaxyl. *Phytiatrie-Phytopharmacie*, **30**, 235–44.

Davies, A. R. & Marriott, M. S. (1981). Inhibitory effects of imidazole antifungals on the yeast-mycelial transformation in *Candida albicans*. *Microbios Letters*, **17**, 155–8.

Despreaux, D., Fritz, R. & Leroux, P. (1981). Mode d'action biochimique du cymoxanil. *Phytiatrie-Phytopharmacie*, **30**, 245–55.

de Waard, M. A. & van Nistelrooy, J. G. M. (1982). Toxicity of fenpropimorph to fenarimol-resistant isolates of *Penicillium italicum*. *Netherlands Journal of Plant Pathology*, **88**, 231–6.

Dimond, A. E. (1963). The selective control of plant pathogens. *World Review of Pest Control*, **2** (4), 7–17.

Dimond, A. E. & Rich, S. (1977). Effects on physiology of the host and on host pathogen interactions. *Systemic fungicides*, 2nd edn, ed R. W. Marsh, pp. 115–30. London: Longman.

Edgington, L. V. (1981). Structural requirements of systemic fungicides. *Annual Review of Phytopathology*, **19**, 107–24.

Elbein, A. D. (1981). The tunicamycins – useful tools for studies on glycoproteins. *Trends in Biochemical Sciences*, **6**, 219–21.

Farih, A., Tsao, P. H. & Menge, J. A. (1981*a*). Fungitoxic activity of Efosite Aluminium on growth, sporulation and germination of *Phytophthora parasitica*. *Phytopathology*, **71**, 934–6.

Farih, A., Tsao, P. H. & Menge, J. A. (1981*b*). *In vitro* effects of metalaxyl on growth, sporulation and germination of *Phytophthora parasitica* and *P. citrophthora*. *Plant Disease*, **65**, 651–4.

Friend, J. & Threlfall, D. R. (eds) (1976). *Biochemical Aspects of Plant–Parasite Relationships*. Annual proceedings of the Phytochemical Society, **13**. London: Academic Press.

Fuchs, A., Davidse, L. C., de Waard, M. A. & de Wit, P. J. G. M. (1983). Contemplations and speculations on novel approaches in the control of fungal plant diseases. *Pesticide Science*, **14**, 272–93.

Gadher, P., Mercer, E. I., Baldwin, B. C. & Wiggins, T. E. (1983). A comparison of the potency of some fungicides as inhibitors of sterol 14-demethylation. *Pesticide Biochemistry and Physiology*, **19**, 1–10.

Geissbühler, H., Brenneisen, P. & Fischer, H.-P. (1982). Frontiers in crop production: chemical research objectives. *Science*, **217**, 505–10.

Gottlieb, D. & Shaw, P. D. (1970). Mechanism of action of antifungal antibiotics. *Annual Review of Phytopathology*, **8**, 371–402.

Grivell, A. R. & Jackson, J. F. (1968). Thymidine kinase: evidence for its absence from *Neurospora crassa* and some other microorganisms and the relevance of this to the specific labelling of deoxyribonucleic acid. *Journal of General Microbiology*, **54**, 307–17.

Grohmann, U. & Hoffmann, G. M. (1982). Light- and electromicroscopic studies about the effect of metalaxyl on species of *Pythium* and *Phytophthora*. *Zeitschrift für Pflanzenkrankheiten und Pflanzenschutz*, **89**, 435–46.

Hancock, J. G. & Huisman, O. C. (1981). Nutrient movement in host pathogen systems. *Annual Review of Phytopathology*, **19**, 309&31.

Harold, F. M. (1966). Inorganic polyphosphates in biology: structure, metabolism and functions. *Bacteriological Reviews*, **30**, 772–94.

Hedin, P. A. (1982). New concepts and trends in pesticide chemistry. *Journal of Agricultural and Food Chemistry*, **30**, 201–15.

Heitefuss, R. & Williams, P. H. (1976). Physiological plant pathology. *Encyclopedia of Plant Physiology, New Series*, **4**. Berlin: Springer Verlag.

Hendrix, J. W. (1974). Physiology and biochemistry of growth and reproduction in *Pythium*. *Proceedings of the American Society of Phytopathology*, **1**, 207–10.

Hock, W. K. & Sisler, H. D. (1969). Specificity and mechanism of action of chloroneb. *Phytopathology*, **59**, 627–32.

Hohl, H. R. (1983). Nutrition of *Phytophthora*. *Phytophthora, its Biology, Taxonomy, Ecology and Pathology*, ed. D. C. Erwin, S. Bartnicki-Garcia & P. H. Tsao, pp. 41–54. St Paul: American Phytopathological Society.

Hollomon, D. W. (1970). Ribonucleic acid synthesis during fungal spore germination. *Journal of General Microbiology*, **62**, 75–87.

Hori, M., Kakiki, K. & Misato, T. (1974). Interaction between polyoxins and active centre of chitin synthetase. *Agricultural and Biological Chemistry*, **38**, 699–705.

Horsfall, J. G. (1972). Selective chemicals for plant disease control. *Pest Control: Strategies for the Future*, Agricultural Board, Division of Biology and Agriculture, National Research Council, pp. 216–225. Washington, D.C.: National Academy of Sciences.

Horsfall, J. G. (1977). Fungicides – past, present and future. *Pesticide Chemistry in the 20th Century*, ed. J. R. Plimmer, pp. 113–22. Washington, D.C.: American Chemical Society.

Horsfall, J. G. & Lukens, R. J. (1969). The strategy of finding fungicides. *Residue Reviews*, **25**, 81–91.

Howell, C. R. & Stipanovic, R. D. (1980). Suppression of *Pythium ultimum* – induced damping off of cotton seedlings by *Pseudomanas fluorescens* and its antibiotic, pyoluteorin. *Phytopathology*, **70**, 712–15.

Jimenez, A., Tipper, D. J. & Davies, J. (1973). Mode of action of thiolutin, an inhibitor of macromolecular synthesis in *Saccharomyces cerevisiae*. *Antimicrobial Agents and Chemotherapy*, **3**, 729–38.

Koser, S. A. (1968). *Vitamin Requirements of Bacteria and Yeasts*. Springfield: Charles C. Thomas.

Langcake, P. (1981). Alternative chemical agents for controlling plant disease. *Philosophical Transactions of the Royal Society of London, B*, **295**, 83–101.

Leighton, T., Marks, E. & Leighton, F. (1981). Pesticides: insecticides and fungicides are chitin synthesis inhibitors. *Science*, **213**, 905–7.

LéJohn, H. B. (1971). Enzyme regulation, lysine pathways and cell wall structures as indicators of major lines of evolution in fungi. *Nature*, **231**, 164–8.

McCorkindale, N. J., Hutchinson, S. S., Pursey, B. A., Scott, W. T. & Wheeler, R. (1969). A comparison of the types of sterol found in some species of the Saprolegniales and Leptomitales with those found in other Phycomycetes. *Phytochemistry*, **8**, 861–7.

Manners, J. G. (1982). *Principles of Plant Pathology*, p. 106. Cambridge University Press.

Manners, J. M. & Gay, J. L. (1982). Accumulation of systemic fungicides and other solutes by haustorial complexes isolated from *Pisum sativum* infected with *Erysiphe pisi*. *Pesticide Science*, **13**, 195–203.

Marchington, A. F. (1984). Application of computer graphics to the design of azolylmethane fungicides. *Pesticide Science* (in press).

Martin, T. J. & Ellingboe, A. H. (1978). Genetic control of ^{32}P transfer from wheat to *Erysiphe graminis* f.sp. *tritici* during primary infection. *Physiological Plant Pathology*, **13**, 1–11.

Misato, T. (1982). Present status and future prospects of agricultural antibiotics. *Journal of Pesticide Science*, **7**, 301–5.

Nes, W. D. & Stafford, A. E. (1983). Evidence for metabolic and functional discrimination of sterols by *Phytophthora cactorum*. *Proceedings of the National Academy of Sciences, USA*, **80**, 3227–31.

Ponchet, J. (1982). Realities and prospects for biological control of plant diseases. *Agronomie*, **2**, 305–14.

Rathmell, W. G. (1983). The discovery of new methods of chemical disease control: current developments, future prospects and the role of biochemical and physiological research. *Advances in Plant Pathology*, **2**, 259–88.

Rathmell, W. G. & Smith, D. A. (1980). Lack of activity of selected isoflavonoid phytoalexins as protectant fungicides. *Pesticide Science*, **11**, 568–72.

Raynal, G., Ravise, A. & Bompeix, G. (1980). Action of aluminium tris-*O*-ethylphosphonate on *Plasmopora viticola* pathogenicity and on stimulation of defense reactions of grapevine. *Annales de Phytopathologie*, **12**, 163–75.

Ridings, W. H., Gallegly, M. E. & Lilly, V. G. (1969). Thiamine requirements helpful in distinguishing isolates of *Pythium* from those of *Phytophthora*. *Phytopathology*, **59**, 737–42.

Ryley, J. F., Wilson, R. G., Gravestock, M. B. & Poyser, J. P. (1981). Experimental approaches to antifungal chemotherapy. *Advances in Pharmacology and Chemotherapy*, **18**, 49–176.

Sbragia, R. J. (1975). Chemical control of plant diseases: an exciting future. *Annual Review of Phytopathology*, **13**, 257–69.

Scheffer, R. J. (1983). Biological control of Dutch elm disease by *Pseudomonas* species. *Annals of Applied Biology*, **103**, 21–30.

Schroth, M. N. & Hancock, J. G. (1981). Selected topics in biological control. *Annual Review of Microbiology*, **35**, 453–76.

Schwinn, F. J. (1983). New developments in chemical control of *Phytophthora*.

Phytophthora, its Biology, Taxonomy, Ecology and Pathology, ed. D. C. Erwin, S. Bartnicki-Garcia & P. H. Tsao, pp. 327–34. St. Paul: American Phytopathological Society.

Shigeura, H. T. & Gordon, C. N. (1962*a*). Hadacidin, a new inhibitor of purine biosynthesis. *Journal of Biological Chemistry*, **237**, 1932–6.

Shigeura, H. T. & Gordon, C. N. (1962*b*). The mechanism of action of hadacidin. *Journal of Biological Chemistry*, **237**, 1937–40.

Siegal, M. R. & Sisler, H. D. (1977). *Antifungal Compounds*, vol. 1 & 2. New York: Marcel Dekker Inc.

Sisler, H. D. (1977). Fungicides; problems and perspectives. *Antifungal Compounds*, ed. M. R. Siegal & H. D. Sisler, pp. 531–47. New York: Marcel Dekker Inc.

Smedegaard-Petersen, V. & Stølen, O. (1981). Effect of energy-requiring defense reactions on yield and grain quality in a powdery mildew-resistant barley cultivar. *Phytopathology*, **71**, 396–9.

Somers, E. (1969). Fungicide selectivity and specificity. *World Review of Pest Control*, **8** (2), 95–100.

Sowa, B. A. & Marks, E. P. (1975). An *in vitro* system for the quantitative measurement of chitin synthesis in the cockroach: inhibition by TH 6040 and polyoxin D. *Insect Biochemistry*, **5**, 855–9.

Tillman, R. W. & Ferguson, M. W. (1980). Toxicity of pyroxychlor to *Pythium aphanidermatum*. *Phytopathology*, **70**, 441–6.

Tuleen, D. M. & Frederiksen, R. A. (1982). Evaluating a crop loss mode for head smut of sorghum. *Phytopathology*, **72**, 1278–80.

Uramoto, M., Kobinata, K., Isono, K., Jenkins, E. E., McCloskey, J. A., Higashijima, T. & Miyazawa, T. (1980). Neopolyoxins A, B and C: new inhibitors of fungal cell wall chitin synthetase. *Nucleic Acid Symposium Series*, **8**, 69–71.

van Andel, O. M. (1958). Investigations on plant chemotherapy. II. Influence of amino acids on the relation of the plant to the pathogen. *Tidjschrift over Plantenziekten*, **64**, 307–27.

Van den Bossche, H., Willemsens, G., Cools, W., Cornelissen, F., Lauwers, W. F. & Van Cutsen, J. M. (1980). *In vitro* and *in vivo* effects of the antimycotic drug ketoconazole on sterol biosynthesis. *Antimicrobial Agents and Chemotherapy*, **17**, 922–8.

van der Kerk, G. J. M. (1963). Fungicides – retrospects and prospects. *World Review of Pest Control*, **2** (3), 29–36.

Vanderplank, J. E. (1978). *Genetic and Molecular Basis of Plant Pathogenesis*. Berlin: Springer Verlag.

VanEtten, H. D. & Puekke, S. G. (1976). Isoflavonoid phytoalexins. *Biochemical Aspects of Plant – Parasite Relationships*, ed. J. Friend & D. R. Threlfall, pp. 239–89. London: Academic Press.

Wade, M., Thomson, D. M. & Miflin, B. J. (1980). Saccharopine: an intermediate of L-lysine biosynthesis and degradation in *Pyricularia oryzae*. *Journal of General Microbiology*, **120**, 11–20.

Wiggins, T. E. & Baldwin, B. C. (1984). Binding of azole fungicides related to diclobutrazol to cytochrome P-450. *Pesticide Science*, **15**, 206–9.

Wolkow, P. M., Sisler, H. D. & Vigil, E. L. (1983). Effect of inhibitors of melanin biosynthesis on structure and function of appressoria of *Colletotrichum lindemuthianum*. *Physiological Plant Pathology*, **22**, 55–71.

Woloshuk, C. P. & Sisler, H. D. (1982). Tricyclazole, pyroquilon, tetrachlorophthalide, PCBA, coumarin and related compounds inhibit melanization and epidermal penetration by *Pyricularia oryzae*. *Journal of Pesticide Science*, **7**, 161–6.

Woloshuk, C. P., Sisler, H. D. & Vigil, E. L. (1983). Action of the antipenetrant, tricyclazole, on appressoria of *Pyricularia oryzae*. *Physiological Plant Pathology*, **22**, 245–59.

Woloshuck, C. P., Sisler, H. D., Tokousbalides, M. C. & Dutky, S. R. (1980). Melanin biosynthesis in *Pyricularia oryzae*: site of tricyclazole inhibition and pathogenicity of melanin deficient mutants. *Pesticide Biochemistry and Physiology*, **14**, 256–64.

Wood, R. K. S. (1967). *Physiological Plant Pathology*. Oxford: Blackwell.

Woods, A. M. & Gay, J. L. (1983). Evidence for a neckband delimiting structural and physiological regions of the host plasma membrane associated with haustoria of *Albugo candida*. *Physiological Plant Pathology*, **23**, 73–88.

Yamaguchi, I. (1982). Fungicides for control of rice blast disease. *Journal of Pesticide Science* **7**, 307–16.

Zhang, Y. & Smith, D. A. (1983). Concurrent metabolism of the phytoalexins phaseollin, kievetone and phaseollinisoflavan by *Fusarium solani* f.sp. *phaseoli*. *Physiological Plant Pathology*, **23**, 89–100.

4

Discovery of antifungal agents: *in vitro* and *in vivo* testing

J. F. RYLEY and W. G. RATHMELL

ICI Pharmaceuticals Division, Mereside, Alderley Park, Macclesfield, Cheshire SK10 4TG and ICI Plant Protection Division, Jealott's Hill Research Station, Bracknell, Berkshire RG12 6EY, UK

Introduction

There are two approaches to the discovery of new antifungal agents: the rational and the empirical. It would be intellectually very satisfying to study the biochemistry or physiology of the parasite, identify ways in which it differs from the host, and then design and synthesise a molecule which would selectively exploit that difference and kill the fungus without in any way harming the host. Indeed mycologists have identified a number of potential targets – cell wall synthesis, membrane sterol biosynthesis, nucleic acid synthesis and nuclear division – and have discovered antifungal agents whose mode of action can be explained by interference at these target sites (Fig. 4.1). Yet the discovery of the polyoxins which interfere with chitin synthesis, the echinocandins and papulocandins which interfere with glucan synthesis, the azoles which interfere with ergosterol biosynthesis, or griseofulvin and the benzimidazoles which interfere with microtubule-associated protein and tubulin polymerisation respectively was, in each case, purely empirical and in no way rational; first came the discovery of antifungal activity and only then an explanation for that activity.

Alternatively it might be possible, following a study of the biochemistry of the interaction between host and parasite, to design molecules that interfere with the establishment of the parasite, and thus prevent infection. In this area, plant physiologists have identified potentiation of the host's natural defence mechanisms as a possible method of preventing fungal disease of crop plants; fosetyl aluminium is an agricultural antifungal agent which may act in this way. Again the discovery was empirical. The vast effort currently being expended in numerous

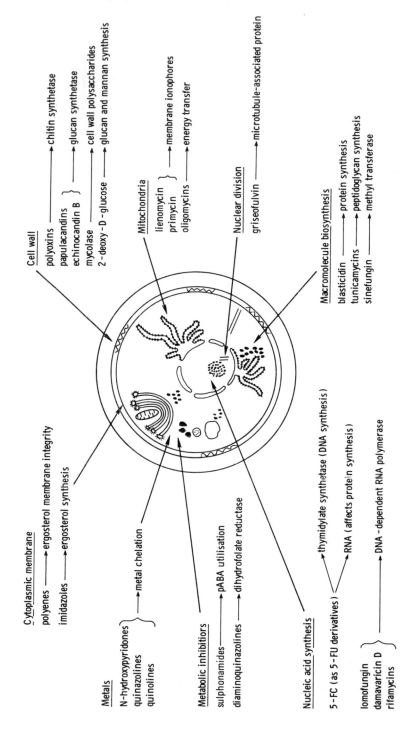

Fig. 4.1. Postulated modes of action of some antifungal agents (from Ryley *et al.*, 1981).

commercial laboratories towards the discovery of more effective molecules to antagonise the ergosterol biosynthetic pathway is eloquent testimony to the value of and need for the empirical approach. Although we can identify in a broad sense potential targets for selective inhibition – such as ergosterol biosynthesis – we cannot as yet design molecules to selectively hit these targets with any degree of optimism that we will succeed at the first, or even at the hundredth attempt. Indeed, by screening culture filtrates and natural product samples, we may well discover that nature is more efficient than our motivated chemists in synthesising antifungal compounds! And so whether our approach is to be rationally empirical, or purely empirical, or both, we need to have systems which will identify antifungal activity among very large numbers of compounds or samples. This is called 'screening'.

The purpose of a screen is simply to provide a yes/no answer to the question 'is there sufficient interaction between this compound or culture filtrate and the fungus to warrant more detailed investigation?' If the answer is 'yes', then a more detailed investigation of that interaction can be carried out. This is called 'evaluation'. The essentials of an efficient screen include factors such as high sample throughput, simplicity and economy of operation, small sample requirements, and unambiguity of interpretation. Provided overall numbers of compounds of interest are reduced considerably by the screen, it matters little that some compounds come through which will be rejected at a later stage of evaluation – false positives. It is important though that the screen does not reject compounds which would be of real interest in the final disease situation – false negatives.

A laboratory model cannot faithfully reproduce the situation in the field or the clinic. With phytopathogens, we may approximate the field situation by using both the fungus and the host of eventual interest in the screen, although there will be numerous differences – such as age of host, method of inoculation, climate and husbandry conditions – which will influence the chemical/disease interaction to some extent. Some of these aspects may be studied by further laboratory evaluation or during field screening, prior to larger scale testing in the field. With medical pathogens, the need to use experimental animals and/or in-vitro models will obviously give greater scope for variation in behaviour between the laboratory and the clinic. The purpose of this review is to discuss the most realistic approaches that we can make in the laboratory in order to solve problems in the field and in the clinic.

Animal mycopathogens

Target selection

For better or for worse, the economic importance of a disease condition is the prime consideration in selecting antifungal targets. The severity of the disease and the efficacy and availability of current treatments are important – particularly to the patient – but, without a vast and viable market potential, research and particularly development costs preclude research investment. Thus although diseases such as coccidioidomycosis, histoplasmosis and cryptococcosis attract a lot of attention in meetings and the scientific literature, and although the treatment of these potentially fatal diseases leaves a great deal to be desired, the number of cases annually is small; it is conditions such as vaginal candidosis and ringworm infections of the skin which underwrite antifungal research and which must be our prime targets. Ideally, however, we are looking for broad-spectrum agents which will be effective in conditions of economic importance – thus paying for the research which led to their discovery – and which will also be effective in the life-threatening systemic mycoses where the economic factor is absent. That such a target is a realistic possibility, though not yet fully attained, is illustrated by the recent development of ketoconazole.

In a laboratory situation, particularly at the screening level, other factors such as ease of handling and safety to the operator are important. Thus, working with *Candida albicans* presents no hazard since up to 50% of the population carries the organism anyway; common dermatophytes present only a minor hazard since they are not lethal, and accidental infections can readily be treated. Not so pathogens such as *Coccidioides immitis*. Because the fungus is potentially lethal, elaborate precautions have to be taken to protect the operator and, consequently, the amount of work which can be done is strictly limited.

Systemic fungal infections obviously require systemic treatment. If, however, a fungus is confined to the skin or mucous membranes, then we can either apply a topical treatment to the affected area or alternatively attack the parasite through the skin by means of a systemic treatment. A woman with a superficial or vaginal infection may feel that with a topical treatment she is doing something useful and relevant by applying medication to the affected area; alternatively she may feel that there is enough mess and discharge already without adding to it! One of the advantages of systemic medication for superficial infections is that it will deal with unrecognised as well as recognised lesions, and will cope with fungi away from the site of the primary lesion, thus reducing the opportunities for reinvasion and reinfection. The discovery of the orally active metronidazole some 25 years ago revolutionised the treatment of vaginal trichomoniasis by virtue of its ability to reach extravaginal protozoa in the patient and her sexual partner, thus achieving radical cure; previous topical treatments could never do more than suppress the vaginal infection until reinvasion took place. We want to do for vaginal candidosis what metronidazole did for vaginal trichomoniasis long ago.

With these considerations in mind, we have set ourselves the target of an orally active compound for the treatment of vaginal candidosis and the treatment of yeast and dermatophyte infections of the skin and nails; our hope is that such a compound would also be effective in treating the systemic mycoses. Accordingly we restrict our laboratory screening and evaluation, both *in vitro* and *in vivo*, to a number of yeasts and dermatophytes. When we find a compound of real interest and activity, then we enlist the help of a number of specialists for evaluation against the more pathogenic and more exotic species. We are not interested in discovering compounds which have only topical activity since many products are already available for this purpose, and it is difficult to see where room for improvement by this route lies; we would, however, consider developing a topical formulation of an orally active product to meet the needs of those who are not convinced of the desirability of systemic treatment of superficial infections. In the discussion which follows we will not therefore consider assessment of topical potential *in vivo* at the screening stage.

In-vitro *versus* in-vivo *testing*

If a compound is going to be an effective antifungal in a clinical situation, then a lot more is required of it than for it just to interact with the fungus. It has to be absorbed from the site of administration – oral or otherwise – and, without being metabolically inactivated, has to reach the site of fungal infection before it can have any hope of showing therapeutic activity. Not every orally administered compound is absorbed from the gut, and few of those which are find their way unchanged to the skin or vaginal secretion. *In vitro* testing will demonstrate potential for interaction between the compound and the fungus, while testing in experimental animals will indicate whether the compound has any hope of reaching the site of infection where it can interact with the fungus. It must be remembered, however, that the pharmacokinetics of a compound in mice or guinea pigs may be very different from pharmacokinetics in humans, and that a compound may consequently work well in the laboratory but not in the clinic. Of course the reverse may be true and, by relying on experimental models, we may well miss compounds with true therapeutic potential. A compound with high activity *in vitro* may not work *in vivo* because of metabolic inactivation; again the reverse may be true and metabolism to an active form might be necessary – a situation which an in-vitro screen would miss.

Any laboratory screen must be a compromise. Different classes of compounds behave in different ways; this is particularly well illustrated by the azoles. Activity against the yeast phase of *Candida albicans* is generally poor when considered in the light of in-vivo potential, but this degree of activity or inactivity can be influenced considerably by the choice of media and conditions of assay. The mycelial phase of *C. albicans*, on the other hand, is very sensitive to the azoles – hence we include this in our screen – but this difference in sensitivity between the two phases is not seen with many other classes of compound. If one is to have a high throughput of samples, and sample size is in many cases limited, then an in-vitro pre-screen is the only realistic option. In-vivo screening and evaluation is then restricted to compounds showing sufficient activity *in vitro* – a difficult level to fix – and to compounds which, for chemical reasons, seem worth testing. The cut-off point for the decision whether to test *in vivo* or not may be decided by the relative capacities of the two screens. Thus, in a 4 year period we have screened 34 316 compounds *in vitro*, 1345 of which went on to in-vivo testing.

Screening in vitro

Having considered all the possible variables and alternatives (Ryley *et al.*, 1981), we devised the following screen: compounds are dissolved or ground up in dimethyl sulphoxide (DMSO) at a concentration of $10\,\text{mg}\,\text{ml}^{-1}$, and a 1 in 4 dilution is made in water. In the light of the activity of known antifungal agents in our screen, we have decided on an arbitrary basis to screen compounds initially at an incorporation rate of $100\,\mu\text{g}\,\text{ml}^{-1}$ for yeasts and $25\,\mu\text{g}\,\text{ml}^{-1}$ for dermatophytes. $50\,\mu\text{l}$ aliquots of the DMSO solution are therefore incorporated into 5 ml amounts of molten yeast-morphology agar (YMA) and $50\,\mu\text{l}$ aliquots of the dilution into Sabouraud's dextrose agar (SDA) in tubes held in a water bath at $50\,°\text{C}$. Following mixing on a Whirlimixer, the two agar samples are poured into either half of a two-compartment plastic Petri dish and allowed to set. Plates are then inoculated using a Denley

Table 4.1. *Animal mycopathogens used for testing in vitro*

Yeasts	Dermatophytes
Candida albicans serotype A (NCPF 3153)	*Trichophyton mentagrophytes*
C. albicans serotype B (NCPF 3156)	*Trichophyton quinckeanum*
C. albicans Janssen B2630	*Trichophyton interdigitale*
C. albicans amphotericin-B-resistant	*Trichophyton rubrum*
C. albicans 5-fluorocytosine resistant	*Trichophyton tonsurans sulphureum*
Candida tropicalis	*Microsporum canis*
Candida parapsilosis Leeds	*Microsporum gypseum*
C. parapsilosis WL	*Epidermophyton floccosum*
	Prototheca zopfii
	Scopulariopsis brevicaulis
	Hendersonula toruloidea

multipoint inoculator, delivering approximately $2\,\mu l$ per inoculation site. The YMA half of the plate is inoculated with 8 different yeasts, the SDA half with 11 different dermatophytes and other organisms; these are listed in Table 4.1. As we have been screening natural product samples as well as synthetic compounds, and because we do not wish to discover any more polyenes, an amphotericin B resistant strain of *Candida albicans* is included in the screen to identify such compounds. Yeast inocula are made up in water at 10^5 colony forming units (cfu) per ml from overnight cultures on YMA, and dermatophyte inocula from 2 week old cultures on SDA; the latter inocula are kept at room temperature and used over a period of 4 weeks. Plates are incubated at $28\,^\circ$C for 4 d and then examined for growth. When activity is found, the test is repeated using 4-fold dilutions of the compound in the agar until the minimum inhibitory concentration (MIC) is found – the lowest concentration of compound which just prevents visible growth.

Side-by-side with this agar-dilution assay we test the susceptibility of the mycelial phase of one strain of *Candida albicans* in Eagle's MEM containing 0.15% $NaHCO_3$ and 1% foetal calf serum. 2 ml quantities of medium containing a mycelial inoculum are dispensed into the wells of Sterilin repli-plates and $10\,\mu l$ aliquots of diluted compound are added ($12.5\,\mu g\,ml^{-1}$ final concentration); the plates are incubated in an atmosphere of 5% CO_2:95% air for 24 h at $37\,^\circ$C and examined for mycelial development with an inverted microscope at $\times 100$ magnification. With active compounds, the test is repeated using four 10-fold dilutions to determine the lowest concentration of compound which just prevents mycelial development.

Some compounds are soluble in water, others can be brought into solution using acid or alkali, while still others are soluble in ethanol or acetone. When handling a large number of compounds in limited quantity there is neither the time nor compound to spare to determine individual solubilities. We find very few compounds which fail to dissolve or form an acceptable dispersion in DMSO. Incubation conditions for 19 different organisms must of necessity be a compromise; yeasts would grow overnight at 37 °C, but dermatophytes prefer a lower temperature and need a longer period of incubation. As a result, there is a marked difference in the extent of growth of some organisms.

The yeast and mycelial phases of *Candida albicans* have been found to be more or less equisensitive to a number of established antifungal agents and to a wide variety of experimental chemicals. With azoles (imidazoles and triazoles) however, this generalisation does not hold. As Table 4.2 indicates, the mycelial phase is more sensitive than the yeast phase to miconazole, clotrimazole and particularly ketoconazole – in the latter case, by several orders of magnitude depending on culture conditions. In the presence of low concentrations of ketoconazole in Eagle's MEM + serum, the fungus grows, but in the yeast rather than the mycelial form. It is because of our interest in azoles and the marked difference in susceptibility of the two forms that we include this mycelial test in our initial screen.

Not only may there be differences in sensitivity between different phases of the same organism growing in different media; there may be markedly different sensitivities between the same phases of the same organism growing in different media. Shadomy, Dixon & May (1982), in a comparison of bifonazole and clotrimazole, have recently found lower MICs on Kimmig's agar than on other media. Table 4.3 compares MICs found with a number of antifungal agents using a modified Kimmig's agar compared with YMA and SDA. In the majority of cases, MICs were lower for both yeasts and dermatophytes when Kimmig's agar was used, although there were some exceptions. The most notable exception was 5-fluorocytosine (5-FC) which was active against *C. albicans* on the defined/synthetic YMA, but inactive on peptone-rich Kimmig's agar. It was in order to detect metabolic antagonists such as 5-FC that we have been using YMA for yeasts in our standard screen. We have recently decided, as a result of the investigation summarised in Table 4.3, to screen against all organisms on Kimmig's agar with an initial testing level of $50 \mu g \, ml^{-1}$. We are thus sacrificing the chance of finding compounds like 5-FC for increased sensitivity to other compounds and a simplification in setting up the test.

Table 4.2. *Sensitivities of yeast and mycelial phases of* Candida spp. *to different antifungal agents*[a]

Compound	C. albicans (serotype A; NCPF 3153)		C. albicans (serotype B; NCPF 3156)		C. tropicalis	
	Yeast	Mycelial	Yeast	Mycelial	Yeast	Mycelial
Amphotericin B	0.4	0.1	0.1	0.1	0.4	0.1
Candicidin	≤0.1	≤0.1	≤0.1	≤0.1	≤0.1	≤0.1
Natamycin	6.2	10	6.2	10	1.6	10
Nystatin	25	10	25	100	25	10
5-Fluorocytosine	1.6	0.1	1.6	0.1	>100	10
Clotrimazole	25	0.01	25	0.01	6.2	0.1
Miconazole	25	0.1	25	0.1	25	1
Ketoconazole	100	0.01–0.001	100	0.01–0.001	25	0.01–0.001

[a]Table gives MICs ($\mu g\,ml^{-1}$) for a variety of drugs when tested against the yeast and mycelial forms of two strains of *C. albicans* and one strain of *C. tropicalis*.

Table 4.3. *Comparison of peptone-rich Kimmig's agar with YMA (48 h incubation) and SDA (96 h incubation) for determining antifungal sensitivity*[a]

Compound	C. albicans (5)		C. parapsilosis (2)		Trichophyton sp. (4)		Microsporum sp. (2)	
	YMA	Kimmig	YMA	Kimmig	SDA	Kimmig	SDA	Kimmig
Amphotericin B	0.2	≤0.1	12.5–0.8	≤0.1	3.1–0.4	≤0.1	12.5–1.6	≤0.1
Nystatin	1.6–0.8	0.4	1.6	0.2	1.6–0.4	0.4–0.2	1.6–0.8	0.2
5-Fluorocytosine	12.5–1.6	>100	>50–0.8	>100	>25	>25	>25	>25
Griseofulvin	>100	>100	>100	>100	6.2–0.2	0.4–≤0.1	6.2–1.6	0.4–0.2
Naftifine	>100	>100	100–25	12.5–6.2	0.4–0.2	≤0.1	25–0.4	≤0.1
Ro 14–4767	25–0.4	6.2–0.1	3.1	1.6	≤0.1	≤0.1	≤0.1	≤0.1
Tridemorph	100	>100–25	>100	>100	12.5–≤0.1	0.8–≤0.1	12.5–6.2	0.8
Miconazole	12.5–3.1	12.5–6.2	1.6–0.4	12.5	0.8–0.4	0.8	1.6–0.4	0.8
Ketoconazole	>100–50	50	0.8–0.4	25–0.4	25–1.6	1.6–0.8	25	1.6–0.8
ICI 153 066	100–50	50–25	0.4–0.2	25	6.2–0.4	3.1	12.5–3.1	3.1–1.6
BAY N 7133	100–25	50–25	3.1	6.2	6.2–0.4	3.1	6.2–0.4	1.6

[a]Figures are MICs in $\mu g\,ml^{-1}$; number of strains tested indicated in brackets.

The agar-dilution system with multipoint inoculation of plates is simpler to use when large numbers of compounds are involved than agar-diffusion on plates seeded with a single organism or broth-dilution using tubes or multiwell plates and microscopical observation or turbidity determinations. In this system we can differentiate between 5-FC sensitive and resistant strains of *Candida albicans* in the presence of 5-FC (when we use YMA), and between amphotericin B sensitive and resistant strains in the presence of polyenes. We have recently shown, however, that two apparently ketoconazole-resistant strains of *C. albicans* cannot be differentiated from sensitive isolates using this agar-dilution assay, but are well differentiated in four other systems *in vitro*, three animal models of infection and several biochemical assays (Ryley, Wilson & Barrett-Bee, 1984).

Screening in vivo

At the screening stage, the amount of compound available will usually be limited; it is only when useful activity *in vivo* has been demonstrated that chemists can be prevailed upon to synthesise realistic quantities of compound for testing! Some workers have tested for activity against *Candida albicans* in the vagina of the rat and for activity against ringworm infections in guinea pigs (Heeres *et al.*, 1979; Thienpont *et al.*, 1979), in each case, appreciable amounts of compound will be required if a realistic dose is to be used. For screening purposes we have chosen the mouse and, to conserve compound, study both infections in the same animal at the same time. Female mice weighing 25–35 g are injected subcutaneously on a Friday with 0.5 mg oestradiol benzoate as a ball-milled suspension in 0.5% (v/v) Tween 80. The following Monday, their backs are clipped and they are dosed orally in groups of five with compound prepared as a ball-milled suspension in 0.5% (v/v) Tween 80. They are inoculated (day 0) intravaginally with a suspension of an overnight culture of *C. albicans* at 10^8 ml^{-1} in saline until run-out using a blunt-ended Pasteur pipette and, following scarification of the back with a few light sideways strokes using a piercing saw fitted with a No. 1 blade (Eclipse), are inoculated with a suspension of *Trichophyton quinckeanum* at 7×10^7 ml^{-1} using a small brush. A further oral dose of compound is given on the same Monday, and then once daily for the following 4 d. Ringworm lesions are scored visually on an arbitrary scale of 0–4 on the Sunday (day 6), and a specimen obtained from the vagina with a wire loop is plated out on BiGGY agar (Difco) on the following Monday (day 7); growth of *C. albicans* is scored

Table 4.4. *Treatment of yeast and dermatophyte infections in animal models*[a]

	Candida albicans (vaginal)				Candida albicans (intravenous)	Trichophyton quinckeanum (cutaneous)	Trichophyton mentagrophytes (cutaneous)
	Mouse (×6)	Mouse (×1)	Rat (×5)	Rat (×1)	Mouse (×10)	Mouse (×6)	Guinea pig (×12)
Griseofulvin	NA 500[b]	—	—	—	—	250	25
Amphotericin B	SI 100[b]	—	—	—	5	25	—
5-Fluorocytosine	—	—	—	50	5	—	75
Ketoconazole	25	50	10	50	50	25	75
ICI 153 066	1–2.5	25–50	0.1	0.25	1	2.5	5
BAY N 7133	100	—	100	—	—	100	—

[a] Figures are minimal daily doses, in mg kg^{-1}, which give a 'satisfactory' and comparable response to treatment. Number of doses is given in parentheses.
[b] NA: not active; Sl: slightly active.

visually on a scale of 0–4 after 2 d incubation at 37 °C. Griseofulvin controls the *T. quinckeanum* infection at a daily dose of 250 mg kg^{-1}, while ketoconazole controls both infections with 25 mg kg^{-1} d^{-1}. In the light of this, we usually start testing novel compounds at 250 mg kg^{-1} – which requires 250 mg of compound. Testing of active compounds is continued until the minimum effective dose has been determined. Table 4.4 indicates the minimum effective doses of several compounds in this screen.

Evaluation in vitro

If a compound shows potentially useful activity *in vivo*, then further work is in order to evaluate the extent of that activity. The nature of this work will vary from compound to compound in the light of observations made at the screening stage and in the light of deficiencies of other compounds on the clinical scene. Experiments *in vitro* will utilise both liquid and solid media, using broth-dilution and agar-diffusion to study different aspects of activity. Of concern may be determining whether the compound has fungicidal or fungistatic activity, studying the speed of response and rate of killing, checking out the effects of pH, serum etc. on antifungal activity, and checking the response of large numbers of recent clinical isolates to see if they behave in the same way as the laboratory strains used in screening. Depending on the nature of the compound, cross-resistance studies and attempts to develop drug-resistant lines by repeated sub-culture in the presence of sub-inhibitory concentrations of drug may be in order. Considerable experience in the azole field has jaundiced our regard for in-vitro studies in that entirely different answers can be obtained as culture conditions are varied, and individual studies often end up asking more questions than they were originally designed to answer.

Evaluation in vivo

Having discovered useful activity *in vivo* at the screening stage, evaluation experiments will have two broad aims in view: studying the same conditions in different species of host, realising that extrapolation to the human situation is the eventual goal, and studying other fungal conditions in relevant animal models to determine the spectrum of activity of the compound. Vaginal infections with *Candida albicans* may be readily produced in rats following removal of the ovaries and treatment every 2 weeks with 0.1 mg oestradiol undecylate (Progynon: Schering) given intra-muscularly; we have been unable, however, to establish vaginal infections in guinea pigs, rabbits, dogs or marmosets following a variety of surgical and chemical manipulations. With the more active azoles – e.g. ICI 153066 – potency in the rat is about 10 times that in the mouse ($0.1\,mg\,kg^{-1}$ as opposed to $1\,mg\,kg^{-1}$), but with less active ones – e.g. BAY N 7133 – this improvement of activity in the rat does not seem to occur (active in both at only $100\,mg\,kg^{-1}$). The mouse and rat models may be used to compare treatment of established infections (therapy) as opposed to treatment from the time of inoculation (prophylaxis), and to compare the relationship between the size of dose and number of doses required to eliminate infection; a single shot treatment of vaginal candidosis must be the ultimate aim. Table 4.4 gives minimal effective doses for single or multidose treatment of vaginal candidosis in mice and rats. These models may also be used to investigate the efficacy of topical formulations of the compound. We formulate in a mixture of polyethylene glycols for initial studies, and treat established infections twice daily for 5 d before looking at shorter regimens of treatment. Rats and mice behaving as they do, we arc never, however, sure we are studying topical rather than systemic activity; it is not clear just how much compound they ingest by licking themselves and each other!

Ringworm infections can be studied in normal guinea pigs, rabbits, dogs and monkeys. We scarify and inoculate in one operation using a piece of wire file-cleaning brush $2.5\,cm^2$, dipped in the spore suspension. Some strains of *Trichophyton mentagrophytes* and most strains of *Microsporum canis* seem to take well in these animals, although lesions with the latter species in the guinea pigs are usually not confluent and are more difficult to score. Although rats and marmosets may be infected, the severity of the infection is insufficient for clear-cut studies with drugs. *Trichophyton rubrum*, currently the most common dermatophyte in humans, will not produce infections in experimental animals in our hands. *T. mentagrophytes* in the guinea pig is something like 10 times more susceptible to griseofulvin than *Trichophyton quinckeanum* in the mouse; the difference is partly due to a greater basic susceptibility of *T. mentagrophytes* and partly due to the different pharmacokinetic behaviour of griseofulvin in the two hosts. *T. mentagrophytes* in the guinea pig is, if anything, less susceptible to ketoconazole than *T. quinckeanum* in the mouse (Table 4.4).

Ringworm lesions in the guinea pig may be used for studies of topical activity using compound formulated in a polyethylene glycol mixture applied twice daily, and following the rate of healing of lesions compared to animals treated with vehicle only. Again there is the problem of the animals licking themselves and ingesting compound, and the added complication that rubbing even unmedicated vehicle into the lesions twice daily modifies the appearance of the lesion somewhat and makes scoring more difficult.

Experimental ringworm infections in all species are self-limiting, recovery varying from 2 weeks or less with *Trichophyton quinckeanum* in the mouse to 6 or 7 weeks with *Trichophyton mentagrophytes* in the monkey. Drug-evaluation studies are not therefore concerned with a cure/no cure situation, but rather a modification in the rate of spontaneous healing. In general, the effectiveness of a compound in speeding up the rate of healing depends on how early in the course of infection treatment is started. With initial studies in guinea pigs, we start treatment three days after inoculation – at a time when hyphal growth in the keratin layers of the skin is well established and migration into the hair follicles is starting, but before macroscopic signs of infection are evident. Figure 4.2 illustrates the effects of daily doses of 5 mg kg^{-1} ICI 153 066 or 75 mg kg^{-1} ketoconazole on the course of infection with *T. mentagrophytes*, in guinea pigs when treatment was started on days 0, 3, 5 or 7 of the infection; the figures given in Table 4.4 are doses which give a comparable and reasonable degree of control of the infection.

Mice inoculated intravenously with *Candida albicans* provide a model – but only a model – of systemic infection. Shortly after inoculation, organisms can be recovered from a variety of internal organs but they only persist for any length of time and multiply in the kidneys – where they may produce abcesses and may cause death. A high inoculum will kill mice overnight, whereas a low inoculum may produce no mortality over a period of many weeks. Although response to medication can be

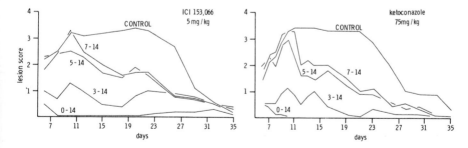

Fig. 4.2. Treatment of *Trichophyton mentagrophytes* infections of guinea pigs with ICI 153 066 and ketoconazole. Figures on graphs indicate period (days) of treatment.

followed by sacrificing animals and determining the number of organisms in the kidney by homogenisation, dilution and plating out (or alternatively by chitin assay), the easiest parameter to follow is mortality. In order to get meaningful results it is essential to adjust the inoculum carefully, by preliminary titration experiments, so that the majority of untreated animals die during the period 2–8 d after inoculation. It is also necessary to start treatment within 1–2 h of inoculation, or even to give the first dose before inoculation. Treatment may be given daily for 10 d or longer, and animals should be kept to follow mortality patterns for at least 5 weeks. Ten daily oral doses of 250 mg kg^{-1} 5-FC or 25 mg kg^{-1} amphotericin B will keep most mice alive for at least 6 weeks, whereas lower or fewer doses prolong survival, but do not give complete protection. Azole drugs such as ketoconazole or ICI 153 066 will extend survival significantly in this model, but will not give permanent protection even after as many as 30 daily doses at the maximum tolerated level. The figures given in Table 4.4 are minimal daily doses which will keep all mice alive during the 10 d treatment period. At these doses most mice will die over the next 3–4 weeks, but at least the figures provide a basis for comparison between compounds. The difficulty with models such as this is to relate experimental findings to the likely utility of the drug in human situations.

Some of the models for other deep-seated systemic mycoses such as cryptococcosis, histoplasmosis, aspergillosis, blastomycosis and paracoccidioidomycosis are based on similar intravenous infections in mice, while models for coccidioidomycosis utilise intranasal inoculation of mice with arthrospores; it is beyond the scope of our experience to discuss these further.

Phytopathogens

Target selection

As in the case of pharmaceutical products, targets for research on agricultural fungicides are determined by economic issues. This means that priority is given to looking for compounds to control diseases that affect crops grown in, or exported to, Western Europe, Japan and the USA. These are generally diseases of the aerial parts of cereals and rice, vegetables, apples and vines. Fortunately the compounds that are effective against the wealthy farmer's foliage/fruit disease problems are usually active against those of the poor farmer. Therefore the technology that is available to the former will for the most part be readily transferable to the latter, given the political and economic will.

A number of important diseases in agriculture have proved difficult to control with chemicals. In general these are caused either by fungi which attack parts of the plant (roots or vascular tissue) remote from the foliage, by fungi that are soil-borne, or by fungi that are too deep-seated to be contacted even by systemic fungicides that may have intrinsic activity against them. Examples are mal-secco of citrus (*Phoma tracheiphila*), *Verticillium* wilt of crops such as cotton, witches' broom of cocoa (*Crinipellis perniciosus*) and root-rot caused by *Armillaria* spp. Partly because of the dearth of active compounds, these diseases are not recognised as major targets and do not often appear in agricultural screens. There are practical reasons for this too. Screening for agricultural fungicides is generally carried out on the plant since the throughput limitations of in-vivo screening in animals do not apply. Producing and caring for large numbers of experimental plants is relatively easy, and offers no ethical problems. Nevertheless, simple tests are required for high-throughput screening, and the ones for foliage diseases are easier to carry out than a test for, say, witches' broom, or eyespot of cereals (*Pseudocercosporella herpotrichoides*). Thus a primary screen is likely to resemble our own at Jealott's Hill, comprising:

(i) a powdery mildew (Ascomycete): *Erysiphe graminis* on barley – closely related to *Podosphaera leucotricha* (apple powdery mildew), *Uncinula necator* (grapevine powdery mildew) and *Sphaerotheca* spp. (vegetable powdery mildew),

(ii) a rust (Basidiomycete): *Puccinia recondita* on wheat – closely related to *Puccinia* spp. (other cereal rusts) and *Hemileia vastatrix* (coffee rust) and also related to *Tilletia* and *Ustilago* spp. (bunts and smuts of cereals and maize etc.);

(iii) a cercospora leaf spot (Deuteromycete): *Cercospora arachidicola* on peanut – related to *Cercospora beticola* (sugar-beet leaf-spot) and *Mycosphaerella musicola* (banana sigatoka);

(iv) an oomycete (Mastigomycete): *Plasmopara viticola* on grapevine – related to *Phytophthora infestans* (potato blight);

(v) rice blast (Deuteromycete): *Pyricularia oryzae*;

(vi) apple scab (Ascomycete): *Venturia inaequalis*;

(vii) grey mould (Deuteromycete): *Botrytis cinerea* on grapes.

The screen also contains representative seed and soil-borne fungi and a phytopathogenic bacterium.

Hence each major subdivision of the phytopathogenic fungi is represented, usually twice, in the screen by pathogens of economic importance. The first four organisms in the list are representative of groups of diseases which usually show a similar response to chemicals; a compound that is active against one powdery mildew is likely to be active against all powdery mildews. The representative chosen is usually the most economically important member of its group. The last three organisms in the list lie outside these clear 'affinity groups', but are included in the screen because of their economic importance. This is illustrated by the rice blast, *Pyricularia oryzae*: this pathogen causes the greatest economic loss of any disease in agriculture, but the chemicals that are now available or in development for its control are practically ineffective against other diseases. *Pyricularia* is therefore an indispensable component of any screen designed to detect novel toxophores.

Screening methods

Examples of the basic methods used to test the efficacy of candidate-agricultural fungicides against foliage diseases are now given. Many of these methods were developed by M. C. Shephard and A. M. Cole at Jealott's Hill.

Test plants are grown from seed in 4 cm diameter plant pots and maintained either in a greenhouse with additional artificial light or in a growth room (16 h day; day temperature 24–26 °C, night temperature 18–20 °C). When they have reached the age indicated in Table 4.5 they are treated with a solution or suspension of the test chemical obtained by shaking the compound in water containing 2% Dispersol T surfactant (ICI Organics Division, Blackley, Manchester M9 3DA) with glass beads for 20 min on a paint shaker. The suspension is diluted to the required concentration and immediately sprayed onto the foliage of the test plants with a DeVilbiss spray gun. Spray is applied to maximum leaf retention, i.e. until just before run-off; the weight of spray solution retained by the leaves of test plants under these conditions is at least equal to the leaf weight. Tween 20 surfactant at a final concentration of 0.5% is added to the suspension when cereal and rice plants are sprayed, in order to aid retention. Chemicals may also be applied to the roots via the soil in order to test their ability to move from roots to leaves in the xylem. This testing is done by standing the pots containing the plants in a solution or suspension of the chemical of appropriate strength.

Suspensions of spores of fungal pathogens are made from cultures grown on stock plants or *in vitro* as appropriate. *Pyricularia oryzae* and *Botrytis cinerea* are grown on 5% malt agar at 26°C and 19°C respectively, and *Cercospora arachidicola* on a peanut-leaf-extract/glucose/yeast-extract agar at 26 °C; other fungal pathogens are maintained on stock plants. Propagules of the pathogens are suspended in water, counted in a haemocytometer, and the suspensions adjusted to give the concentrations shown in Table 4.5. Spores of *Erysiphe graminis* are suspended at $1 \, mg \, ml^{-1}$ in perfluorotributylamine (3M Commercial Chemicals, St Paul, Minnesota 55101, USA) as described by Bushnell & Rowell (1967). Plants are inoculated, generally 1–2 d after chemical treatment, by spraying suspensions of fungal spores from a DeVilbiss spray gun onto the foliage. After a period of up to 24 h in a humidity cabinet (100% relative humidity, 18 °C; this step is omitted for *E. graminis*) the plants are returned to the greenhouse or growth room to await development of symptoms. After sufficient time for this (as indicated in Table 4.5) the area of leaf covered by lesions is estimated as a percentage.

In the initial screen, the test substance is applied to both foliage and roots at $100 \, \mu g \, ml^{-1}$. This level is considerably higher, for most diseases, than that required by contemporary agricultural fungicides. Diclobutrazol, for example, requires $5–10 \, \mu g \, ml^{-1}$ to control the wheat rust *Puccinia recondita*. If good activity (95% disease control or greater) is detected at $100 \, \mu g \, ml^{-1}$, root-drench and foliage-spray tests are carried out separately at a dose rate close to the highest commercial dose (HCD) for that disease. The HCD is the level at which an inexpensive compound would need to be active to be as cost-effective as the best treatment currently available to the farmer. If such a compound requires a higher dose than the HCD to be active, it is unlikely to be competitive. For many diseases the HCD is about $10 \, \mu g \, ml^{-1}$. However, for diseases where few treatments are as yet available to the farmer (soil diseases) or where the existing compounds are not very active (*Pyricularia oryzae, Botrytis cinerea*) the HCD rate is set at $100 \, \mu g \, ml^{-1}$. If activity is confirmed on re-test at the HCD, the compound passes on to evaluation tests, which use broadly similar methods to those described above, but which have somewhat different purposes.

The level of activity of the compound can be determined by testing it at ever-decreasing doses until the degree of control falls below 95% ('breakpoint'). Comparison of the breakpoint with that of pesticides of known field performance enables an appropriate dose of the candidate chemical for field trials against a particular disease to be estimated. Before a decision is made to field test a compound, however, its systemic properties are evaluated in detail in order to characterise its likely performance and novelty. Thus, if the compound is active in

Table 4.5. *Summary of experimental procedures for screening against phytopathogenic fungi*

Host plant	Age at treatment (d)	Pathogen	Inoculum concentration (number ml^{-1})	Time from inoculation to assessment (d)
Apple	7	*Venturia inaequalis*	2×10^5	12
Wheat	9	*Puccinia recondita*	2×10^5	9
Barley	6	*Erysiphe graminis*	$(1 \, \mathrm{mg} \, \mathrm{ml}^{-1})$	6
Rice	11	*Pyricularia oryzae*	1×10^5	6
Tomato	17	*Botrytis cinerea*[a]	1×10^{5} [b]	5
Peanut	9	*Cercospora arachidicola*	3×10^5	12
Grapevine	28	*Plasmopara viticola*	5×10^4	7

[a] Test now superseded by a grape berry test
[b] 1% sucrose added to spore suspension

root-drench tests, it must have the capacity to move through the xylem to give control of a foliage disease. Tests for eradicant action are then carried out, using methods similar to those described above, but with inoculation preceding chemical treatment by up to 6 d or more. Tests for phloem mobility may also be carried out. These are more time-consuming and can only be done rigorously on a minority of chemicals. All the tests are used to build up a profile of the activity and systemic properties of the compound compared with materials already available to the farmer, and help to make decisions on whether field testing is worthwhile.

Evaluation – a search for novelty and technical improvement

Many of the diseases that are included in the screen can be controlled, often well controlled, by compounds already available to the farmer. There is, however, a need to discover new materials in order to overcome problems that beset the use of compounds presently available (Rathmell, 1983). Apart from low efficacy against some diseases, these problems include unwanted side effects, fungicide resistance and technical difficulties associated with spraying. Resistance amongst phytopathogenic fungi, especially to systemic fungicides, makes it necessary that new compounds with different biochemical modes-of-action are made available to the farmer. Alternation of the different fungicides will then delay, or at least ameliorate the effects of pesticide resistance. Spraying, the method by which most agricultural fungicides are applied, is an inefficient method of bringing compounds into contact with target fungi (Graham-Bryce, 1977), and compounds having properties that either reduce or eliminate the need for application by this means, or enable its efficiency to be increased, are well worth finding (Rathmell & Skidmore, 1982). High levels of activity are likely to give rise, especially in those diseases where large doses of compound are currently needed (*Botrytis cinerea* and *Pyricularia oryzae*), to improved cost-efficiency. High potency also allows the use of advanced spraying systems (Rathmell & Skidmore, 1982), such as the electrodynamic one (Coffee, 1981).

Enhanced ability to enter and move within plant tissues will lead to compounds that can be used more efficiently, either as sprays or as seed treatments for the control of foliage diseases (Rathmell, 1983). For example, if a chemical has a curative action against apple scab (*Venturia inaequalis*), because the fungicide penetrates the plant's tissues and reaches the fungal mycelium growing within, this provides the grower with a particularly flexible method of disease control. The chemical does not have to be applied prophylactically; meteorological data (Mills' periods) can be used as a guide to spray timing, so that fewer applications are needed (French, McClellan & Shephard, 1983). Ability to move in the phloem (symplast) of plants, rarely found even amongst the systemic fungicides, could also reduce the number of treatments needed in orchards and plantations by permitting transport of pesticide from one leaf to another. The pesticide would move from the leaves present at the time of spraying to the new growth (which is often most susceptible to disease); the period of control would thereby be prolonged. Phloem-mobile fungicides applied to the foliage might also move to the crown or roots of infected plants and be exuded in the soil, thus controlling vascular, root-infecting and soil-borne pathogens for which few chemical control measures are as yet available.

The requirement in evaluation, then, is to take toxophores found active against the important diseases in the screen and find which of them are more advanced than, or different from, existing materials in respect of level and spectrum of activity, mode of action, and/or systemic properties. The technical advances being sought are not always easy to detect in high-throughput screens; an attempt is therefore made to get an early indication during evaluation that a toxophore, as well as being active, has some novelty. This information can then be given to fungicide chemists to aid them in their structure–activity studies. Hence, candidate compounds are applied in these tests, both as root-drenches – to detect systemic movement to the leaves – and as foliage-sprays. Low-rate testing is done as quickly as possible after activity has first been detected, in order to determine whether the compound has higher activity than materials known so far. Testing is often done with strains of fungi resistant to known toxophores, and thus insensitive (usually) to compounds with the same mode of action. Some tests inevitably take longer, and cannot be incorporated in the early screens; the tests for phloem mobility or for usefulness as a cereal-seed dressing belong to this category. However a rapid, but relatively imprecise, phloem test has been devised for use immediately after activity of a compound has been detected on *Plasmopara viticola* and *Erysiphe graminis*. Standard root-drench tests give a rough indication of the likely activity of a candidate chemical as a seed-dressing for the control of foliage diseases.

Field screening

The methods described above for artificially inoculating test plants may be adapted for field use on some crops for the first year's outside testing of a candidate compound. In the early stages of field testing, supplies of compound are too scarce for large-plot trials on farms, so small-plot trials are carried out, with inoculation to ensure adequate disease challenge and therefore a guaranteed result. P. N. French and R. Abbott at Jealott's Hill have developed methods for obtaining disease with a high level of reliability in cereal trials.

The trial is sited in a sheltered field – the trial area may be surrounded by a closely webbed fence – and drilled to a high plant density with a variety susceptible to the chosen disease. Nitrogen fertiliser is applied in excess of normal amounts (230–80 units over the growing season). These practices ensure that the environment within the crop is as stable as possible and that there is a large amount of actively growing soft-leaf tissue which is at a suitable stage for infection. Trial plots are interspersed by areas of untreated crop ('guard areas') to ensure that epidemic build-up is as even as possible throughout the trial. Situations of water stress are avoided by using well-drained soils to prevent waterlogging, and by irrigating in times of drought. Unwanted diseases are controlled by selective fungicides wherever possible. Fungal inoculum is prepared by large-scale versions of methods already described and applied when weather conditions are most favourable for the disease. Application may be by spraying or by placing infected plants amongst the plots in the 'guard areas'. Control of the humidity of the microclimate within the crop is achieved by extending natural dew periods with overhead irrigation or misting with small water droplets. Most diseases require some humidity to develop and some require free water on the leaf surface. Accurate application of small quantities of chemicals in the field may be achieved by hand-held boom sprayers or by mechanised plot sprayers such as the one described by French (1980). If a compound continues to show promising activity in such field screens it will go on to field evaluation, a topic beyond the scope of this chapter.

Conclusions

Although something can be done to prevent or treat the majority of fungal diseases of man and the crops on which he relies for food and a livelihood, there is always room for improvement. In both the medical and agricultural fields, compounds with a greater spectrum of activity simplify matters from the user's point of view, compounds with greater potency can simplify the method of administration, while compounds with systemic rather than topical only activity add a new dimension to disease control. We need new antifungal agents, however, not just because control of a few pathogens is poor or methods of application are not ideal; fungi, especially in the agricultural field, are learning to live with the present-day fungicides, and a constant supply of new agents with new modes of action is needed to deal with the problems of resistance.

The rational design of new antifungal agents is fraught with difficulties. Not only is it necessary to devise chemicals which exploit some subtle metabolic difference between host and parasite to achieve selective toxicity, consideration has also to be given to factors governing absorption, distribution and persistence – or lack of it – in the host, and the avoidance of more subtle, long-term toxicological and environmental problems such as teratogenicity and liver toxicity (for pharmaceutical products), and contamination of and persistence in the soil (for agricultural agents). This being the case, rational drug design will be modulated with empiricism, and we must expect to have to synthesise, test and evaluate large numbers of compounds before real success is achieved. We have described high-throughput screens to detect antifungal activity and support such a programme of chemical synthesis. Simplifications, short-cuts and compromises characterise such screens in order to make possible the testing of large numbers of compounds with the minimum of time, effort and cost. Once having detected antifungal activity, there is considerable scope for flair in evaluating that activity, and in detecting unique properties in a particular compound which may be exploited in novelty of usage, application and type of antifungal control. We strive for perfection – but we live in a real world. Although our sights may change as new competitive products emerge, fungicide resistance and the possibility of doing even better will spur us on for a long while yet.

References

Bushnell, W. R. & Rowell, J. B. (1967). Fluorochemical liquid as a carrier for spores of *Erysiphe graminis* and *Puccinia graminis*. *Plant Disease Reporter*, **51**, 447–8.
Coffee, R. A. (1981). Electrodynamic crop spraying. *Outlook on Agriculture*, **10**, 350–6.
French, P. N. (1980). A mechanised field sprayer for small-plot pesticide trials. In *Proceedings of the Fifth International Conference on Mechanisation of Field Experiments*, ed. J. Drijkstra & A. van Santen, pp. 135–9. Wageningen, Netherlands: PUDOC.
French, P. N., McClellan, W. D. & Shephard, M. C. (1983). Evaluation of an eradicant fungicide against apple scab and coffee rust. In *Proceedings of the 10th International Congress of Plant Protection*, Brighton, UK, Nov. 1983, p. 1008. London: British Crop Protection Council.

Graham-Bryce, I. J. (1977). Crop protection: a consideration of the effectiveness and disadvantages of current methods and of the scope for improvement. *Philosophical Transactions of the Royal Society*, **281B**, 163–79.

Heeres, J., Backx, L. J. J., Mostmans, J. H. & Van Cutsem, J. (1979). Antimycotic imidazoles. Part 4. Synthesis and antifungal activity of ketoconazole, a new potent orally active broad-spectrum antifungal agent. *Journal of Medicinal Chemistry*, **22**, 1003–5.

Rathmell, W. G. (1983). The discovery of new methods of chemical disease control: current developments, future prospects and the role of biochemical and physiological research. *Advances in Plant Pathology*, **2**, 259–88.

Rathmell, W. G. & Skidmore, A. M. (1982). Recent advances in the chemical control of cereal rust diseases. *Outlook on Agriculture*, **11**, 350–6.

Ryley, J. F., Wilson, R. G. & Barrett-Bee, K. J. (1984). Azole resistance in *Candida albicans*. *Sabouraudia*, **22**, 53–63.

Ryley, J. F., Wilson, R. G., Gravestock, M. B. & Poyser, J. P. (1981). Experimental approaches to antifungal chemotherapy. *Advances in Pharmacology and Chemotherapy*, **18**, 49–176.

Shadomy, S., Dixon, D. M. & May, Y. (1982). A comparison of bifonazole (Bay H 4502) with clotrimazole *in vitro*. *Sabouraudia*, **20**, 313–23.

Thienpont, D., Van Cutsem, J., Van Gerven, F., Heeres, J. & Janssen, P. A. J. (1979). Ketoconazole – a new broad spectrum orally active antimycotic. *Experientia*, **35**, 606–7.

5

Development of resistance to antifungal agents

J. DEKKER

Laboratory of Phytopathology, Agricultural University, Wageningen, The Netherlands

Introduction

When a fungus which is sensitive to a particular fungicide becomes less sensitive due to stable heritable changes occurring following exposure to the compound, this is called fungicide resistance. Development of fungicide resistance in plant pathogens causes considerable problems in agriculture.

Cases of disease-control failure due to fungicide resistance occurred particularly after the introduction of systemic fungicides about 15 years ago. In contrast, hardly any such problems have arisen during a century of application of conventional fungicides like copper compounds and dithiocarbamates. This difference seems to be related to the modes of action of these two groups of fungicide. Most conventional fungicides react with thiol groups present in enzymes and they interfere with numerous metabolic processes. These so-called multi-site inhibitors are therefore general plasma-toxicants and they are only suitable as surface protectants; fungicides which react at many sites may be phytotoxic when they penetrate plant cells and may hamper transport within the plant. Systemic fungicides, however, appear to react with one site, or only a few specific sites, in the cell. They may inhibit the synthesis of protein, ergosterol or chitin, or they may block mitosis, or they may interfere with respiration, membrane function, etc. Although all systemic fungicides are presumably site-specific inhibitors, the reverse is not always true since some site-specific fungicides are not taken up by plants.

There is not only a difference between multi-site and site-specific inhibitors with respect to development of fungicide resistance but, within the latter category, there is considerable variation in the risk of development of fungicide resistance.

This chapter discusses if a relationship exists between the mode of action of a fungicide and the mechanism of resistance. Attention will be paid to factors which determine whether or not the emergence of fungicide resistance leads to a failure of disease control, with emphasis on the fitness of resistant strains. Finally, the phenomena of positively and negatively correlated cross resistance will be considered.

Mechanisms of resistance
General

Fungi may become resistant to fungicides by the following mechanisms (Dekker, 1977; Georgopoulos, 1977):

  (i) a change at the site of inhibitor action which results in a decreased affinity to the fungicide;
 (ii) decreased uptake or decreased accumulation of the fungicide in the fungus;
(iii) detoxification of the fungicide before the site of action has been reached or lack of conversion of a compound into thc fungitoxic principle;
 (iv) compensation for the inhibitory effect, e.g. by an increased production of an inhibited enzyme;
  (v) circumvention of the blocked site by the operation of an alternate pathway.

Often such changes may be brought about spontaneously by the mutation of a single gene. In laboratory experiments, ultra-violet radiation or mutagenic chemicals may be used to increase the mutation frequency.

In the case of site-specific fungicides, a slight change at the target site may result in reduced affinity to the fungicide, and, as a consequence, decreased sensitivity. It is obvious that resistance to multi-site inhibitors cannot arise in this manner, as it would require changes at many sites. Resistance mechanisms (ii) and (iii) above may, in principle, effect sensitivity to systemic as well as to conventional fungicides. However, the resistance obtained in this way to conventional fungicides is usually not very high, and has only rarely led to problems in practice. An example of such a problem is the resistance of *Pyrenophora avenae* to organic mercury; this developed after many years of treating oat seeds with this fungicide (Noble *et al.*, 1966). Resistance mechanisms (iv) and (v) are seldom found.

Our present knowledge about the mechanisms of fungal resistance to different fungicides will be discussed.

Carboxamides

Of this group of systemic fungicides, carboxin and oxycarboxin have been most studied (Kuhn, Chapter 8, this volume). Carboxin is used as a seed treatment for the control of loose smut of cereals (which is caused by a pathogen present in the interior of the seed). Oxycarboxin is used as a spray treatment against rust diseases. These compounds mainly affect Basidiomycetes, and the site of action lies somewhere between succinate and coenzyme Q in the complex II region of the respiratory chain. Carboxin binds to the nonhaem–iron–sulphur protein associated with this enzyme complex (Lyr, 1977).

In laboratory experiments, carboxin-resistant strains of various *Ustilago* species have been obtained. Resistance in *U. maydis* brought about by two mutations at the *oxr-1* locus is thought to be caused by a slight change at the target site in the succinate dehydrogenase complex (Georgopoulos, 1982*b*). In this fungus, a second type of resistant mutant has been found; these lack the alternative electron transport system and, as a consequence, they are more sensitive to cyanide and antimycin A than the wild type strain (Georgopoulos & Sisler, 1970; Fig. 5.1).

Resistant strains with a mutation at the *oxr-1* locus are also more or less resistant to related compounds, such as oxycarboxin and pyracarbolid. Exceptions do occur, however, and with some carboxin analogues even negatively correlated cross-resistance has been observed

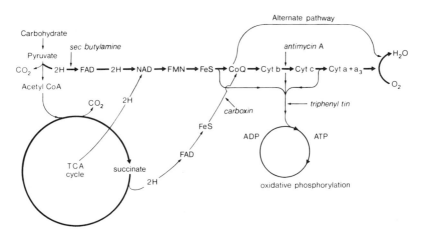

Fig. 5.1. Target sites of the respiration inhibitors carboxin, sec butylamine, triphenyl tin and antimycin A.

Table 5.1. *Carbendazim-binding activity in mycelial extracts of Aspergillus nidulans and Penicillium brevicompactum[a]*

		Binding
A. nidulans 003	(wild type)	5640
A. nidulans R	(resistant)	320
A. nidulans 186	(extra sensitive)	12.200
P. brevicompactur	(sensitive)	.4140
P. brevicompactu	(resistant)	220

After Davidse,
[a]Figures repre sent dpm ^{14}C-carbendazim per ml extr act.

(Georgopoulos, 198 c). Oxycarboxin-resistant strains of *Puccinia chrysanthemi* have been found in chrysanthemum green houses.

Benzimidazoles and thiophanates

These fungicides have a rather wide spectrum of antifungal activity, but are not active against Oomycetes. Both benomyl and thiophanate methyl are converted into the actual toxic principle, carbendazim. Davidse (1982) demonstrated that carbendazim prevents mitosis in *Aspergillus nidulans* by binding to tubulin subunits of the mitotic spindle.

In laboratory experiments with *Aspergillus nidulans*, carbendazim-resistant mutants were easily obtained, either spontaneously or after mutagenic treatment (Van Tuyl, 1977). Van Tuyl (1977) found high- and low-level resistance. The former was based on mutation of gene *ben A* on linkage group (chromosome) VIII, and the latter was based on mutations of at least two genes, namely *ben B* and *ben C* on linkage groups II and VIII respectively. The mechanism of resistance has been studied in *A. nidulans* and *Penicillium* species. A positive correlation was found between carbendazim sensitivity and affinity of carbendazim to the tubulin protein of fungi (Davidse, 1982; Table 5.1). The mode of action of these fungicides is discussed by Burland & Gull (Chapter 14, this volume).

Strains of numerous pathogens, which have become resistant to benzimidazoles and thiophanates have been isolated from the field. As a rule there is cross-resistance among fungicides belonging to this group.

Hydroxypyrimidines

These fungicides are only active against powdery mildews, so that no in-vitro studies can be carried out. Ethirimol has been reported to interfere with adenosine deaminase in *Erysiphe graminis* during the first phase of infection of barley plants (Hollomon, Chapter 9, this volume).

Strains with decreased ethirimol sensitivity have been found in barley fields, where the seeds or the plants had previously been treated with this fungicide. The adenosine deaminase of these strains did not differ from that of sensitive strains (Hollomon, 1979), and the mechanism of resistance has not yet been elucidated. Resistance is probably not controlled by a single gene, but by a complex, hereditable system in which the effects of the genes are thought to be additive (Hollomon, 1981).

Acylalanines

These fungicides are selectively active against Oomycetes. Davidse (Chapter 11, this volume) found that metalaxyl acts as a specific inhibitor of one or more of the RNA polymerases of *Phytophthora megasperma* f.sp. *medicaginis*. After irradiation of zoospores with ultra-violet (UV) or treatment with *N*-methyl-*N'*-nitro-*N*-nitrosoguanidine, a considerable number of metalaxyl-resistant strains were obtained (Davidse, 1981). Bruin & Edgington (1982) also obtained metalaxyl-resistant strains of *Phytophthora capsici* and *Pythium splendens* after UV-treatment of zoospores. Metalaxyl-resistant strains of various Oomycetes have also been obtained from various field crops. Davidse (Chapter 11, this volume) has obtained evidence that resistance is due to a change at the target site.

Dicarboximides

Fungicides belonging to this group are primarily active against diseases caused by species of *Botrytis*, *Sclerotinia* and *Monilia*. The systemic properties of these fungicides, when present, are not very pronounced. The dicarboximides cause mitotic instability in diploids of *Aspergillus nidulans*, but their mode of action is still a subject of discussion (Beever & Byrde, 1982; Leroux & Fritz, Chapter 10, this volume).

Strains resistant to iprodione, vinclozolin or procymidone were easily obtained in laboratory experiments with *Botrytis cinerea*, *Sclerotinia fructicola* and several other fungi, and strains of *B. cinerea* with decreased fungicide sensitivity have also been isolated from vineyards. The mechanism of resistance to dicarboximides is still obscure.

According to Pappas, Cooke & Jordan, (1979), resistance cannot be attributed to reduced uptake of the fungicide by the pathogen.

Aromatic hydrocarbons

Fungicides considered to belong to this structurally rather variable group (Leroux & Fritz, Chapter 11, this volume) include quintozene, *O*-phenylphenol, diphenyl, hexachlorobenzene, dicloran and chloroneb, most of which are not systemic and are active against various plant pathogens. Their mode of action is not yet clear, and it is not even known if it is similar for all members of the group. As aromatic hydrocarbons cause mitotic instability, and since cross-resistance has been observed between some hydrocarbons and dicarboximides, there might be some similarity between the modes of action of these two groups of fungicide.

Mutants resistant to aromatic hydrocarbons have frequently been obtained in laboratory experiments and have also given problems in the field (Georgopoulos, 1977). The mechanism of resistance to these fungicides has not yet been elucidated. No difference in uptake or breakdown of chloroneb was observed between sensitive and resistant strains of *Ustilago maydis* (Tillman & Sisler, 1973).

Organophosphates

The compounds edifenphos and 5-benzyl *O,O*-diisopropyl phosphorothiolate (IBP) are used to control rice blast caused by *Pyricularia oryzae*. They interfere with the biosynthesis of phosphatidylcholine, an important constituent of the cell membrane (Kodama, Yamashita & Akatsuka, 1980). Isoprothiolane, which lacks phosphorous in the molecule, seems to act in the same way. Mutants of *P. oryzae*, resistant to the three fungicides mentioned above, were found in laboratory experiments; after several years of fungicide use, these could also be isolated from the field. Strains were obtained which had moderate or a relatively high level of resistance to IBP. The former type of resistance is probably due to detoxification of the fungicide by cleavage of the S–C bond (Uesugi & Sisler, 1978). The highly resistant strains, however, did not detoxify IBP any more than did the wild type strain, which means that another, as yet unknown mechanism of resistance is involved.

Among the organophosphate fungicides, pyrazophos is used against powdery mildew diseases, although a few other plant pathogens are also sensitive to this fungicide. In studies with *Pyricularia oryzae* De Waard

(1974) observed that a metabolite of pyrazophos is the fungitoxic principle, namely 2-hydroxy-5-methyl-6-ethoxycarbonylpyrazolo-(1,5α)pyrimidine (PP). The latter compound, but not pyrazophos, affects respiration, nucleic acid and protein synthesis, but the primary site of action is not yet known. In laboratory experiments, strains of *P. oryzae* were obtained which were resistant to pyrazophos. These strains were unable to convert pyrazophos into PP and this was presumed to be the basis of the resistance mechanism (De Waard & Van Nistelrooy, 1980*b*). Isolates of *Sphaerotheca* with reduced sensitivity to pyrazophos have been obtained from cucumber greenhouses (Dekker & Gielink, 1979*b*).

Inhibitors of sterol biosynthesis

Several groups of structurally unrelated compounds interfere with fungal growth by inhibiting sterol biosynthesis. These fungicides include imidazoles, triazoles, pyrimidines, a piperazine and a pyridine, all of which inhibit C-14 demethylation, and morpholines, which inhibit Δ^8–Δ^7 isomerisation in the pathway of ergosterol biosynthesis (Sisler & Ragsdale, Chapter 12, this volume). The spectrum of the antifungal activity of these compounds varies from narrow to wide, but all are inactive against Oomycetes since these fungi lack ergosterol. Some compounds are only active against powdery mildews, but others also inhibit many other fungi.

In laboratory experiments, strains of various fungi have been obtained which have reduced sensitivity to the piperazine triforine, to imidazoles, triazoles and pyrimidines. Strains with increased fungicide resistance have also been obtained in the field, but only few cases of disease control failure have been reported.

The mechanism of resistance has been studied extensively by De Waard & Van Nistelrooy (1979, 1980*a*), using fenarimol and *Aspergillus nidulans*. There was no evidence that resistance involves a change at the target site. However, after incubating mycelia, in a solution of ^{14}C-labelled fenarimol, there was less accumulation of the fungicide in resistant than in the sensitive strain; this difference did not appear to be due to reduced permeability of the resistant strain. It was shown that accumulation of fenarimol in the fungus resulted from the net difference between passive influx and energy-dependent efflux of the fungicide. In sensitive cells, the efflux mechanism was not constitutive but was gradually induced after addition of the fungicide. This resulted in a rather large accumulation of fenarimol in the fungus during the first half hour of the experiment. In the resistant cells, however, the energy-dependent efflux mechanism is constitutive and, thus, fully operative from the start of the experiment. In this strain, the rate of fenarimol efflux equals the rate of influx and therefore fenarimol does not accumulate to any appreciable extent (Fig. 5.2). That the efflux mechanism is energy dependent is shown by the immediate steep increase in fenarimol accumulation when a respiration inhibitor, such as oligomycin, is added. Lack of accumulation of fenarimol may be responsible for resistance as, presumably, a high level of the fungicide is necessary to saturate the target site. Similar results were obtained with fenarimol-sensitive and -resistant strains of *Penicillium italicum* (M. A. De Waard, personal communication).

Studies by Siegel & Solel (1981) of imazalil resistance in *Aspergillus nidulans* indicate that binding of this fungicide in the cell involves a small number of high-affinity sites, probably sterol carrier proteins. Evidence was obtained that in this case increased efflux may also be responsible, at least in part, for resistance.

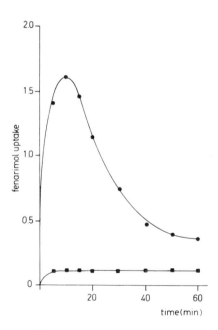

Fig. 5.2. Accumulation of fenarimol (30 μM) by mycelium of a wild type strain (filled circles), and a fenarimol-resistant mutant (filled squares) of *Aspergillus nidulans*. Uptake in nmol mg^{-1} dry weight (after De Waard & Van Nistelrooy, 1980*a*).

It still has not yet been established if the phenomenon of energy-dependent efflux is the only mechanism of resistance to fungicides which inhibit sterol biosynthesis.

Antibiotics

Several antifungal antibiotics are used to control plant diseases. Interesting data have been obtained about the development of resistance to these compounds, and about the mechanism of resistance.

Polyoxins. This group of antibiotics interferes with chitin synthesis in fungal cell walls, by inhibiting chitin synthase. They are inactive against non-chitinous fungi, such as the Oomycetes. Polyoxin B is used for control of black spot of Japanese pear, caused by *Alternaria kikuchiana*, and polyoxin D against rice sheath blight, caused by *Pellicularia sasakii*.

Strains of *Alternaria kikuchiana*, resistant to polyoxin B, were obtained in the laboratory and from orchards treated with the antibiotic. Chitin synthesis was much less inhibited by polyoxin B in the resistant strain than the sensitive strain (Nishimura, Kohmoto & Udagawa, 1973). Resistance is not due to a change at the target site, as the chitin synthase of resistant and sensitive strains was equally inhibited by the antibiotic. Resistance appeared to be correlated with a strongly reduced uptake of polyoxin B by the pathogen (Misato, Kakiki & Hori, 1977).

Griseofulvin. This antibiotic is active against powdery mildews and diseases caused by *Botrytis* species. In studies with *Basidiobolus ranarum*, Gull & Trinci (1973) observed that daughter nuclei did not separate during cell division and, shortly after treatment of *Aspergillus nidulans* with the antibiotic, spindle distortion was observed (Crackower, 1972). In view of these observations, griseofulvin is now considered to be an inhibitor of mitosis. Although it interferes with the functioning of microtubules, the site of action of griseofulvin is not identical with that of carbendazim and this conclusion is supported by the lack of cross-resistance.

Development of resistance to griseofulvin has been reported only in the dermatophyte, *Arthrodermi simii*. It depends on a single gene mutation (Darbord *et al.*, 1974), but the mechanism of resistance is still unknown. There are no reports of the development of griseofulvin resistance amongst plant pathogens. However, this antibiotic has been used to only a limited extent in plant protection.

Kasugamycin. This antibiotic is used against rice blast caused by *Pyricularia oryzae*. Kasugamycin inhibits protein synthesis in this fungus, by preventing binding of aminoacyl-tRNA to the ribosome (Siegel, 1977). Kasugamycin-resistant isolates have been obtained in laboratory experiments and from the field. In a cell-free system, protein synthesis was inhibited by the antibiotic when ribosomes from a sensitive strain were used, but not with those of a resistant strain. This result indicates that resistance is due to a change at the target site in the ribosome, which presumably decreases its affinity to the antibiotic (Siegel, 1977). Genetic analysis using the perfect stage of the rice blast pathogen, *Magnaporthe grisea*, showed that resistance is due to a single gene mutation, and that at least three loci for resistance exist (Taga *et al.*, 1978).

Blasticidin-S. This antibiotic, which inhibits growth of bacteria and a few fungi, has been used for control of rice blast. It has now been largely replaced by kasugamycin. Blasticidin-S is an inhibitor of protein synthesis which binds to the 50-S ribosomal subunit in bacteria and to the 60-S subunit in fungi. It inhibits peptidyl-transferase activity associated with chain elongation (Siegel, 1977).

In cell-free systems obtained from resistant strains of *Pyricularia oryzae*, protein synthesis was inhibited by the antibiotic to the same degree as in similar systems obtained from sensitive strains. Resistance is thus not caused by a change at the target site, but probably by reduced permeability of the cell membrane for the antibiotic (Misato, 1967).

Cycloheximide. This antibiotic is active against many fungi. It inhibits protein synthesis by interfering with the transfer of amino acids from aminoacyl-tRNAs to the nascent polypeptide chain. The 60-S subunit of the ribosome contains the binding site for the antibiotic. In cycloheximide-resistant mutants of *Saccharomyces cerevisiae*, eight loci for resistance were detected (Wilkie & Lee, 1965). Roa & Grollman (1967) showed that resistance to the antibiotic is based on a change in a single protein component of the 60-S unit of the ribosome. No resistance in plant pathogens has been reported, but practical use of this antibiotic has been very limited because of its phytotoxicity.

Other fungicides

Sec-butylamine. This fungicide, also known as 2-aminobutane is used for control of *Penicillium* rot of citrus fruits. Its primary site of action is pyruvate dehydrogenase (Fig. 5.1). Within a few years after its introduction, failure of disease control occurred due to the development of resistant strains (Eckert, 1982). It has not yet been established whether or not resistance depends on a change at the target site.

Organotin compounds. Triphenyl tin compounds are non-systemic compounds used to control a broad range of fungal diseases of plants. Their antifungal activity seems primarily due to specific interference with oxidative phosphorylation (Kaars Sijpesteijn, 1970 and Chapter 7, this volume; Fig. 5.1). Only in a single case has unsatisfactory disease control been attributed to decreased sensitivity of the pathogen to the fungicide, namely with *Cercospora* leaf spot of beets in Greece (Georgopoulos, 1982a). The mechanism of fungicide resistance has not been studied.

Non-fungicidal compounds. A few new compounds have been introduced recently, which are not fungicidal *in vitro*, but which are still active against fungal plant diseases, either by increasing host-plant resistance (e.g. probenazole) or by decreasing fungal pathogenicity (e.g. tricyclazole). It will be interesting to see if 'resistance' eventually develops to these 'fungicides'. This may depend on whether such compounds act at one or several target sites in the host or parasite. These compounds are discussed by Wade (Chapter 13, this volume).

Fitness of fungicide-resistant strains
General

Failure of disease control due to development of fungicide-resistance will only occur when most of the population of the pathogen has become resistant. Whether or not this will happen depends upon various factors including:

(i) fitness of the resistant mutants;

(ii) the type of pathogen, its speed of reproduction and nature of the disease;

(iii) disease control management.

Attention will be focused primarily on the first factor, since this is of predominant importance. Moreover, the other factors fall outside the scope of this chapter.

A mutational change responsible for resistance to a fungicide, may or may not result in decreased fitness of the fungus. This will depend on the mechanism of resistance which is often, but not always, related to the mode of action of the fungicide.

The build-up of resistance in the population of a pathogen will be inhibited when, in absence of the fungicide, the level of resistance is inversely related to the degree of fitness of the fungus. This might indeed happen with some fungicides, for example pimaricin; this antibiotic exerts its fungitoxicity by complexing with ergosterol and causing cell leakage. In pimaricin-resistant mutants of yeasts, ergosterol is replaced in the membrane by a precursor, which shows less affinity for the antibiotic. However, the precursor functions less efficiently in the fungal membrane than ergosterol and results in a decrease in cell vigour. Pimaricin-resistant strains of *Cladosporium cucumerinum* on cucumber and *Fusarium oxysporum* f.sp. *narcissi* on tulips, consequently show a considerable decrease in pathogenicity (Dekker & Gielink, 1979*a*).

Only those groups of site-specific fungicides will be considered which have been used extensively in agriculture since it is only with these that evidence is available about the build-up of resistant strains in the population of the pathogen.

Carboxamides

Strains of *Ustilago maydis* resistant to carboxin developed on maize (Georgopoulos, Chrysayi & White, 1975). A slight change in the succinate dehydrogenase complex which causes reduced affinity to the fungicide is apparently not detrimental to the fitness of the pathogen. In practice, however, fungicide resistance has not caused problems; the fungicide is used as a seed treatment. The build-up of resistant strains in the population may be inhibited by the pathogen's low apparent infection rate, or by the type of treatment employed. The closely related oxycarboxin, which is used as a spray treatment against rust diseases, has encountered fungicide-resistance problems in *Puccinia chrysanthemi* in France (Grouet, Montfort & Leroux, 1981) and in The Netherlands (Dirkse *et al.*, 1982). The repeating infection cycle of the pathogen, together with a continuous selection of pressure imposed by frequent fungicide applications, may have favoured the build-up of resistant strains in the pathogen population. Isolates of this pathogen, obtained from chrysanthemum greenhouses in The Netherlands, appeared as pathogenic as sensitive strains, even one year after application of oxycarboxin had been discontinued (author's unpublished results).

Benzimidazoles and thiophanates

Mutational changes in the tubulin subunits of fungal micro-tubules which are responsible for reduced affinity to carbendazim may affect the fitness of the fungus. In the absence of the fungicide, some resistant strains may be less competitive than the parental strain (Dekker, 1982*a*), but other resistant strains appear to be as fit as the sensitive strain. In the field a selection in favour of resistant strains will take place in the presence of the fungicide and this may eventually lead to the establishment of resistant strains in the population and possibly to disease-control failure. Indeed this has already happened with various diseases and, in some cases, even more rapidly than was anticipated. Moreover, resistance may persist for a long time in the pathogen population, even after application of the fungicide has been discontinued. This occurred, for example, with benomyl-resistant strains of *Cercospora beticola* on sugar beets (Georgopoulos, 1982a) and with resistant strains of *Sphaerotheca fuliginea* in cucumber greenhouses in The Netherlands (H. T. A. M. Schepers, unpublished results). It thus appears that although, in some species, mutation towards fungicide resistance may be accompanied by lower fitness, there is no absolute correlation between fungicide resistance and reduced fitness. Apparently a slight change in the tubulin subunits of fungal microtubules does not necessarily affect the normal functioning of the mitotic spindle. This means that benzimidazoles and thiophanates have to be considered as risky with respect to development of fungicide resistance.

Hydroxypyrimidines

Although strains of *Erysiphe graminis* with decreased sensitivity to ethirimol have been isolated from barley fields treated with the fungicide, no failure of disease control has yet been observed. Most isolates appeared to have a relatively low level of resistance to ethirimol, and could be controlled by the normal dose of fungicide. Shepherd *et al.* (1975) observed that, under selection pressure induced by the presence of the fungicide, the proportion of strains with a low level of resistance increased, in contrast to the decrease observed in the proportion of strains with a relatively high level of resistance. They suggested that this result was due to the lower fitness of the latter strains. Indeed, Walmsley-Woodward, Laws & Whittington (1979) found that, in the absence of the fungicide, ethirimol-resistant isolates of barley powdery mildew compete poorly with sensitive isolates.

It should be appreciated, however, that under conditions which are particularly favourable for the development of fungicide resistance, practical problems may indeed occur. This happened when dimethirimol was used to control powdery mildew in cucumber greenhouses in The Netherlands (Bent *et al.*, 1971). In this case, a highly sporulating pathogen was subjected to a continuous selection pressure in a closed environment (the fungicide was injected into the soil at the stem base of each plant to allow continuous uptake). However, only a moderate level of resistance developed, indicating a reduced fitness of strains which have a high level of fungicide resistance. Nevertheless, the pathogenicity of the resistant strains was high enough to cause practical problems.

It seems that a high level of resistance to hydroxypyrimidines is correlated with a somewhat reduced fitness, so these compounds should be classified as moderately risky with respect to the development of fungicide resistance. The biochemical basis of this resistance is not yet clear.

Acylalanines

Among the metalaxyl-resistant strains of *Phytophthora megasperma* f.sp. *medicaginis* obtained by Davidse (1981 and Chapter 11, this volume), several strains appeared to be as virulent as the wild type strain, suggesting that this fungicide has a high-resistance risk. Shortly afterwards, resistance problems did indeed suddenly appear in potato fields (Davidse *et al.*, 1981) which had been treated with metalaxyl to control *Phytophthora infestans*. The development of these resistant strains resulted in considerable crop losses; the fungicide was finally withdrawn from use against this disease in The Netherlands. Factors which favour rapid build-up of fungicide resistance are:

   (i) the presence of resistant strains with normal pathogenicity;

  (ii) the well-known high reproduction rate of this pathogen;

 (iii) the high selection pressure induced by this extremely efficient systemic fungicide.

Development of resistance has also been reported in various populations of downy mildews on cucumber, tobacco and vine. Metalaxyl and related compounds must therefore be considered as risky with respect to the development of fungicide resistance. With several other less risky types of disease, application of this fungicide has not yet met with problems.

Dicarboximides

Strains of *Botrytis cinerea*, which had become resistant to dicarboximides have been isolated from vine fields (Leroux & Fritz,

Chapter 10, this volume). Although strains of *B. cinerea* with increased resistance to these fungicides have been isolated repeatedly, it was several years before cases of inadequate disease control were first reported (Beever & Byrde, 1982). Insufficient control of grey mould of cyclamen (Pappas, 1982) and of grey mould of cucumber and tomato (Katan, 1982) has been reported. In most cases, however, even when strains with increased fungicide resistance were isolated, disease control remained satisfactory, probably because most strains only developed a moderate degree of resistance. Moreover, although resistant strains are usually less virulent than sensitive strains, the range of virulence shown by both groups is broad and can overlap (Beever & Byrde, 1982). In addition, strains with high resistance to the fungicide are osmotically sensitive and this may result in reduced virulence in the field; the high content of soluble sugar in grape plants may be unfavourable for such strains (Beever & Byrde, 1982).

Aromatic hydrocarbons

Sodium-*O*-phenylphenate and diphenyl are applied to harvested citrus fruits to prevent infection by *Penicillium digitatum* and *Penicillium italicum*. Spontaneous mutations give rise to strains of these fungi which are resistant to these fungicides. A large proportion of these resistant strains are still virulent. As a result of selection pressure during post-harvest fungicide treatments, the resistant strains may become dominant in the packing houses, and this may result in poor control of citrus rot and heavy losses during shipment (Eckert, 1982).

Other fungicide-resistance problems have also been reported, e.g. with hexachlorobenzene seed disinfection of wheat for control of bunt (Kuiper, 1965) and with dicloran treatment of onion for control of white rot (Locke, 1969). The fitness of resistant strains is apparently sufficient to cause problems when there is a continued selection pressure by these fungicides.

Organophosphates

Although strains of *Pyricularia oryzae*, resistant to phosphorothiolates and isoprothiolane were easily obtained *in vitro*, no failure of disease control due to fungicide resistance was reported during the 10 years which followed their introduction. When inadequate control was observed after that period, field isolates were obtained which were more resistant to the fungicide than were the wild type strain. A few of these isolates were highly resistant to the fungicide, but most had only a low level of resistance (Uesugi, 1982). The former strains appeared to have a lower fitness than the wild type strain, which explains their rare occurrence. To a lesser extent this may also be the case for the strains with a low resistance, since the proportion of these strains in the population increased only slowly under the selection pressure imposed by the application of the fungicide.

Isolates of cucumber powdery mildew, obtained from greenhouses where pyrazophos had given poor control, appeared to be more resistant to this fungicide. In competition tests, these isolates appeared somewhat less fit than the pyrazophos-sensitive strain (Dekker & Gielink, 1979*b*). These results indicate that the organophosphates can be considered to be only moderately risky with respect to fungicide resistance.

Inhibitors of sterol biosynthesis

The piperazine triforine has been used for almost a decade to control powdery mildews and other diseases and, during this time, no problems of fungicide resistance have developed. Triforine-resistant strains of *Cladosporium cucumerinum*, obtained in laboratory experiments, grew and sporulated poorly on agar and were only weakly pathogenic on cucumber (Fuchs *et al.*, 1977). These authors suggest that fungicide resistance is correlated with reduced fitness. Moreover this fungicide is not very persistent in plants, reducing the selection pressure exerted by this fungicide. Both factors counteract the rapid build-up of a population of resistent pathogens.

The imidazole imazalil and the pyrimidine fenarimol are more persistent; application of these fungicides may, in view of cross-resistance, also increase the number of triforine-resistant strains. The situation concerning the influence of resistance on fitness in the absence of the fungicide is not yet clear. Fenarimol-resistant mutants of *Aspergillus nidulans* showed reduced fitness as judged by growth characteristics *in vitro* (De Waard & Gieskes, 1977). The question arises as to whether the constitutive ability of resistant strains for energy-dependent fungicide efflux is disadvantageous for the fungus under fungicide-free conditions; the continuous operation of such a pump would certainly require extra energy. In *Penicillium italicum*, the pathogenicity of fungicide-resistant strains was not demonstrably less than that of the parental strain. But, in a competition experiment with oranges infected with a mixed inoculum from resistant and sensitive strains, the proportion of resistant conidia isolated gradually decreased, indicating the somewhat lower fitness of the resistant strain (De Waard

& van Nistelrooy, 1982). Fitness, however, seems sufficiently high to cause problems. Fenarimol-resistance has already been reported to have caused failure of powdery mildew control on cucumber in Israel (E. Koren, personal communication).

The behaviour of triazole fungicides, which have become important tools in the control of many serious plant diseases, is being watched with great care. The application of triadimefon to control barley powdery mildew has already resulted in an increase in the frequency of isolation of fungicide-resistant strains, but Wolfe & Fletcher (1981) report that this has not yet led to failure of disease control. Very recently, however, indications have been obtained from The Netherlands and other countries that the efficacy of triadimefon is declining. Surveys are in progress to check whether this is true, and if so, whether it is due to increased fungicide resistance.

Although, in comparison with other groups of site-specific fungicides, the inhibitors of sterol biosynthesis may be classified in the low-risk category with respect to fungicide resistance, careful fungicide management is nevertheless advised.

Antibiotics

Polyoxins. Four years after the introduction of polyoxin B for black spot on Japanese pears, unsatisfactory disease control was observed in several places due to fungicide resistance. After use of the antibiotic was discontinued, the proportion of fungicide-resistant strains in the pathogen population dropped markedly in the following years (Nishimura, Kohmoto & Udagawa, 1976), which indicates the reduced fitness of these resistant strains. This was confirmed in experiments on agar and in competition tests on plants. Apparently the change in membrane permeability, presumed to be the basis of resistance, presents a slight disadvantage for the fungus in the absence of the fungicide.

Kasugamycin. Failure of disease control was observed in certain rice-growing areas in Japan, where this antibiotic had been used exclusively for a number of years (Miura, Ito & Takahashi, 1975). A large proportion of isolates from these locations appeared to be kasugamycin-resistant. After application of the antibiotic was stopped, the proportion of fungicide-resistant strains dropped in the following years. No differences in pathogenicity were detected when infection by resistant strains was compared with that by sensitive strains, but in competition test using a mixture of fungicide-resistant and -sensitive conidia as inoculum, the former gave more lesions than the latter. This suggests a somewhat lower competitive ability of the resistant pathogen (Uesugi, 1982).

Other fungicides
Sec-butylamine. The competitive ability of resistant isolates of *Penicillium digitatum* has been tested by inoculating lemons with 1:1 mixture of fungicide-resistant and -sensitive spores. The results suggest that mutation towards resistance is not linked to reduced fitness of the pathogen (Smilanick & Eckert, 1983).

Organotin compounds. In competition experiments with a mixture of fungicide-resistant and -sensitive spores of *Cercospora beticola* on sugar beet plants, the percentage of resistant cells decreased with time, indicating a reduced fitness of the resistant strain. The risk of resistance to these fungicides developing seems to be rather limited, and no resistance problems have been encountered outside Greece (Georgopoulos, 1982*a*).

Positively or negatively correlated cross-resistance
When emergence of resistance to one chemical automatically implies resistance to another chemical, this is called cross-resistance. Proof of cross-resistance may be obtained when genetic analysis shows that resistance to both chemicals is controlled by the same gene. If resistance to these chemicals depends on different genes, it is called multiple resistance. The phenomenon of cross-resistance is of practical importance; e.g. when failure of disease control occurs, and a farmer has to choose a second fungicide, he should always consider the possibility of cross-resistance of the pathogen to the first and the second fungicides. Thus, he should not choose a new chemical of which the existence of cross-resistance with the first chemical is known.

Cross-resistance usually exists between chemicals which have related structures and the same mechanism of action, but there are exceptions. Sometimes cross-resistance occurs with chemically unrelated fungicides but is not observed with closely related chemicals.

The phenomenon of negatively correlated cross-resistance is particularly interesting. In this case resistance to one fungicide is correlated with sensitivity to a second fungicide and vice versa. This may even happen with closely related fungicides. Although cross-resistance between benomyl and thiabendazole is the rule, Van Tuyl, Davidse

& Dekker (1974) obtained a few thiabendazole-resistant strains of *Aspergillus nidulans* which appeared more sensitive to benomyl than the wild type strain. This is not as strange as it may first seem. Affinity of an organism for a fungicide may be reduced by a small change at the receptor site, in this case the tubulin subunit of microtubules. This protein may be modified in many ways. It is conceivable that most of these small modifications influence affinity to both fungicides in the same direction, but some modifications may increase affinity to one chemical, but reduce it to the other one. A negatively correlated cross-resistance has been found in *Ustilago maydis* for structural analogues of carboxin (Georgopoulos, 1982c), and in *Penicillium italicum* for the sterol biosynthesis inhibitors, fenarimol and fenpropimorph (De Waard & van Nistelrooy, 1982). Uesugi, Katagiri & Noda (1974) discovered that strains of *Pyricularia oryzae* with a high resistance to phosphorothiolates (such as IBP) were extra sensitive to phosphoramidates which are not active against the wild type strain. In a population of such strains, mutants may emerge which are again resistant to phosphoramidates, but sensitive to phosphorothiolates. It appears that high resistance to phosphorothiolates is correlated to a slower metabolism of phosphoramidates, resulting in sensitivity to the latter compounds and vice versa (Uesugi, 1982). This, however, did not hold for the strains with a low-level type of fungicide resistance.

Development of fungicide resistance in the field might in principle be counteracted by application of a mixture of two fungicides with negatively correlated cross-resistance. This, however, will only be possible in practice if all strains of the pathogen, without exception, respond in this way to a combination of such chemicals. De Waard & van Nistelrooy (1983) suggest that this might be true with respect to dodine- and fenarimol-sensitivity of various fungi.

Conclusions

Fungicide resistance has become one of the most important problems in the control of fungal disease of plants. Knowledge about emergence and behaviour of fungicide-resistant strains is essential for devising strategies to counteract the development of resistance in practice.

Fungicide resistance may emerge to virtually all site-specific inhibitors whether or not they are systemic in plants. The most common resistance mechanism involves a change at the target site which reduces its affinity to the fungicide or results in a decrease in uptake of the fungicide.

Resistance is often brought about by a single gene mutation. Development of resistance problems to conventional multi-site inhibitors which, in principle, may be caused by detoxification or reduced uptake, rarely occurs.

Emergence of fungicide resistance *in vitro*, however, does not necessarily imply that resistance problems will also be encountered in the field. This will only happen after a considerable part of the pathogen population has become resistant. Among the factors which influence the build-up of fungicide resistance in the field, the fitness of resistant strains plays a predominant role. With some fungicides, resistance is correlated with a more or less reduced fitness; with others, however, this is not so. On this basis, fungicides may be classified as risky (e.g. benomyl), moderately risky (e.g. kasugamycin), or as involving little risk (e.g. triforine). However, even risky compounds may continue to be used, at least against particular diseases, when proper attention is given to sound fungicide management in practice (Dekker, 1982*b*).

References

Beever, R. E. & Byrde, R. J. W. (1982). Resistance to the dicarboximide fungicides. In *Fungicide Resistance in Crop Protection*, ed. J. Dekker & S. G. Georgopoulos, pp. 101–17. Wageningen: Pudoc.

Bent, K. J., Cole, A. M., Turner, J. A. W. & Woolner, M. (1971). Resistance of cucumber powdery mildew to dimethirimol. In *Proceedings British Insecticide Fungicide Conference Brighton*, vol. 1, 274–82. Brighton: British Crop Protection Council.

Bruin, C. G. A. & Edgington, L. V. (1982). Induction of fungal resistance to metalaxyl by ultraviolet irradiation. *Phytopathology*, **72**, 476–80.

Crackower, S. H. B. (1972). The effect of griseofulvin on mitosis in *Aspergillus nidulans*. *Canadian Journal of Microbiology*, **18**, 683–7.

Darbord, J. C., Vidon, D., van Nguyen, H. & Desvignes, A. (1974). Etude génétique de mutants résistants à la griseofulvine chez un dermatophyte, *Arthroderma simii*. *Annals of Microbiology* (Institut Pasteur), **125A**, 17–25.

Davidse, L. C. (1981). Resistance to acylalanine fungicides in *Phytophthora megasperma* f.sp. *medicaginis*. *Netherlands Journal of Plant Pathology*, **87**, 11–24.

Davidse, L. C. (1982). Benzimidazole compounds: selectivity and resistance. In *Fungicide Resistance in Crop Protection*, ed. J. Dekker & S. G. Georgopoulos, pp. 60–70. Wageningen: Pudoc.

Davidse, L. C., Looyen, D., Turkensteen, L. J. & van der Wal, D. (1981). Occurrence of metalaxyl resistant strains of *Phytophthora infestans* in Dutch potato fields. *Netherlands Journal of Plant Pathology*, **87**, 65–8.

Dekker, J. (1977). Resistance. In *Systemic Fungicides*, ed. R. W. Marsh, pp. 176–9. London & New York: Longman.

Dekker, J. (1982*a*). Can we estimate the fungicide-resistance hazard in the field from laboratory and greenhouse tests? In *Fungicide Resistance in Crop Protection*, ed. J. Dekker & S. G. Georgopolous, pp. 128–38. Wageningen: Pudoc.

Dekker, J. (1982b). Counter measures for avoiding fungicide resistance. In *Fungicide Resistance in Crop Protection*, ed. J. Dekker & S. G. Georgopolous, pp. 177–86. Wageningen: Pudoc.

Dekker, J. & Gielink, A. J. (1979a). Acquired resistance to pimaricin in *Cladosporium cucumerinum* and *Fusarium oxysporum* f.sp. *narcissi* associated with decreased virulence. *Netherlands Journal of Plant Pathology*, **85**, 67–73.

Dekker, J. & Gielink, A. J. (1979b). Decreased sensitivity to pyrazophos of cucumber and gherkin powdery mildew. *Netherlands Journal of Plant Pathology*, **85**, 137–42.

De Waard, M. A. (1974). Mechanism of action of the organophosphorus fungicide pyrazophos. *Mededelingen Landbouwhogeschool Wageningen*, 74–14, 98p.

De Waard, M. A. & Gieskes, S. A. (1977). Characterisation of fenarimol-resistant mutants of *Aspergillus nidulans*. *Netherlands Journal of Plant Pathology*, **83**, Suppl. 1, 177–88.

De Waard, M. A., Groeneweg, H. & van Nistelrooy, J. G. M. (1982). Laboratory resistance to fungicides which inhibit ergosterol biosynthesis in *Penicillium italicum*. *Netherlands Journal of Plant Pathology*, **88**, 99–112.

De Waard, M. A. & van Nistelrooy, J. G. M. (1979). Mechanism of resistance to fenarimol in *Aspergillus nidulans*. *Pesticide Biochemistry and Physiology*, **10**, 219–29.

De Waard, M. A. & van Nistelrooy, J. G. M. (1980a). An energy-dependent efflux mechanism for fenarimol in a wild-type strain and fenarimol-resistant mutants of *Aspergillus nidulans*. *Pesticide Biochemistry and Physiology*, **13**, 255–66.

De Waard, M. A. & van Nistelrooy, J. G. M. (1980b). Mechanism of resistance to pyrazophos in *Pyricularia oryzae*. *Netherlands Journal of Plant Pathology*, **86**, 251–8.

De Waard, M. A. & van Nistelrooy, J. G. M. (1982). Toxicity of fenpropimorph to fenarimol-resistant isolates of *Penicillium italicum*. *Netherlands Journal of Plant Pathology*, **88**, 231–6.

De Waard, M. A. & van Nistelrooy, J. G. M. (1983). Negatively correlated cross resistance to dodine in fenarimol resistant isolates of various fungi. *Netherlands Journal of Plant Pathology*, **89**, 67–73.

Dirkse, F. B., Dil, M., Linders, R. & Rietstra, I. (1982). Resistance in white rust (*Puccinia horiana* P. Hennings) of chrysanthemum to oxycarboxin and benodanil in The Netherlands. *Mededelingen Faculteit Landbouwweten-schappen Rijksuniversiteit Gent*, **47**, 793–800.

Eckert, J. W. (1982). *Penicillium* decay of citrus fruits. In *Fungicide Resistance in Crop Protection*, ed. J. Dekker & S. G. Georgopoulos, pp. 231–50. Wageningen: Pudoc.

Fuchs, A., de Ruig, S. P., van Tuyl, J. M. & de Vries, F. W. (1977). Resistance to triforine: a non existent problem? *Netherlands Journal of Plant Pathology*, **83**, Suppl. 1, 189–205.

Georgopoulos, S. G. (1977). Development of fungal resistance to fungicides. In *Antifungal Compounds*, ed. M. R. Siegel & H. D. Sisler, vol. 2, pp. 409–95. New York & Basel: Marcel Dekker Inc.

Georgopoulos, S. G. (1982a). *Cercospora beticola* of sugarbeets. In *Fungicide Resistance in Crop Protection*, ed. J. Dekker & S. G. Georgopoulos, pp. 187–94. Wageningen: Pudoc.

Georgopoulos, S. G. (1982b). Genetical and biochemical background of fungicide resistance. In *Fungicide Resistance in Crop Protection*, ed. J. Dekker & S. G. Georgopoulos, pp. 46–52. Wageningen: Pudoc.

Georgopoulos, S. G. (1982c). Cross resistance. In *Fungicide Resistance in Crop Protection*, ed. J. Dekker & S. G. Georgopoulos, pp. 53–59. Wageningen: Pudoc.

Georgopoulos, S. G., Chrysayi, M. & White, G. A. (1975). Carboxin resistance in the haploid, the heterozygous diploid and the plant parasitic dicaryotic phase of *Ustilago maydis*. *Pesticide Biochemistry and Physiology*, **5**, 543–51.

Georgopoulos, S. G. & Sisler, H. D. (1970). A gene mutation eliminating antimycin A tolerant electron transport in *Ustilago maydis*. *Journal of Bacteriology*, **103**, 745–50.

Grouet, D., Montfort, F. & Leroux, P. (1981). Mise en evidence, en France, d'une souche de *Puccinia horiana* resistente à l'oxycarboxin. *Phytiatrie-Phytopharmacie*, **30**, 3–12.

Gull, K. & Trinci, A. J. P. (1973). Griseofulvin inhibits fungal mitosis. *Nature*, **244**, 292–4.

Hollomon, D. W. (1979). Evidence that ethirimol may interfere with adenine metabolism during primary infection of barley powdery mildew. *Pesticide Biochemistry and Physiology*, **10**, 181–9.

Hollomon, D. W. (1981). Genetic control of ethirimol resistance in a natural population of barley powdery mildew. *Phytopathology*, **71**, 536–40.

Kaars Sijpesteijn, A. (1970). Biochemical modes of action of fungicides. *World Review of Pest Control*, **9**, 85–93.

Katan, T. (1982). Resistance to dicarboximide fungicides in the grey mould pathogen *Botrytis cinerea* on protected crops. *Plant Pathology*, **31**, 133–41.

Kodama, O., Yamashita, K. & Akatsuka, T. (1980). Edifenphos, inhibitor of phosphatidylcholine biosynthesis in *Pyricularia oryzae*. *Agricultural Biological Chemistry*, **44**, 1015–21.

Kuiper, J. (1965). Failure of hexachlorobenzene to control common bunt of wheat. *Nature*, **206**, 1219–20.

Locke, S. B. (1969). Botran tolerance of *Sclerotium cepivorum* isolates from fields with different Botran treatment histories. *Phytopathology*, **59**, 13 (abstract).

Lyr, H. (1977). Effect of fungicides in energy production and intermediary metabolism. In *Antifungal Compounds*, vol. 2, ed. M. R. Siegel & H. D. Sisler, pp. 301–32. New York & Basel: Marcel Dekker Inc.

Misato, T. (1967). Blasticidin S. In *Antibiotics*, ed. D. Gottlieb & P. D. Shaw, pp. 434–9. Berlin: Springer Verlag.

Misato, T., Kakiki, K. & Hori, M. (1977). Mechanism of polyoxin resistance. *Netherlands Journal of Plant Pathology*, **83**, Suppl. 1, 253–60.

Miura, H., Ito, H. & Takahashi, S. (1975). Occurrence of resistant strains of *Pyricularia oryzae* to kasugamycin as a cause of the diminished fungicidal activity to rice blast. *Annals Phytopathological Society Japan*, **41**, 415–17.

Nishimura, S., Kohmoto, K. & Udagawa, H. (1973). Field emergence of fungicide-tolerant strains in *Alternaria kikuchiana* Tanaki. *Report Tottori Mycological Institute, Japan*, **10**, 677–86.

Nishimura, S., Kohmoto, K. & Udagawa, H. (1976). Tolerance to polyoxin *Alternaria kikuchiana* Tanaka, causing black spot of Japanese pear. *Review of Plant Protection Research, Tokyo*, **9**, 47–57.

Noble, M., MacGarvie, Q. D., Hams, A. F. & Leafe, L. L. (1966). Resistance to mercury of *Pyrenophora avenae* in Scottish seed oats. *Plant Pathology*, **15**, 23–8.

Pappas, A. C. (1982). Inadequate control of grey mould on cyclamen by dicarboximide fungicides. Greece. *Zeitschrift für Pflanzenkrankheiten und Pflanzenschutz*, **89**, 52–8.

Pappas, A. C., Cooke, B. K. & Jordan, V. W. L. (1979). Insensitivity of *Botrytis cinerea* to iprodione, procymidone and vinclozolin and their uptake by the fungus. *Plant Pathology*, **28**, 71–6.

Roa, S. S. & Grollman, A. (1967). Cycloheximide resistance in yeast: a property of the 60S ribosomal subunit. *Biochemical and Biophysical Research Communications*, **29**, 696–704.

Shephard, M. C., Bent, K. J., Woolner, M. & Cole, A. M. (1975). Sensitivity to ethirimol of powdery mildew from UK barley crops. In *Proceedings British Insecticide Fungicide Conference, Brighton*, vol. 1, 59–65. Brighton: British Crop Protection Council.

Siegel, M. R. (1977). Effect of fungicides on protein synthesis. In *Antifungal Compounds*, ed. M. R. Siegel & H. D. Sisler, vol. 2, pp. 399–438. New York & Basel: Marcel Dekker Inc.

Siegel, M. R. & Solel, Z. (1981). Effects of imazalil on a wild-type and fungicide-resistant strain of *Aspergillus nidulans*. *Pesticide Biochemistry and Physiology*, **15**, 221–3.

Smilanick, J. L. & Eckert, J. W. (1983). Cultural characteristics and competitive ability of citrus green mold (*Penicillium digitatum* Sacc.) resistant to the fungicide Sec-butylamine. *Phytopathology*, **73**, 802 (abstract).

Taga, M., Nakagawa, H., Tsuda, M. & Ueyama, A. (1978). Ascospore analysis of kasugamycin resistance in the perfect stage of *Pyricularia oryzae*. *Phytopathology*, **68**, 815–17.

Tillman, R. W. & Sisler, H. D. (1973). Effect of chloroneb on the metabolism and growth of *Ustilago maydis*. *Phytopathology*, **63**, 219–25.

Uesugi, Y. (1982). *Pyricularia oryzae* in rice. In *Fungicide Resistance in Crop Protection*, ed. J. Dekker & S. G. Georgopoulos, pp. 207–18. Wageningen: Pudoc.

Uesugi, Y. & Sisler, H. D. (1978). Metabolism of a phosphoramidate by *Pyricularia oryzae* in relation to tolerance and synergism by a phosphorothiolate and isoprothiolane. *Pesticide Biochemistry and Physiology*, **9**, 247–54.

Uesugi, Y., Katagiri, M. & Noda, O. (1974). Negatively correlated cross-resistance and synergism between phosphoramidates and phosphorothiolates in their fungicidal actions on rice blast fungi. *Agricultural and Biological Chemistry*, **38**, 907–12.

Van Tuyl, J. M. (1977). Genetics of fungal resistance to systemic fungicides. *Mededelingen Landbouwhogeschool Wageningen*, **77–2**, 137p.

Van Tuyl, J. M., Davidse, L. C. & Dekker, J. (1974). Lack of cross resistance to benomyl and thiabenzole in some strains of *Aspergillus nidulans*. *Netherlands Journal of Plant Pathology*, **80**, 165–8.

Walmsley-Woodward, D. J., Laws, F. A. & Whittington, W. J. (1979). The characteristics of isolates of *Erysiphe graminis* f.sp. *hordei* varying in response to tridemorph and ethirimol. *Annals of Applied Biology*, **92**, 211–19.

Wilkie, D. & Lee, B. K. (1965). Genetic analysis of actidione resistance in *Saccharomyces cerevisiae*. *Genetical Research*, **6**, 130–8.

Wolfe, M. S. & Fletcher, J. T. (1981). Insensitivity of *Erysiphe graminis* f.sp. *hordei* to triadimefon. *Netherlands Journal of Plant Pathology*, **87**, 239 (abstract).

6

Antifungal agents which affect hyphal extension and hyphal branching

A. P. J. TRINCI

Botany Department, University of Manchester, Manchester M13 9PL, UK

Introduction

Antifungal agents are usually either fungistatic, e.g. griseofulvin and the benzimidazole carbamates, or fungicidal, e.g. nystatin and 5-fluorocytosine. However, it is possible to control certain fungal diseases of plants using compounds which neither kill the pathogen nor inhibit its growth. The general properties of this type of antifungal are given in Table 6.1, and these or similar compounds have sometimes been referred to as 'paramorphogens' (Tatum, Barratt & Cutter, 1949). Antifungal agents belonging to the group include sclareol (Bailey, Vincent & Burden, 1974; Bailey *et al.*, 1975), L-sorbose (Trinci & Collinge, 1973; Howell, 1978), 3-O-methylglucose (Galpin, Jennings & Thornton, 1977), cellobiose (Wilson & Niedepruem, 1967), α-amino-isobutyric acid (S. C. Watkinson, personal communication), and validamycin A (Wakae & Matsuura, 1975). Validamycin A, widely used in Japan to control rice sheath blight caused by *Rhizoctonia solani* (=*Pellicularia sasakii*), prevents the pathogen spreading from the lower to the upper portions of the plant, where it causes a reduction in crop yield. Compounds which have the characteristics listed in Table 6.1 may be effective in controlling fungal diseases of plants either because they reduce the rate of spread of the pathogen in the host, or because they inhibit penetration of the host. Validamycin A inhibits the formation of infection cushions by *R. solani* and therefore inhibits host penetration (Wakae & Matsuura, 1975). In addition, any compound like validamycin A which reduces the rate at which a pathogen spreads in the host will increase the time available to the plant to mobilise its defence mechanisms. (See Wade, Chapter 13).

Antifungal agents of the validamycin A type are likely to be less

Table 6.1. *Characteristics of antifungal compounds which are neither fungistatic nor fungicidal*

1. The compound has little or no effect on the organism's maximum specific growth rate (μ_{max}).
2. The compound has no effect on the organism's growth yield (Y_E) from a normal carbon source.
3. The compound causes a decrease in the maximum (E_{max}) and mean (E) rates of hyphal extension.
4. The compound causes a decrease in hyphal growth unit length (G).
5. The compound induces the formation of colonies which are denser and expand more slowly than control colonies.

phytotoxic than most antifungal compounds since they do not even inhibit the growth (μ_{max}) of the fungus. However, because of their particular mode of action, they may be effective in controlling only a limited range of fungal pathogens. In addition, they are unlikely to be of value as antifungal drugs since they lack fungicidal or fungistatic activity.

The kinetics of mycelial growth are first described, and this is followed by an account of the effects of various 'paramorphogen' type compounds on mycelial growth and morphology.

Kinetics of mycelial growth

When a fungal mycelium is cultured on a medium which contains an excess of all nutrients and lacks inhibitory substances, its total hyphal length and the number of branches increase exponentially at the same specific rate (Trinci, 1974). It follows from this that the ratio between these two parameters will be a constant, and that therefore mycelial growth can be considered in terms of the duplication of a hypothetical growth unit consisting of a tip and a strain-specific length of hypha (G, the hyphal growth unit; Bull & Trinci, 1977). If the mean hyphal diameter of the mycelium remains constant as it increases in size, the hyphal growth unit will be of constant volume as well as constant length (Trinci, 1984).

Trinci (1974) used the following equation to determine the mean rate of hyphal extension (E) of mycelia growing exponentially under unrestricted conditions:

$$E = 2 \frac{(H_t - H_0)}{B_0 + B_t},\tag{1}$$

where $H_0 = $ total hyphal length of the mycelium at zero time, $H_t = $ total

Table 6.2. *Effect of* L-*sorbose on the colony radial growth rate of Rhizoctonia cerealis and on the morphology of leading hyphae; the culture medium contained* 5 mM *glucose*

L-sorbose concentration (mM)	Colony radial growth rate (K_r, μm h^{-1})	Intercalary compartment length[a] (μm)	Hyphal radius (μm)
0	278	186	1.62
2	223	144	1.66
4	200	111	1.72
8	123	86	1.71
16	77	54	2.13
32	34	45	2.20

[a]Distance between adjacent septa.

hyphal length 1 h later, $B_0 =$ number of tips at zero time, and $B_t =$ number of tips 1 h later. Steele & Trinci (1975) showed that E was a function of the length of the organism's hyphal growth unit (G) and its specific growth rate (μ). Thus

$$E = G\mu. \tag{2}$$

This relationship has also been demonstrated for mycelia of *Streptomyces hygroscopicus* (Schuhmann & Bergter, 1976) and *Candida albicans* (Gow & Gooday, 1982), and thus appears to reflect a fundamental property of microbial mycelia. It follows from Eqn (2) that, in the absence of a change in hyphal radius (see below), compounds which cause a decrease in hyphal extension without altering the organism's specific growth rate will induce mycelia to branch more profusely.

Eqn (2) can be re-written (Trinci, 1984):

$$E = \left(\frac{V_g}{\pi r^2}\right)\mu, \tag{3}$$

where $V_g =$ hyphal growth unit volume and $r =$ hyphal radius. Eqn (3) predicts that compounds or conditions which alter hyphal radius but not the organism's specific growth rate, will cause a change in both hyphal growth unit length and hyphal extension, e.g. an increase in hyphal radius will result in a decrease in hyphal growth unit length and a decrease in the mean rate of hyphal extension. Thus, G is affected by conditions which either alter the ratio E/μ or alter hyphal radius.

Effect of L-sorbose on mycelial growth and morphology

L-sorbose is used by geneticists to restrict the growth of colonies of *Neurospora crassa*; Fig. 6.1 and Table 6.2 show the effects of this sugar on the growth and morphology of *Rhizoctonia cerealis*. However, contrary to popular belief, growth of *N. crassa* is not inhibited by L-sorbose. For example, Crocken & Tatum (1968) and Trinci & Collinge (1973) showed that L-sorbose has no appreciable effect on the specific growth rate of *N. crassa* cultured under unrestricted conditions on solid (Table 6.3) or liquid media. However, L-sorbose causes a dramatic reduction in the length of the hyphal growth unit of *N. crassa* when mycelia are grown on solid media (Fig. 6.2a, Table 6.3). Trinci & Collinge (1973) suggested that this effect of L-sorbose on hyphal branching eventually leads to the formation of dense colonies which have narrower peripheral growth zones than control colonies and expand in radius more slowly than control colonies (Table 6.3). The inhibitory effect of L-sorbose on colony radial-growth rate is probably in part due to a direct effect on hyphal extension (L-sorbose inhibits the activity of glycogen synthetase and β-1,3-glucan synthetase (Mishra & Tatum, 1972), and in part due to an indirect effect on peripheral

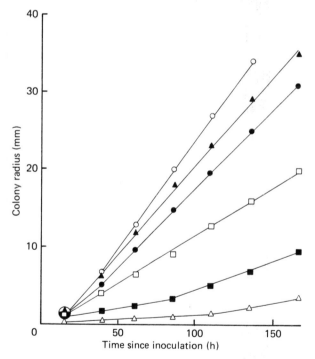

Fig. 6.1. Effect of 0 mM (open circles), 2 mM (closed triangles), 4 mM (closed circles), 8 mM (open squares), 16 mM (closed squares), and 32 mM (open triangles) L-sorbose on the radial growth rate of colonies of *Rhizoctonia cerealis* cultured on medium containing 5 mM glucose.

Table 6.3. *Influence of L-sorbose on the growth and morphology of Neurospora crassa spco 9 cultured on solid medium containing 10 g l⁻¹* sucrose

	Mycelia grown on medium lacking L-sorbose	Mycelia grown on medium containing L-sorbose $(20\,\mathrm{g}\,l^{-1})$
Specific growth rate (μ, h⁻¹)	0.375	0.324
Hyphal growth unit length (μm)	323	40
Colony radial growth rate (K_r, μm h⁻¹)	1001	100
Peripheral growth zone width (w, μm)	3930	525

From Trinci & Collinge, 1973.

growth-zone width (Trinci, 1971). In dense colonies, adverse conditions (nutrient limitation, accumulation of staling products, pH changes etc.) are established much closer to the colony margin than is normal.

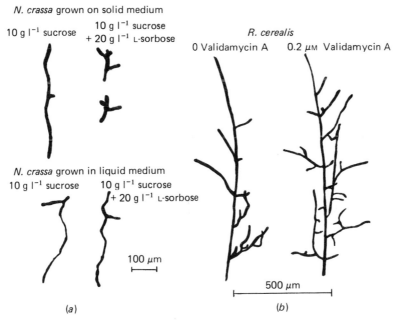

Fig. 6.2. (a) Effect of L-sorbose on the morphology of mycelia of *Neurospora crassa* cultured on solid and liquid media (from Trinci & Collinge, 1973). (b) Effect of validamycin A on the morphology of leading hyphae of colonies of *Rhizoctonia cerealis* grown on media containing 0.5 mM glucose.

Howell (1978) showed that L-sorbose (10 mg seed^{-1}) could be used to control damping off of cotton seedlings by *Rhizoctonia solani*; prior to planting, cotton seeds were treated with a slurry of cellulose acetate and L-sorbose. He correlated the protective effect of L-sorbose with its effects on hyphal extension and hyphal branching.

'Sclareol' and α-aminoisobutyric acid

'Sclareol' (Fig. 6.3), an epimeric mixture of the diterpenes sclareol and 13-epi-sclareol, was isolated by Bailey *et al.* (1974) from the leaves of *Nicotiana glutinosa*. This compound induces mycelia of *Alternaria longipes* and *Cladosporium cucumerinum* to branch more profusely than control mycelia, and causes a reduction in colony radial-growth rate (Bailey *et al.*, 1974). However, 'sclareol' has no affect on growth yield. Thus, the present evidence suggests that 'sclareol' affects *A. longipes* and *C. cucumerinum* in a manner similar to the effect of L-sorbose on *Neurospora crassa*.

S. C. Watkinson (personal communication) has shown that α-aminoisobutyric acid, an analogue of alanine, is taken up but not metabolised by *Serpula lacrimans* (the dry rot fungus). Results obtained by Watkinson suggest that α-aminoisobutyric acid inhibits hyphal extension and increases hyphal branching. The analogue is translocated within hyphae, and it may therefore be possible to use it to control the fungus by inhibiting strand elongation.

Validamycin A

Like the compounds considered above, validamycin A has no effect on growth yield but reduces hyphal extension and decreases hyphal growth unit length (Nioh & Mizushima, 1974). The widespread use of this antibiotic in Japan to control rice sheath blight illustrates the potential value of compounds with this particular mode of action.

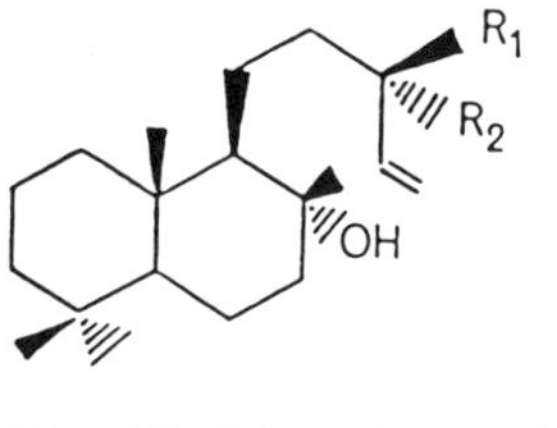

Fig. 6.3. Sclareol: an epimeric mixture of sclareol (R_1:OH; R_2:Me) and 13-epi-sclareol (R_1:Me; R_2:OH). From Bailey *et al.* (1975), reprinted by permission from *Nature*, **255**, No. 5506, pp. 328–9. Copyright (©) 1975, Macmillan Journals Limited.

Development of validamycin A

Various changes in agricultural practices (the introduction of early maturing rice cultivars, early rice transplantation, the use of high concentrations of fertilisers, and the cultivation of dense crop populations) led, in Japan, to an increase in the severity of rice sheath blight. Up to the middle of the 1960s this disease was controlled by organoarsenic compounds, but the use of these fungicides caused considerable environmental pollution. Validamycin A was therefore developed by Takeda Chemical Industries as an alternative antifungal and was eventually registered in Japan (1972), Korea (1976), Taiwan (1977) and Colombia (1977) as a controlling agent for rice sheath blight. In addition, the antibiotic has been registered in Japan (1975) and The Netherlands (1977) as a controlling agent for black scurf of potatoes. Approximately 12 tons of validamycin A (active ingredient) was produced in Japan in 1974 (Misato, Ko & Yamaguchi, 1977) and production of the antibiotic had increased to about 100 tons a year by 1978; about a third of that year's production was exported.

Production and structure of validamycin A

Six validamycins (designated A–F) have been isolated from the culture filtrate of *Streptomyces hygroscopicus* var. *limoneus* (Iwasa *et al.*, 1971c; Horii, Kameda & Kawahara, 1972). The revised (Suami, Ogawa & Chida, 1980) structure of validamycin A is shown in Fig. 6.4; it has a molecular weight of 515.5 daltons and is a water-soluble, weakly basic compound. Validamycin A has been classified as an aminoglycosidic antibiotic, but its sugar moiety lacks an amino group and it contains unique hydroxylmethyl branched cyclitol moieties. Validamycin analogues differ in the aglycone part of the molecule (validoxylamine A or validoxylamine B), the configuration of the anomeric centre of the glucoside, the position of the glucosidic linkage, and the number of D-glucose residues. Validamycins A, C, D, E and F contain validoxylamine A, and validamycin B contains validoxylamine B. Validamycins A, B and D contain one D-glucose residue and validamycins C, E and F contain two glucose residues.

Bioassays for validamycin activity

Validamycin activity is assessed *in vitro* by determining the minimum antibiotic concentration which induces *Rhizoctonia solani* to branch abnormally or causes a reduction in colony radial-growth rate. The 'inhibitory' effects of validamycin A are reversed by glucose (see

Fig. 6.8), fructose and peptone (Shibata, Uyeda & Mori, 1981), and therefore in-vitro bioassays are best carried out using water agar or a culture medium which contains only a low concentration of carbon source. The history of the discovery of validamycin A illustrates the value of using in-vivo systems to screen for novel fungicides; the activity of validamycin A was originally detected by in-vivo but not by in-vitro screening procedures, presumably because the culture medium employed in the latter contained too high a concentration of carbon source.

The 'dish' method (Fig. 6.5*a*), which was eventually developed to assay validamycin activity, consists of a small dish containing culture medium inoculated with *Rhizoctonia solani* which is embedded in a

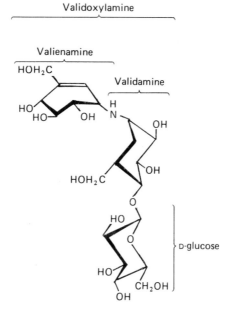

Fig. 6.4. Structure of (*a*) validamycin A as revised by Suami *et al.* (1980), and (*b*) meso-inositol.

Petri dish of water agar which contains the antibiotic (Iwasa, Higashide & Shibata, 1971*a*); hyphae grow radially outwards from the inoculum into the agar which contains the validamycin. The 'dish' method was later simplified by placing the inoculum plug on a glass disk (Fig. 6.5*b*); this is usually referred to as the 'dendroid' method. In both these bioassays, the validamycin titre is established by determining the highest

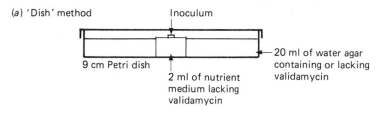

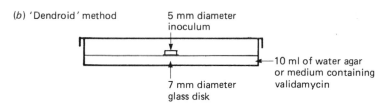

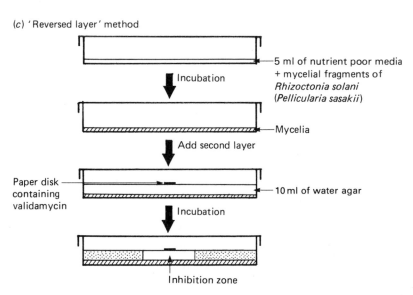

Fig. 6.5. Methods used to assess validamycin A activity *in vitro* (from Iwasa *et al.*, 1971*a*).

dilution capable of inducing *R. solani* to branch abnormally. Iwasa *et al.* (1971*a*) also developed an inhibition zone method for assaying validamycin; in this 'reversed layer' method, *R. solani* hyphae suspended in 5 ml of nutrient-poor medium are added to a 9 cm Petri dish and the culture is incubated at 27 °C for about 2 d. The plate is then overlaid with water agar and a paper disk containing validamycin is positioned centrally on the plate; Iwasa *et al.* (1971*a*) observed a linear relationship between antibiotic concentration (between 62.5 and 1000 μg ml^{-1}) and the diameter of the inhibition zone which developed.

Biological activity of validamycin A

Iwasa *et al.* (1971*b*) detected no antimicrobial activity of validamycin A against nearly 3000 strains of fungi and bacteria, and the antibiotic had no affect on the microbial flora in soil or on plant surfaces (Wakae & Matsuura, 1975). Validamycin A (1 mg ml^{-1}) is phytotoxic to mulberry but not to 150 other species of plant, and the antibiotic has very low acute and subacute toxicities to mammals (Wakae & Matsuura, 1975; Misato *et al.*, 1977). Most of the fungi which are sensitive to validamycin A are Basidiomycetes belonging to the Corticiaceae. However, four members of the Fungi Imperfecti and one Ascomycete are also sensitive to the antibiotic (Table 6.4); Oomycetes and Zygomycetes are apparently insensitive.

Rhizoctonia solani isolates have been separated into four anastomosis groups on the basis of the hyphal fusions which occurs between isolates from the same group (Parmeter, Sherwood & Platt, 1969; Anderson, 1982). Isolates of *R. solani* (= *Pellicularia sasakii*) which cause rice sheath blight belong to anastomosis group I and isolates which cause black scurf of potatoes belong to anastomosis group IV. The characteristics which Boerema & Verhoeven (1977) and Hollins, Jellis & Scott (1983) used to distinguish between *R. solani* and *Rhizoctonia cerealis* are listed in Table 6.5; both species are sensitive to validamycin A.

The capacity of validamycins and their constituent moieties to induce *Rhizoctonia solani* to branch abnormally is related to the ED$_{84}$ values of these compounds (Table 6.6). It should be noted that validoxylamine A is much less active than validamycin A, and that validamine and valienamine show even lower activity *in vivo* than validoxylamine A (Table 6.6). These observations may either reflect differences in the biological activity of validamycin A and its constituent moieties, or they may reflect differences in the relative efficiency of uptake of the various compounds.

Table 6.4. *Fungi[a] which are sensitive and insensitive to validamycin A*

Sensitive fungi	Insensitive fungi	
Acrocylindrium sp.[c]	*Alternaria alternata*[e]	*Mortierella hyalina*[e]
Chaetomium globosum[e]	*Alternaria kikuchiana*[b]	*Mucor hiemalis*[e]
Coprinus macrorhizus[d]	*Aspergillus oryzae*[c]	*Penicillium notatum*[c]
Corticium rolfsii[ce]	*Aspergillus niger*[d]	*Penicillium verrucosum*[e]
Corticium gramineum[c]	*Botrytis cinerea*[b,c,d,e]	*Petriellidium boydii*[e]
Fusarium culmorum[e]	*Broomella acuta*[e]	*Phytophthora infestans*[b,c,d]
Helicobasidium mompa[d]	*Cochliobolus miyabeanus*[c,d]	*Pyrenophora graminea*[c]
Helminthosporium sigmoideum[c,d]	*Corticium rolfsii*[b]	*Pythium aphanidermatum*[b,c]
Lentinus edodes[c]	*Cunninghamella elegans*[e]	*Pythium dissotocum*[e]
Pellicularia praticola[b]	*Epicoccum nigrum*[e]	*Rhizopus nigricans*[c]
Rhizoctonia cerealis[f]	*Flammulina velutipes*[d]	*Rhizopus oryzae*[d]
Rhizoctonia solani (=*Pellicularia sasakii*)[b,e]	*Fusarium oxysporum*[b]	*Sclerotinia sclerotiorum*[b,d]
Sclerotium hydrophilum[c,d]	*Gaeumannomyces graminis*[e]	*Scopulariopsis brevicaulis*[e]
Stereum roseum[c,d]	*Geotrichum candidum*[e]	*Stereum fasciatum*[b]
	Gerlachia nivalis[e]	*Thelebolus polysporus*[e]
	Gliocladium roseum[e]	*Trichoderma viride*[e]
	Glomerella cingulata[b]	*Zygorhynchus moelleri*[e]

[a] The specific names are those given by the authors concerned.
[b] Iwasa et al. (1971b): activity assessed using the 'dish' method (Fig. 6.5a) and 0.1 to 5.0 μg ml^{-1} validamycin A.
[c] Nioh & Mizushima (1974): activity probably assessed using the 'dendroid' method (Fig. 6.5b) and 0.1 μg ml^{-1} validamycin A.
[d] Wakae & Matsuura (1975): method unknown; 10 μg ml^{-1} validamycin A.
[e] Gams & Van Laar (1982): activity assessed using the conventional plate method and 100 μg ml^{-1} validamycin A.
[f] Trinci: activity assessed using the conventional plate method and 0.2 μM validamycin A.

Table 6.5. *Characteristics used by Boerema & Verhoeven (1977) and Hollins et al. (1983) to distinguish between Rhizoctonia solani and Rhizoctonia cerealis*

Characteristic	R. solani	R. cerealis
Number of nuclei per intercalary compartment	2–33	2
Radius of leading hyphae cultured on potato dextrose agar (μm)	3–6	2–3
Pigmentation of sclerotia	Dark coloured	White, turning yellow
Colony radial growth rate	Relatively fast	Relatively slow
Host plant	Rice, potatoes etc. but not cereals	Cereals
Perfect state	*Thanatephorus cucumeris* (Frank) Donk	*Ceratobasidium* sp.

Table 6.6. *ED_{84} values for validamycins, validoxylamines, validamine and valienamine against rice sheath blight, and the minimum concentrations of validamycins and validoxylamines which induce Rhizoctonia solani to branch abnormally*

	ED_{84} value[a] (μM)	Minimum concentration[b] which induces abnormal branching (μg ml^{-1})
Validamycin A	0.7	0.01
Validamycin B	1.7	0.5
Validamycin C	8.5	10.0
Validamycin D	13.2	100.0
Validamycin E	1.0	0.13
Validamycin F	1.4	0.13
Validoxylamine A	8.0	10.0
Validoxylamine B	84.0	>100.0
Validamine	>300.0	
Valienamine	>300.0	

[a]From Wakae & Matsuura, 1975.
[b]From Horii *et al.*, 1972.

Soil microorganisms degrade validamycin A to validoxylamine A and D-glucose, and the former is further degraded to valienamine and validamine (Wakae & Matsuura, 1975); validamycin A has a half-life in soil of less than 4 h (Fig. 6.6), and therefore does not accumulate in the environment.

Kameda, Asano & Hashimoto (1978) and Kameda *et al.* (1980) used *Rhodotorula marina* and *Rhodotorula lactosa* to prepare various glucosidic analogues of validamycin; the β-glucoside analogues were more active than the α-glucoside analogues, but the activity *in vitro* of these new compounds was lower than that of validamycin A.

Effect of validamycin A on growth yield, hyphal extension and hyphal branching

Validamycin A, at concentrations of up to $1\,mg\,ml^{-1}$, has no appreciable effect on the yield of biomass (Wakae & Matsuura, 1975; Shibata *et al.*, 1981), RNA or protein (Nioh & Mizushima, 1974) from cultures of *Rhizoctonia solani*. Other experiments (A. P. J. Trinci, unpublished) suggest that the antibiotic (10 μM) also does not affect the specific growth rate of *Rhizoctonia cerealis* grown on media containing 50 mM glucose. However, validamycin A, like L-sorbose and the other compounds discussed above, reduces the maximum rate of extension of hyphae of *R. solani* (Nioh & Mizushima, 1974) and *R. cerealis* (A. P. J. Trinci, unpublished) and increases hyphal branching (Fig. 6.2*b*). As a result, the antibiotic induces sensitive fungi to form colonies which are denser than control colonies and which expand in diameter more slowly than control colonies. Nioh & Mizushima (1974) showed that the density of hyphae (total hyphal length per unit area of substrate) at the margin of a validamycin A treated colony of *R. solani* was about three times higher than the density of hyphae at the margin of a control

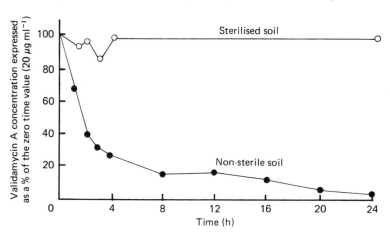

Fig. 6.6. Degradation of validamycin A in soil. From a figure in *Validacin: Humans and Nature First*, published by Takeda Chemical Industries Ltd, Agricultural Chemicals Division, Tokyo, Japan.

colony; there is probably an inverse relationship between hyphal density and peripheral growth-zone width (Bull & Trinci, 1977).

Figure 6.7 shows the effect of validamycin A on the radial growth rate of colonies of *Rhizoctonia cerealis* grown on a medium containing 5 mM glucose. Results similar to those shown in Fig. 6.7 have also been observed for *Rhizoctonia solani* (A. P. J. Trinci, unpublished); the characteristics used to distinguish between these two species are listed in Table 6.5. A deceleration in the radial growth rate of colonies of *R. cerealis* induced by validamycin A was always correlated with an increase in hyphal density at the colony margins. The response of *R. cerealis* to various concentrations of L-sorbose (Fig. 6.1 and Table 6.2) is quite different to its response to various concentrations of validamycin A (Fig. 6.7, Table 6.7). Fig. 6.7 and Table 6.7 show that the lag between inoculation and the onset of a deceleration in the radial growth rate of

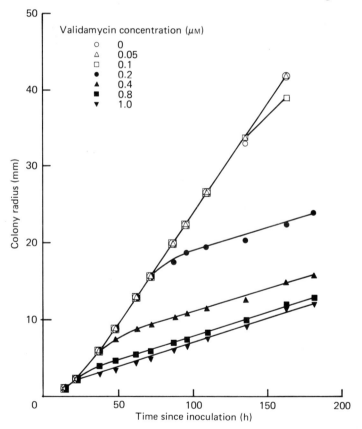

Fig. 6.7. Effect of validamycin A concentration on the radial growth rate of colonies of *Rhizoctonia cerealis* cultured on media containing 5 mM glucose.

Table 6.7. *Effect of validamycin A on colony radial growth rate and hyphal morphology of Rhizoctonia cerealis*[a]

Validamycin A (μM)	Colony radial growth rate (K_r, μm h^{-1})	Morphology of leading hyphae at the colony margin	
		Intercalary compartment length (μm)	Hyphal radius (μm)
0	277	188	1.86
0.05	281	180	1.98
0.10	101	134	2.09
0.20	68	92	1.11
0.40	59	93	1.28
0.80	65	98	1.12
1.00	68	91	1.28

[a]The cultures were grown on media containing 5 mM glucose.

R. cerealis colonies is inversely related to the validamycin concentration, and that the radial growth rate finally attained by validamycin 'inhibited' colonies is independent of antibiotic concentration.

Thus, validamycin A, unlike L-sorbose, apparently induces an 'all or none response'. However, the timing of the onset of the response varies with antibiotic concentration. The deceleration in colony radial growth rate induced by validamycin A is correlated with a decrease in intercalary compartment length (approximately halved) and a decrease in hyphal radius (Table 6.7). The changes in intercalary compartment length and hyphal radius induced by validamycin A (Table 6.7), unlike those induced by L-sorbose (Table 6.2), are not apparently affected by the concentration of the inhibitor; the 0.10 μM validamycin A result presumably reflects the fact that the measurements were made on colonies in the deceleration phase of growth (Fig. 6.7). Thus, validamycin A causes a decrease in the maximum rate of hyphal extension, a decrease in intercalary compartment length, a decrease in hyphal radius and an increase in hyphal branching. The antibiotic has also been shown to increase the angle which branches make with their parent hyphae (Nioh & Mizushima, 1974; Matsuura, 1982). However, it has no affect on the organism's maximum specific growth rate or growth yield.

Antagonism of the activity of validamycin A by meso-inositol and glucose

Meso-inositol (also known as myco- or *i*-inositol) is a cyclic polyalcohol or cyclitol; in plants it apparently acts as a precursor for other isomeric cyclitols. No co-enzyme function has been assigned to meso-inositol in fungi, but the 'vitamin' has been associated with fungal phospholipids, e.g. as phosphatidyl inositol in the mitochondrial membranes of *Neurospora crassa*. It would therefore seem that meso-inositol is a structural component of fungi rather than a true vitamin. Fungi unable to synthesise meso-inositol require its presence in the medium at a concentration of about $1–10\,\mu\text{M}$ (Griffin, 1981).

Wakae & Matsuura (1975) grew *Rhizoctonia solani* in the presence of validamycin A $(0.75–75\,\mu\text{g}\,\text{ml}^{-1})$ and found that the antibiotic reduced the amount of meso-inositol present in the culture filtrate. It has also been shown that when meso-inositol is applied with validamycin A it reduces the effectiveness of the antibiotic in preventing infection of rice by *R. solani* (Wakae & Matsuura, 1975; Akechi & Matsuura, 1982). This antagonism observed between validamycin A and meso-inositol was attributed by Wakae & Matsuura (1975) to the structural similarity between the 'vitamin' and component moieties of validoxylamine A (Fig. 6.4). However, meso-inositol $(400\,\mu\text{M})$ does not antagonise the 'inhibitory' effects of validamycin A $(0.2\,\mu\text{M})$ on *Rhizoctonia cerealis* when this organism is cultured on media containing $0.5\,\text{mM}$ or $5.0\,\text{mM}$ glucose (A. P. J. Trinci, unpublished), and Shibata, Uyeda & Mori (1980*a*, 1981) report that meso-inositol does not stimulate the rate of extension of hyphae of *R. solani* which have been 'inhibited' with validamycin A. Thus, although meso-inositol apparently antagonises the effect of validamycin A on the pathogenicity of *R. solani in vivo* (Wakae & Matsuura, 1975), no antagonism is observed between these two compounds *in vitro*.

Figure 6.8 shows the effects of validamycin A $(0.2\,\mu\text{M})$ on the radial-growth rate and morphology of colonies of *Rhizoctonia cerealis* grown on media containing various concentrations of glucose (1.5–20 mM). The lag between the time of inoculation and the onset of the 'inhibitory' effect (increased hyphal branching and reduced colony radial-growth rate) is directly related to glucose concentration. The antagonism observed between validamycin A and glucose may either result from catabolite repression or from competition for a common uptake mechanism; the latter explanation would seem unlikely in view of the structure of these two compounds.

A factor has been isolated from *Rhizoctonia solani* which stimulates the extension of validamycin A 'inhibited' hyphae (Shibata *et al.*, 1980*a*, 1981); the biological activity of this factor was only revealed after the removal of basic components from the original methanol extract.

Shibata *et al.* (1980*b*) also isolated a factor from *R. solani* which, like validamycin A, inhibited the extension of hyphae of this and other fungi (*Rhizoctonia fragarius*, *Sclerotium hydrophilum* and *Coprinus cinereus* f. *microsporum*). The authors suggest that these two factors may be involved in the endogenous regulation of hyphal extension in *R. solani*.

Translocation of validamycin A and of the antagonistic effect of glucose on validamycin A

The onset of reduction in the radial growth rate of colonies of *Rhizoctonia cerealis*, associated with the presence of validamycin A, is delayed when an inoculum plug lacking the antibiotic is separated from a medium containing validamycin A by a disc of aluminium foil or glass (Fig. 6.9). This result shows that the effect of glucose in antagonising the action of validamycin A can be translocated within hyphae over a considerable distance. The 'inhibitory' effect of validamycin A can also be translocated within hyphae (Fig. 6.10).

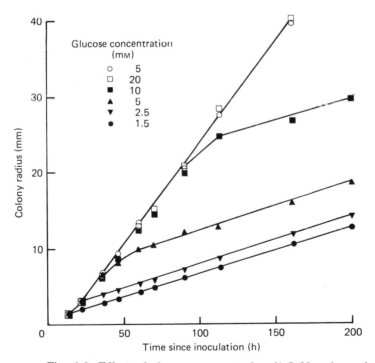

Fig. 6.8. Effect of glucose concentration (1.5–20 mM) on the radial-growth rate of colonies of *Rhizoctonia cerealis* cultured on media containing 0.2 μM validamycin A; the control medium (open circles) contained 5 mM glucose and no validamycin.

Validamycin A as an antifungal

Sclerotia of *Rhizoctonia solani* stick to the sheath joints of the rice plant and form a mycelium which spreads to the upper portions of the plant. Treatment of infected rice plants with validamycin A restricts the pathogen to the lower leaves and reduces loss of crop yield. For various reasons (see Fig. 6.6) validamycin A is more effective as a curative than as a protective antifungal. The development of strains of the pathogen resistant to validamycin A is apparently not a practical problem since no resistance was found amongst 650 isolates of *R. solani* isolated from rice infected with sheath blight (Matsuura, 1982). In

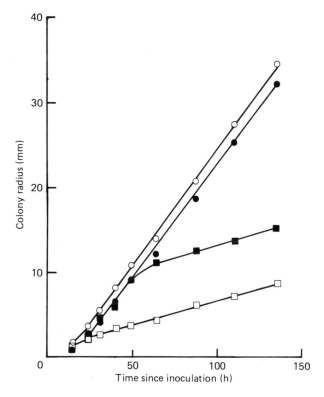

Fig. 6.9. Translocation of the 'glucose' effect within hyphae. Cultures were inoculated with mycelial plugs (15 mm diameter) taken from colonies cultured on a medium containing 0.5 mM glucose; some plugs were placed directly upon the medium (open circles, open squares) whilst others were placed upon glass coverslips (16 mm diameter; filled circles, filled squares). The medium contained 0.5 mM glucose with (open and filled squares) or without (open and filled circles) 0.2 μM validamycin A.

addition, sensitivity of the fungus to the antibiotic did not vary after a strain had been sub-cultured 20 times on medium containing validamycin A. One of the advantages of this type of antifungal may be that there is no obvious selective pressure which favours the development of resistance.

Validamycin A is sold as liquid (3.0% w/v validamycin A) and dust (0.3% w/w validamycin A) formulations. Solacol® was registered in The Netherlands in 1978 as a controlling agent of black scurf of potato; this liquid formulation contains $(g l^{-1})$ validamycin, 30, non-ionic detergent, 4, preservative, 1, antifoam agent, 0.5 and blue dye, 0.12 (Gams & Van Laar, 1982). Results obtained by Gams & Van Laar (1982) using

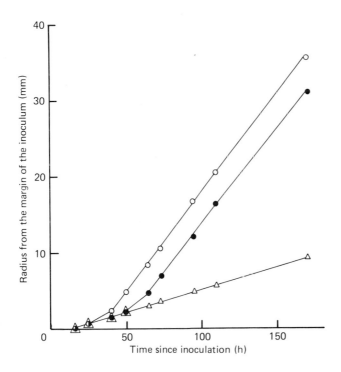

Fig. 6.10. Translocation of the 'inhibitory' effect of validamycin A within hyphae. Cultures were inoculated with plugs (5.4 mm diameter) taken from colonies cultured on a medium containing 0.5 mM glucose and 0.2 μM validamycin A; some plugs were placed directly upon the medium (open circles and triangles), whilst others (filled circles), were placed upon discs of aluminium foil which had a diameter of 5.6 mm. The medium contained 0.5 mM glucose with (triangles) or without (open and filled circles) 0.2 μM validamycin A.

Solacol® show that some of the antifungal activity of the formulation was due to the detergent rather than the antibiotic.

Conclusions

Little is known about the mechanism of hyphal extension although it is thought to involve a delicate balance between primary wall expansion and primary wall rigidification (Saunders & Trinci, 1979). As shown above, it is possible to reduce the maximum rate of hyphal extension (E_{max}) of a fungus without changing its maximum specific growth rate (μ_{max}). Conditions or mutations (Trinci, 1973*a*, *b*) which cause a reduction in E_{max} without altering μ_{max} will change the spatial distribution of the organism's biomass but not its rate of production. The maximum rate of extension of a hypha may be reduced either as the direct result of an inhibition of primary wall expansion or indirectly because of a decrease in the rate of supply of wall precursors etc. to the extending tip.

Antifungals with a validamycin-type mode of action may be effective in controlling certain pathogens of plants either because they have a direct affect on host penetration, or because the pathogen spreads more slowly in the host and therefore allows it more time to mobilise its defence mechanisms. The effectiveness of validamycin A in controlling rice sheath blight is strong evidence of the potential value of antifungals which have this particular mode of action.

Acknowledgement. I thank Mr Colin Dean for his assistance with the *R. cerealis* experiments.

References

Akechi, K. & Matsuura, K. (1982). Abstracts of the 5th International Congress of Pesticide Chemistry, Kyoto, Japan.

Anderson, N. A. (1982). The genetics and pathology of *Rhizoctonia solani*. *Annual Review of Phytopathology*, **20**, 329–47.

Bailey, J. A., Carter, G. A., Burden, R. S. & Wain, R. L. (1975). Control of rust diseases by diterpenes from *Nicotiana glutinosa*. *Nature*, **255**, 328–9.

Bailey, J. A., Vincent, G. G. & Burden, R. S. (1974). Diterpenes from *Nicotiana glutinosa* and their effect on fungal growth. *Journal of Geneal Microbiology*, **85**, 57–64.

Boerema, G. H. & Verhoeven, A. A. (1977). Check list for scientific names of common parasitic fungi. Series 2B Fungi on field crops: cereals and grasses. *Netherlands Journal of Plant Pathology*, **83**, 165–204.

Bull, A. T. & Trinci, A. P. J. (1977). The physiology and metabolic control of fungal growth. *Advances in Microbial Physiology*, **15**, 1–84.

Crocken, B. & Tatum, E. L. (1968). The effect of sorbose on metabolism and morphology of *Neurospora*. *Biochemica et Biophysica Acta*, **156**, 1–8.

Galpin, M. F., Jennings, D. H. & Thornton, J. D. (1977). Hyphal branching in *Dendryphiella salina*: effect of various compounds and further elucidation of the effect of sorbose and the role of cyclic AMP. *Transactions of the British Mycological Society*, **69**, 175–82.

Gams, W. & Van Laar, W. (1982). The use of solacol (validamycin) as a growth retardant in the isolation of soil fungi. *Netherlands Journal of Plant Pathology*, **88**, 39–45.

Gow, N. A. R. & Gooday, G. W. (1982). Growth kinetics and morphology of colonies of the filamentous form of *Candida albicans*. *Journal of General Microbiology*, **128**, 2187–94.

Griffin, D. H. (1981). *Fungal Physiology*. John Wiley & Sons: New York.

Hollins, T. W., Jellis, G. J. & Scott, P. R. (1983). Infection of potato and wheat by isolates of *Rhizoctonia solani* and *Rhizoctonia cerealis*. *Plant Pathology*, **32**, 303–10.

Horii, S., Kameda, Y. & Kawahara, K. (1972). Studies on validamycin, new antibiotics. VIII. Isolation and characterisation of validamycins C, D, E and F. *The Journal of Antibiotics*, **25**, 48–53.

Howell, C. R. (1978). Seed treatment with L-sorbose to control damping-off of cotton seedlings by *Rhizoctonia solani*. *Phytopathology*, **68**, 1096–8.

Iwasa, T., Higashide, E. & Shibata, M. (1971a). Studies on validamycins, new antibiotics. III. Bioassay methods for the determination of validamycin. *The Journal of Antibiotics*, **24**, 114–18.

Iwasa, T., Higashide, E., Yamamoto, H. & Shibata, M. (1971b). Studies on validamycins, new antibiotics. II. Production and biological properties of validamycins A and B. *The Journal of Antibiotics*, **24**, 107–13.

Iwasa, T., Kameda, Y., Asai, M., Horii, S. & Mizuno, K. (1971c). Studies on validamycins, new antibiotics. IV. Isolation and characterisation of validamycins A and B. *The Journal of Antibiotics*, **24**, 119–23.

Kameda, Y., Asano, N. & Hashimoto, T. (1978). Microbial glycosidation of validamycins. *The Journal of Antibiotics*, **31**, 936–8.

Kameda, Y., Asano, N., Wakae, O. & Twasa, T. (1980). Microbial glycosidation of validamycins. II. The preparation of α- and β-D-glucoside analogs of validamycins. *The Journal of Antibiotics*, **33**, 764–6.

Matsuura, K. (1982). Characteristics of validamycin A in controlling *Rhizoctonia* diseases. In *Proceedings of the 5th International Congress of Pesticide Chemistry*. Kyoto, Japan (in press).

Misato, T., Ko, K. & Yamaguchi, I. (1977). Use of antibiotics in agriculture. *Advances in Applied Microbiology*, **21**, 58–88.

Mishra, N. C. & Tatum, E. L. (1972). Effect of L-sorbose on polysaccharide synthetases of *Neurospora crassa*. *Proceedings of the National Academy of Sciences, USA*, **69**, 313–17.

Nioh, T. & Mizushima, S. (1974). Effect of validamycin on the growth and morphology of *Pellicularia sasakii*. *Journal of General and Applied Microbiology*, **20**, 373–83.

Parmeter, J. R., Sherwood, R. T. & Platt, W. D. (1969). Anastomosis grouping among isolates of *Thanatephorus cucumeris*. *Phytopathology*, **59**, 1270–8.

Saunders, P. T. & Trinci, A. P. J. (1979). Determination of tip shape in fungal hyphae. *Journal of General Microbiology*, **110**, 469–73.

Schuhmann, E. & Bergter, F. (1976). Mikroskopische Untersuchungen zur

Wachstumskinetik von *Streptomyces hygroscopicus*. *Zeitschrift für Allgemeine Microbiologie*, **16**, 201–15.

Shibata, M., Uyeda, M. & Mori, K. (1980*a*). Reversal of validamycin inhibition by the hyphal extract of *Rhizoctonia solani*. *The Journal of Antibiotics*, **33**, 679–81.

Shibata, M., Uyeda, M. & Mori, K. (1980*b*). Existence of hyphal extension factor and hyphal extension inhibitor in the hyphae of *Rhizoctonia solani*. *Agricultural and Biological Chemistry*, **44**, 2241–3.

Shibata, M., Uyeda, M. & Mori, K. (1981). Stimulation of the extension of validamycin-inhibited hyphae by the hyphal extension present in *Rhizoctonia solani*. *The Journal of Antibiotics*, **34**, 447–51.

Steele, G. C. & Trinci, A. P. J. (1975). Morphology and growth kinetics of hyphae of differentiated and undifferentiated mycelia of *Neurospora crassa*. *Journal of General Microbiology*, **91**, 362–8.

Suami, T., Ogawa, S. & Chida, N. (1980). The revised structure of validamycin A. *The Journal of Antibiotics*, **33**, 98–9.

Tatum, E. L., Barratt, R. W. & Cutter, V. M. (1949). Chemical induction of colonial paramorphs in *Neurospora* and *Syncephalastrum*. *Science*, **109**, 509–11.

Trinci, A. P. J. (1971). Influence of the peripheral growth zone on the radial growth rate of fungal colonies. *Journal of General Microbiology*, **67**, 325–44.

Trinci, A. P. J. (1973*a*). The hyphal growth unit of wild type and spreading colonial mutants of *Neurospora crassa*. *Archive für Mikrobiologie*, **91**, 127–36.

Trinci, A. P. J. (1973*b*). Growth of wild type and spreading colonial mutants of *Neurospora crassa* in batch culture and on agar medium. *Archive für Mikrobiologie*, **91**, 113–26.

Trinci, A. P. J. (1974). A study of the kinetics of hyphal extension and branch initiation of fungal mycelia. *Journal of General Microbiology*, **81**, 225–36.

Trinci, A. P. J. (1984). Regulation of hyphal branching and hyphal orientation. In *Ecology and Physiology of the Fungal Mycelium*, 6th Symposium of the British Mycological Society, ed. A. D. M. Rayner & J. H. Jennings, pp. 23–52. Cambridge University Press.

Trinci, A. P. J. & Collinge, A. J. (1973). Influence of L-sorbose on the growth and morphology of *Neurospora crassa*. *Journal of General Microbiology*, **78**, 179–92.

Wakae, O. & Matsuura, K. (1975). Characteristics of validamycin as a fungicide for *Rhizoctonia* disease control. *Review of Plant Protection Research*, **8**, 81–92.

Wilson, R. W. & Niederpruem, D. J. (1967). Cellobiose as a paramorphogen in *Schizophyllum commune*. *Canadian Journal of Microbiology*, **16**, 629–34.

7

Mode of action of some traditional fungicides

A. KAARS SIJPESTEIJN

Netherlands Organisation for Applied Scientific Research, Institute of Applied Chemistry, P.O. Box 5009, 3502 JA Utrecht, The Netherlands

Introduction

The following is an account of the modes of action of the protectant fungicides, thiram, bisdithiocarbamates, fentin and dinocap. They are all thought to act on functions involved in energy production but their precise inhibition mechanisms have yet to be elucidated. Since the advent of systemic fungicides, mode of action studies have been largely focused on these new compounds, and many of the older protectant fungicides have not yet received the attention which they undoubtedly deserve.

Thiram

Thiram and sodium dimethyldithiocarbamate

Tetramethylthiuram disulphide, TMTD or thiram, was one of the first fungicides used in agriculture. It has been used to control a variety of fungal diseases of plants. The sodium salt of dimethyl-dithiocarbamate (NaDDC) has often been used in laboratory studies

$$(CH_3)_2N-C-S-S-C-N(CH_3)_2$$
$$\underset{S}{\overset{\parallel}{}}\qquad\underset{S}{\overset{\parallel}{}}$$

Thiram

$$(CH_3)_2N-C-S-Na$$
$$\underset{S}{\overset{\parallel}{}}$$

Na-dimethyldithiocarbamate

since thiram is rather insoluble in water. This reduction product of the disulphide shows the same antifungal properties as thiram itself; in fact, thiram is generally supposed to act after it has been reduced to the dithiocarbamate. The compound is also a potent inhibitor of gram-positive bacteria.

Antifungal spectrum

The sensitivity of a variety of fungi to NaDDC has been studied on glucose mineral salts agar made up with tap water (final pH 7.0); the crude agar used in these experiments had first been washed thoroughly with tap water. Conidia were suspended in the molten agar medium. Characteristic results obtained for NaDDC (thiram gives the same result) are shown in Table 7.1. According to their behaviour, fungi can be roughly divided into three groups.

Group I. The growth of fungi in this group is inhibited completely by *c.* $0.2\,\mu g\,ml^{-1}$ of NaDDC. Examples of these NaDDC-sensitive fungi include *Glomerella cingulata, Botrytis allii, Botrytis cinerea, Ceratostomella ulmi, Glomerella fructigenum, Sclerotinia fructicola* and the yeast *Rhodotorula glutinis*.

Group II. Growth of fungi in this group is inhibited at low concentrations ($0.5–2\,\mu g\,ml^{-1}$) of NaDDC ('first zone of inhibition'), but growth occurs at higher (5, 10 and $20\,\mu g\,ml^{-1}$) NaDDC concentrations ('zone of inversion growth') and growth inhibition again occurs at NaDDC concentrations of $50\,\mu g\,ml^{-1}$ and above ('second zone of inhibition'). We have called fungi showing this unusual inhibition response pattern, 'fungi with inversion growth'. Fungi belonging to this group include *Aspergillus niger* (Fig. 7.1), *Aspergillus oryzae, Aspergillus ruber, Alternaria solani, Alternaria tenuis, Cladosporium cucumerinum, Cladosporium herbarum* and *Penicillium brevicompactum*.

Group III. The growth of fungi in this group is inhibited by NaDDC concentrations of *c.* $50\,\mu g\,ml^{-1}$ and above. These 'insensitive' fungi

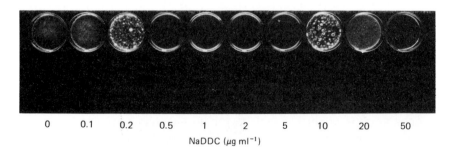

| 0 | 0.1 | 0.2 | 0.5 | 1 | 2 | 5 | 10 | 20 | 50 |

NaDDC ($\mu g\,ml^{-1}$)

Fig. 7.1. *Aspergillus niger* on glucose mineral salts agar made up with tap water; pH 7.0. The plates were photographed 5 d after inoculation with conidia (from Kaars Sijpesteijn & Van der Kerk, 1954*b*).

Table 7.1. *Inhibition of fungi by NaDDC on glucose mineral salts agar made up with tap water, pH 7.0[a]*

	Concentration of NaDDC (μg ml^{-1})										
	0	0.05	0.1	0.2	0.5	1	2	5	10	20	50
Group I *Glomerella cingulata* type	+[b]	+	+	−[c]	−	−	−	−	−	−	−
Group II *Aspergillus niger* type	+	+	+	+	−	−	−	±[d]	+	+	−
Group III *Fusarium oxysporum* type	+	+	+	+	+	+	+	+	+	+	−

From Kaars Sijpesteijn & Janssen, 1959.
[a]Aneurin and biotin were added, if necessary, and the cultures were inoculated with conidia.
[b]Growth observed.
[c]No macroscopically visible growth observed.
[d]Only poor growth observed.

include *Fusarium oxysporum, Aspergillus fumigatus, Aspergillus nidulans, Aspergillus ochraceus, Aspergillus versicolor, Fusarium avenaceum, Fusarium bulbigenum, Fusarium coeruleum, Penicillium citrinum* and *Penicillium expansum*.

Fungi which show behaviour intermediate between groups I and II, and II and III, can also be found, e.g. slow growth may occur in the 'zone of inversion growth', (*Penicillium italicum* and *Saccharomyces cerevisiae*) or in the 'first zone of inhibition'. The peculiar phenomenon of 'inversion growth' was first described by Dimond *et al*. (1941) who studied the effect of TMTD on spore germination of *Macrosporium sarciniforme*; they found a double maximum in the dose-response curve. When mycelium was exposed to NaDDC, growth appeared, especially after a long incubation period, to be somewhat less affected than spore germination (Kaars Sijpesteijn & Van der Kerk, 1952).

With the exception of *Aspergillus niger*, the action of NaDDC on all the fungi studied was independent of pH. *A. niger* behaves like a Group II fungus at pH values above 6.5, and like a Group III fungus at pH values below *c*. 6.5, i.e. the 'first zone of inhibition' is not observed.

Only fungi showing Group I 'sensitivity' can be controlled by thiram, but an understanding of the mode of action of thiram in fact emerged from studies made on Group II fungi.

Effect on Aspergillus niger *(Group II)*

The 'first zone of inhibition' (Fig. 7.1) was observed only on a glucose mineral salts medium (pH 7), but not when the medium was enriched with malt extract, peptone or casamino acids. In these latter cases, the minimal inhibitory concentration was increased to 20–50 μg ml^{-1}. An analysis of these antagonistic effects disclosed that the L-histidine could counteract antifungal activity in the 'first zone of inhibition' but not at higher fungicide concentrations (Fig. 7.2). The amount of L-histidine contained in the antagonistic substances listed above could account for the effects observed (Kaars Sijpesteijn & Van der Kerk, 1952). D-histidine and several other imidazole derivatives also proved antagonistic to the fungicide (Kaars Sijpesteijn & Van der Kerk, 1954*a*). Histidine also antagonised the inhibitory effects of NaDDC on Group I and other Group II fungi, but not on the 'insensitive' Group III fungi. The compound was more effective as an antagonist at pH 7.0 than at lower pH values (Kaars Sijpesteijn & Van der Kerk, 1952).

The role of metals in the action of NaDDC and thiram had long been suspected, but direct evidence first came from the investigations of Goksøyr (1955*a*, *b*) on the action of NaDDC on *Saccharomyces cerevisiae*. He stressed the importance of the 1:1 and 1:2 complexes of DDC⁻ ions with Cu^{2+} ions.

$$\underset{\text{1:1 Complex}}{\chemfig{}} \qquad \underset{\text{1:2 Complex}}{\chemfig{}}$$

1:1 Complex   1:2 Complex

Traces of Cu^{2+} are probably invariably present on plant surfaces or in culture media. In media, an equilibrium will exist between the 1:1 complex ($CuDDC^+$) and the 1:2 complex ($CuDDC_2$) and this equilibrium will be governed by the relative amounts of Cu^{2+} and DDC⁻ ions present.

$$[CuDDC]^+ \underset{+Cu}{\overset{+DDC}{\rightleftharpoons}} CuDDC_2$$

With an excess of DDC⁻ ions, all Cu^{2+} present is bound to the very insoluble, 1:2 complex; only then can free DDC⁻ ions exist. The 1:1 complex, which is soluble in water, is only present when excess Cu^{2+} ions are available.

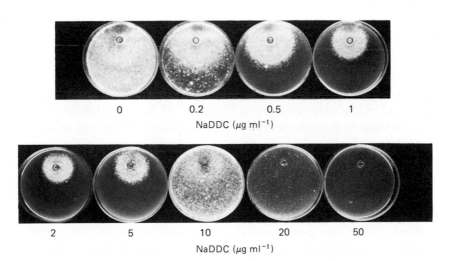

0   0.2   0.5   1

NaDDC (μg ml⁻¹)

2   5   10   20   50

NaDDC (μg ml⁻¹)

Fig. 7.2. *Aspergillus niger* on glucose mineral salts agar made up with tap water; pH 7.0. The cup contained 2% L-histidine and the plates were photographed 5 d after inoculation with conidia.

Table 7.2. *Effect of NaDDC and copper sulphate on growth of Aspergillus niger cultured in glucose mineral salts solution at pH 7.0*

Concentration of $CuSO_4 \cdot 5H_2O$ ($\mu g\,ml^{-1}$)	Concentration of NaDDC ($\mu g\,ml^{-1}$)								
	0	0.2	0.5	1	2	5	10	20	50
0	+[b]	+	+	+	+	+	+	+	−[c]
3	+	+	+	+	+	+	+	+	−
10	+	+	−	−	−	±[d]	+	+	−

From Kaars Sijpesteijn & Janssen, 1959.
[a]The basal medium was 'copper free', cultures were inoculated with conidia and incubated for 3 d in shake culture.
[b]Growth observed.
[c]No macroscopically visible growth observed.
[d]Only poor growth observed.

Goksøyr suggested that, for *Saccharomyces cerevisiae*, the 'first zone of inhibition' coincides with the presence of the 1:1 complex and he correlated the 'zone of inversion growth' with the presence of the 1:2 complex. However, he did not carry out growth experiments in which he used different concentrations of Cu^{2+}. His hypothesis predicts there should be no growth inhibition of *Aspergillus niger* in the 'first zone' if the medium was Cu^{2+}-free. As Table 7.2 shows, this was indeed the case; the flasks used in the experiment were cleaned with steam, the culture medium was made up with double-distilled water and the total Cu^{2+} content did not exceed $0.01\,\mu g\,ml^{-1}$. The addition of $3\,\mu g\,ml^{-1}$ of copper sulphate to this medium was insufficient to restore the inhibitory effect of the fungicide, but at $10\,\mu g\,ml^{-1}$ the 'first zone of inhibition' reappeared.

Figure 7.3 gives the amounts of the complexes and of free DDC^- ions dissolved in the series containing $10\,\mu g\,ml^{-1}$ of copper sulphate. These were calculated from the stability constants for the complexes and their solubilities (Janssen, 1956 and 1957; Janssen & Kaars Sijpesteijn, 1961).

Table 7.2 shows that both the 1:1 complex and free DDC^- ions are toxic when present at certain concentrations. Because of its low solubility in water ($0.01\,\mu g\,ml^{-1}$) the 1:2 complex only attains low concentrations in solution; this complex appears to be non-toxic. Indeed, when the compound $CuDDC_2$ was tested in a Cu^{2+}-free medium, it was non-toxic to *Aspergillus niger*, and the same result was obtained for *Fusarium oxysporum*. In our earlier experiments we always used tap water to prepare media and used agar which had been thoroughly washed with tap water. By doing so we had apparently

Table 7.3. *Effect of NaDDC and copper sulphate on the growth of Aspergillus niger cultured on glucose mineral salts agar at pH 7.0[a]*

Concentration of $CuSO_4 \cdot 5H_2O$ ($\mu g\, ml^{-1}$)	Concentration of NaDDC ($\mu g\, ml^{-1}$)							
	0	0.5	1	2	5	10	20	50
0	+[b]	+	+	+	+	+	+	−[c]
0.1	+	+	+	+	+	+	+	−
0.3	+	+	±[d]	+	+	+	+	−
1	+	±	−	−	+	+	+	−
3	+	+	−	−	−	+	+	−

From Kaars Sijpesteijn, Janssen & Van der Kerk, 1957.
[a]The basal medium was 'copper-free'; cultures were inoculated with conidia and incubated for 3 d.
[b]Growth observed.
[c]No macroscopically visible growth observed.
[d]Only poor growth observed.

introduced considerable amounts of Cu^{2+} into the media (Kaars Sijpesteijn, Janssen & Van der Kerk, 1956). Similar results were obtained in Cu^{2+}-free agar media and liquid cultures, except that (for unknown reasons) Cu^{2+} proved more effective in causing the 'first zone of inhibition' in agar media than in liquid cultures (Table 7.3).

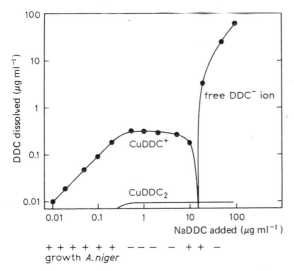

Fig. 7.3. Calculated concentration of dimethyldithiocarbamate ion and its copper complexes compared with growth inhibition of *Aspergillus niger*, pH 7.0, copper sulphate $10\,\mu g\, ml^{-1}$ (from Janssen & Kaars Sijpesteijn, 1961).

Of other heavy metals tested only Co was able to give an effect, although to a far less extent than Cu^{2+} (Kaars Sijpesteijn, Janssen & Van der Kerk, 1957).

Effect on Glomerella cingulata *(Group I)*

Glomerella cingulata is an extremely sensitive fungus which shows no 'inversion growth' i.e. it belongs to Group I. It was thought that 'inversion growth' might not occur if the 1:2 complex was toxic to this fungus. Indeed, in a Cu^{2+}-free medium, $CuDDC_2$ completely inhibited spore germination at a concentration of $0.05\ \mu g\ ml^{-1}$, but did not inhibit it at $0.02\ \mu g\ ml^{-1}$; the solubility of $CuDDC_2$ is only *c.* $0.01\ \mu g\ ml^{-1}$. Probably this concentration is just inhibitory, but the rate of dissolution is slow, therefore somewhat more may be required to quickly build up an inhibitory concentration (Kaars Sijpesteijn & Janssen, 1958).

Tables 7.4*a* and *b* give the growth response of *Glomerella cingulata* in relation to the calculated amounts of copper complexes dissolved (Kaars Sijpesteijn & Janssen, 1959). Cu^{2+}-free ingredients were used in these shake cultures. Growth inhibition in flasks without added Cu^{2+} must be due to complex formation between dithiocarbamate and the Cu^{2+} adhering to the glass or the conidia. When all Cu^{2+} still present was bound by the addition of $0.3-1\ \mu g\ ml^{-1}$ of sodium dibutyldithiocarbamate, the minimal inhibitory concentration of NaDDC increased to $20\ \mu g\ ml^{-1}$ (Kaars Sijpesteijn & Janssen, 1958). The dibutyl derivative is a homologue of NaDDC which is non-toxic to fungi and has much stronger chelating properties.

As shown in Tables 7.4*a* and *b*, the toxic 1:2 complex always accompanies the 1:1 complex; it is, therefore, difficult to assess the toxicity of the latter complex. We assume, however, that it is quite toxic. A strong indication was obtained from the fact that, in a copper-containing medium, the homologue Na-diethyldithiocarbamate causes a toxicity pattern of the Group II type, (Table 7.5), i.e. *Glomerella cingulata* then displays 'inversion growth' and two 'zones of inhibition' (Kaars Sijpesteijn & Janssen, 1958, 1959). This means that the 1:1 complex of this homologue is toxic, whereas the 1:2 complex is too insoluble to cause toxicity.

At higher NaDDC-concentrations both the saturated solution of 1:2 complex and the free DDC^- ions will cause toxicity to *Glomerella cingulata*. Interestingly, inhibition of *Bacillus subtilis* by NaDDC and

Table 7.4. *Response of Glomerella cingulata to NaDDC[a]*
(a) Response in relation to the amount of 1:1 complex (CuDDC[+])
calculated to be present[b]

Concentration of CuSO$_4 \cdot$5H$_2$O (μg ml^{-1})	Concentration of NaDDC (μg ml^{-1})								
	0.02	0.05	0.1	0.2	0.5	1	2	5	10
				c					
0									
0.03	0.01	0	0						
0.1	0.02	0.03	0.02	0					
0.3	0.02	0.04	0.05	0.04	0	0			
1	0.02	0.05	0.1	0.1	0.1	0.06	0		
3	0.02	0.05	0.1	0.2	0.2	0.2	0.1	0	0

(b) Response in relation to calculated amount of 1:2 (CuDDC$_2$) in
solution (μg ml^{-1})

Concentration of CuSO$_4 \cdot$ 5H$_2$O (μg ml^{-1})	Concentration of NaDDC (μg ml^{-1})				
	0.02	0.05	0.1	0.2	0.5
			c		
0					
0.03	S[d]	S	S	S	S
0.1	0.002	S	S	S	S
0.3	0.001	0.004	S	S	S
1		0.001	0.005	S	S
3			0.001	S	S

From Kaars Sijpesteijn & Janssen, 1959.
[a]The organism was grown on 'copper-free' glucose mineral salts solution with various concentrations of copper sulphate; pH 7.0.
[b]The figures indicate the calculated concentrations of 1:1 complex (μg ml^{-1}) in the medium. The shake cultures were inoculated with conidia and incubated for 3 d.
[c]—: inhibition pattern. Left: growth; right: inhibition.
[d]S: saturated concentration of CuDDC$_2$ (c. 0.01 μg ml^{-1}).

the diethyl homologue follow the same pattern as described here for *G. cingulata* (Kaars Sijpesteijn & Janssen, 1959).

One must assume that under field conditions sufficient Cu^{2+} is always present to enable NaDDC to act at very low concentrations on sensitive pathogens.

Biochemical mode of action of NaDDC

In the case of *Glomerella cingulata*, low concentrations of NaDDC act as both the 1:1 and the 1:2 complex with Cu^{2+}, whereas in

Table 7.5. *Action of sodium diethyldithiocarbamate and copper sulphate on growth of Glomerella cingulata in glucose mineral salts solution at pH 7.0*[a]

Concentration of CuSO$_4$ · 5H$_2$O (μg ml^{-1})	Sodium diethyldithiocarbamate concentration (μg ml^{-1})									
	0	0.1	0.2	0.5	1	2	5	10	20	50
0	+[b]	±[c]	±	+	+	+	+	±	−[d]	−
0.1	+	+	±	±	+	+	+	+	−	−
0.3	+	+	−	−	+	+	+	+	−	−
1	+	+	−	−	−	−	±	±	±	−
10	+	+	−	−	−	−	−	−	±	−

From Kaars Sijpesteijn & Janssen, 1958.
[a]The basal medium was 'copper-free'; the shake cultures were inoculated with conidia and incubated for 3 d.
[b]Growth observed.
[c]Only poor growth observed.
[d]No macroscopically visible growth observed.

the case of *Aspergillus niger* only the former complex is toxic. *Fusarium oxysporum* (a Group III fungus) is insensitive to both complexes. All three fungi are, however, inhibited by the free dithiocarbamate ion at higher concentrations.

We regard the 1:1 complex as the primary toxic agent which, in the cell, interferes with some essential function. Toxicity of the 1:2 complex, as in the case of *Glomerella cingulata*, may well be due to formation of a 1:1 complex inside the cell. The same suggestion was also made by Albert, Gibson & Rubbo (1953) for the complexes of oxine, a compound for which we assume that Cu-complexes act in the same way as those of NaDDC (Kaars Sijpesteijn & Janssen, 1958). Oxine also has antibacterial as well as antifungal activity.

The antagonistic effect of histidine and other imidazoles cannot be ascribed to a direct extraction of Cu^{2+} from dithiocarbamate complexes since they have insufficient complexing ability. Instead, we believe that, if present in large excess, they bind to the 1:1 complex and form an imidazole:Cu:DDC complex (Kaars Sijpesteijn *et al.*, 1957). The imidazoles would then compete with the target enzyme for the 1:1 complex. It is of interest that the antagonistic action of the imidazoles appears to run parallel to their affinity for metal ions, especially for copper (Janssen & Kaars Sijpesteijn, 1961).

The question of the actual target of the 1:1 complex in the cell is still unresolved, although it is generally assumed that low concentrations of NaDDC inhibit some function in respiration.

Goksøyr (1955*a*, *b*; 1958) found a strong effect of the 1:1 complex on acetate oxidation by intact cells of *Saccharomyces cerevisiae*, but the results of his experiments on O_2-uptake, with glucose as a substrate, are less clear. Inhibition of O_2-uptake by *Aspergillus niger* in the presence of 500 μg ml^{-1} NaDDC, but not at 200 μg ml^{-1} and below, was observed by Klöpping (1951) and Van der Kerk & Klöpping (1952). They did, however, use a malt medium which contained histidine, and probably had a pH at which *A. niger* does not show a 'first żone of inhibition'.

Working with rather high concentrations of NaDDC or thiram, and sometimes with fungi of Group III, Sisler & Cox (1954 & 1955) reported that respiration was inhibited. This result, however, is relevant to the mode of action of the free DDC$^-$ ions, not the 1:1 complex. The same applies to studies on the action of NaDDC or thiram on enzyme systems from various fungi, plants and animals. A survey of these experiments is given by Thorn & Ludwig (1962) in their review of dithiocarbamates.

In experiments with pellets of *Aspergillus niger* and other fungi, we found some accumulation of pyruvate under conditions of 'first zone of inhibition' (Fig. 7.4; Kaars Sijpesteijn *et al.*, 1957). This observation

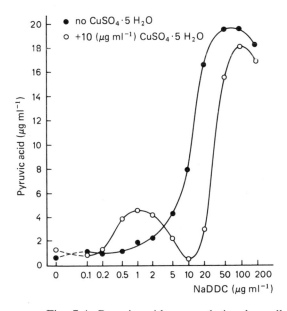

Fig. 7.4. Pyruvic acid accumulation by pellets (dry weight 7 mg) of *Aspergillus niger* in glucose mineral salts solution containing various concentrations of NaDDC in the absence (filled circles) and presence (open circles) of copper sulphate (10 μg ml^{-1}). The pellets were shaken for 4 h in 20 ml of medium (from Kaars Sijpesteijn *et al.*, 1957).

may indicate interference with the pyruvate dehydrogenase system. Experiments with germinating conidia may show a more pronounced effect since they are more sensitive than mycelium to NaDDC.

We have hypothesised that the 1:1 complex may interfere with dithiol groups present in various enzyme systems e.g. in lipoic acid or lipoic acid dehydrogenase. Dithiol groups may well combine with the 1:1 complex (Kaars Sijpesteijn & Janssen, 1959). Moreover, it is known that lipoic acid dehydrogenase from *Saccharomyces cerevisiae* is very sensitive to Cu^{2+} ions (Wren & Massey, 1966). If, however, this inhibition accounts for fungal toxicity, one would also expect to observe some inhibition of respiration.

Recently Yoneyama, Sekido & Misato (1982) observed an effect of thiram on lipid biosynthesis in the gram-negative bacterium, *Xanthomonas campestris* p.v. *oryzae*. However, gram-negative bacteria are rather insensitive to thiram. Further, the concentration (*c.* $120 \mu g\, ml^{-1}$) used in these experiments is not relevant to the inhibitory activity of the 1:1 complex. Recent studies on the mutagenic activity of thiram in the Ames test (Zdzienicka *et al.*, 1979) and on the induction of abnormal chromosome segregation (Upshall & Johnson, 1981) also do not seem to be relevant to its mode of action as a fungicide.

In conclusion we can say that at low NaDDC concentrations the 1:1 complex, and, for very sensitive fungi, the 1:2 complex, are the inhibitory agents. Their biochemical modes of action are thought to involve inhibition of some specific site in respiration. Although this may be true, convincing proof, even for inhibition of respiration, is still lacking.

Activity of NaDDC at somewhat higher concentrations is due to the free DDC^- ions. They almost certainly act by inhibiting respiration; several possible targets have been suggested in the literature. Their inhibitory action has also been ascribed to removal of heavy metals from enzyme systems, but this process presumably only occurs at very high fungicide concentrations.

Thus there is a need for critical investigation and re-evaluation of the mode of action of NaDDC at low concentrations, i.e. of the action of the 1:1 complex ($CuDDC^+$) on sensitive fungi.

Bisdithiocarbamates

The derivatives mancozeb, maneb and zineb still have wide practical application. For laboratory studies, however, the sodium salt nabam is generally used because of its much better solubility. The

Table 7.6. *Antimicrobial activity of nabam, DIDT and ETU*

Microorganisms	Minimum concentration required for complete growth inhibition (μg ml^{-1})		
	nabam	DIDT[a]	ETU[b]
Fungi[c]			
Botrytis allii	2	1	>500
Penicillium italicum	5	0.5	>500
Aspergillus niger	10	2	>500
Cladosporium cucumerinum	2	0.5	>500
Saccharomyces cerevisiae	5	10	>500
Bacteria[d]			
Bacillus subtilis	5	20	>500
Escherichia coli	50	100	>500
Pseudomonas fluorescens	200	100	>500
Mycobacterium phlei	20	10	>500

From Vonk, 1975
[a]DIDT: 5,6-dihydro-3H-imidazo [2,1-c]-1,2,4-dithiazole-3-thione.
[b]ETU: ethylene thiourea.
[c]The glucose mineral salts agar cultures (pH 7.0) were inoculated with conidia and incubated for 3 d.
[d]The peptone glucose agar cultures (pH 7.0) were inoculated and incubated for 2 d.

compound has a broad antifungal spectrum and also possesses some activity against bacteria. Characteristic results are given in Table 7.6 (Vonk, 1975). Nabam inhibits spore germination and mycelial growth. Growth is slightly less affected at pH 7 than at pH 5; cysteine and other thiol compounds strongly antagonise the effect of nabam on sensitive fungi (Kaars Sijpesteijn & Van der Kerk, 1954b).

Because of the presence of a H-atom attached to the N-atom, nabam, in contrast to dialkyldithiocarbamates, can lose H$_2$S during the simultaneous formation of the reactive thiol-reagent, ethylene diisothiocyanate (Fig. 7.5). The observation that this compound is more toxic than nabam itself led Klöpping & Van der Kerk (1951) to postulate that nabam acts after formation *in situ* of the unstable isothiocyanate, which will in turn combine with SH-groups in the fungus. This would also explain the antagonistic effect of thiol compounds.

It was later shown by Pluijgers, Vonk & Thorn (1971) that in aqueous solution nabam forms the compound 5,6-dihydro-3H-imidazo [2,1-c]-1,2,4-dithiazole-3-thione (DIDT). This compound was previously thought to be ethylene thiuram monosulphide. Moreover, nabam and (at a very slow rate) DIDT gave rise to ethylene thiourea (ETU) (Thorn

& Ludwig, 1962; Vonk, 1971, 1975). Under physiological conditions DIDT undergoes reductive ring opening (Kaars Sijpesteijn & Vonk, 1975; Vonk & Kaars Sijpesteijn, 1976). Whereas ETU has no antifungal activity, DIDT is an even more potent antifungal agent than nabam itself (Table 7.6). This activity can be explained on the basis of the formation of ethylene diisothiocyanate from the reduction product of DIDT (Vonk, 1975; Kaars Sijpesteijn & Vonk, 1975).

Thus nabam acts after conversion into ethylene diisothiocyanate and this compound will combine unspecifically with many cell constituents, such as thiol-containing enzymes. Although this would lead to an inhibition of respiration, Klöpping (1951), using the Warburg technique, only observed such inhibition at concentrations well above the growth-inhibitory concentration of nabam. A reason for this may be that some time is needed to convert this compound into the active agent. Moreover, pellets of *Aspergillus niger* were used in these Warburg experiments; spore germination may be more sensitive to the fungicide than mycelial growth.

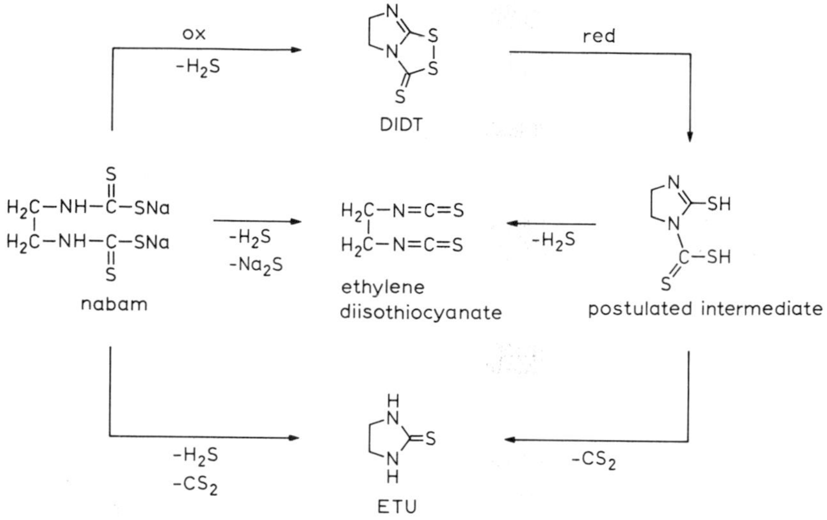

Fig. 7.5. Scheme of the decomposition of nabam in aqueous solution (from Kaars Sijpesteijn & Vonk, 1975).

Fentin

Triphenyltin acetate, or fentin acetate, has a broad antimicrobial spectrum; it is active against many fungi as well as gram-positive bacteria (cf. Kaars Sijpesteijn *et al.*, 1962; Kaars Sijpesteijn, Luijten & Van der Kerk, 1969). The phytotoxicity of fentin has limited its use in crop protection to control of *Phytophthora infestans* on potato and *Cercospora beticola* on sugar beet. The same applies to fentin hydroxide.

Fentin acetate

Fentin acetate inhibits spore germination and mycelial growth equally; pH or composition of the medium do not influence activity and no antagonists of its action are known. The compound affects respiration; it has an immediate effect on O_2 uptake by pellets of *Aspergillus niger* in glucose phosphate buffer at pH 7 (Kaars Sijpesteijn, 1970). Whereas $1\,\mu g\,ml^{-1}$ suppresses growth from conidia almost completely, $3\,\mu g\,ml^{-1}$ depressed respiration of washed mycelial suspensions by about 50%.

In fungi the target was not further elucidated. However, trialkyl- and triphenyltin derivatives are known to inhibit oxidative phosphorylation at low concentrations in rat mitochondria in a manner similar to oligomycin (Aldridge & Cremer, 1955; Aldridge & Rose, 1969; Stockdale, Dawson & Selwyn, 1970). The same will probably apply to fungal cells because, according to Watson & Smith (1967), oxidative phosphorylation proceeds in mitochondria of *Aspergillus niger* much in the same way as in yeast and mammalian mitochondria. This hypothesis is supported by the finding that some mutants of *Saccharomyces cerevisiae* resistant to triethyltin sulphate were also resistant to oligomycin (Lancashire & Griffith, 1971). In his review on the influence of organotin compounds, Aldridge (1976) mentions two different ways in which triorganotins derange mitochondrial function. Thus fentin may be regarded as a specific inhibitor of respiration.

In the field, serious resistance to fentin was found in Greece, where *Cercospora beticola* could no longer be controlled in several areas using this fungicide (Georgopoulos, 1982).

Dinocap

Dinitro-octylphenylcrotonate, or dinocap, has rather specific activity against powdery mildews. The compound inhibits spore germination.

Dinocap

Dinitrophenols are known to act on oxidative phosphorylation in mammalian mitochondria. They uncouple respiration from phosphorylation and result in a stimulation of respiration and a depression of ATP-formation. Presumably, dinocap acts in the same way on mildews, after transformation to a phenol. Because of its high lipophilicity, dinocap probably does not penetrate into other fungi.

General conclusions

All the protectant fungicides dealt with in this chapter probably interfere with energy production and, with the exception of the bis-dithiocarbamates, their target is presumably quite specific. Spore germination is at least as sensitive as mycelial growth. The biochemical mode of action of thiram and its derivatives especially needs further study.

The modes of action of the fungicides treated above seem to differ in many respects from those of most systemic fungicides; the latter compounds act quite frequently on biosynthetic processes and leave energy production unaffected. Also they are usually more effective against the growing mycelium than against spore germination.

References

Albert, A., Gibson, M. I. & Rubbo, S. D. (1953). The influence of chemical constitution on antibacterial activity. Part VI. The bactericidal action of 8-hydroxyquinoline (oxine). *British Journal of Experimental Pathology*, **34**, 119–30.

Aldridge, W. N. (1976). The influence of organotin compounds on mitochondrial functions. In *Advances in Chemistry Series*, No. 157, ed. J. J. Zukerman, pp. 186–96. American Chemical Society.

Aldridge, W. N. & Cremer, J. E. (1955). The biochemistry of organo-tin compounds: diethyltin dichloride and triethyltin sulphate. *Biochemical Journal*, **61**, 408–18.

Aldridge, W. N. & Rose, M. S. (1969). The mechanism of oxidative phosphorylation: a hypothesis derived from studies of trimethyltin and triethyltin compounds. *FEBS Letters*, **4**, 61–8.

Dimond, A. E., Horsfall, J. G., Heuberger, J. W. & Stoddard, E. M. (1941). Role of the dosage-response curve in the evaluation of fungicides. *Connecticut Agricultural Experiment Station Bulletin*, **451**, 635–67.

Georgopoulos, S. G. (1982). Case study 1: *Cercospora beticola* of sugar-beets. In *Fungicide Resistance in Crop Protection*, ed. J. Dekker & S. G. Georgopoulos, pp. 187–94. Wageningen: Pudoc.

Goksøyr, J. (1955a). Reversal of the fungicidal effect of dithiocarbamyl compounds. *Nature*, **175**, 820.

Goksøyr, J. (1955b). The effect of some dithiocarbamyl compounds on the metabolism of fungi. *Physiologia Plantarum*, **8**, 719–835.

Goksøyr, J. (1958). The mechanism of action of dithiocarbamyl compounds on the acetate oxidation pathway in yeast. *Universitetet i Bergen Årbok*, 1958. Naturvitenskapelig rekke, Nr. 3, 1–16.

Janssen, M. J. (1956). The stability constants of metal complexes of some *n*-dialkyldithiocarbamic acids. Part I. Copper complexes in 75% (v/v) ethanol water. *Recueil des Travaux Chimiques des Pays Bas*, **75**, 1411–22.

Janssen, M. J. (1957). The stability constants of metal complexes of some n-dialkyldithiocarbamic acids. Part II. Copper complexes in ethanol/water mixtures of various compositions. *Recueil des Travaux Chimiques des Pays-Bas*, **76**, 827–35.

Janssen, M. J. & Kaars Sijpesteijn, A. (1961). Mode of action of dithiocarbamates. *Fungicides in Agriculture and Horticulture*. S.C.I. monograph No. 15. London: Society Chemical Industry.

Kaars Sijpesteijn, A. (1970). Bichemical modes of action of agricultural fungicides. *World Review Pest Control*, September 1970, 85–93.

Kaars Sijpesteijn, A. & Janssen, M. J. (1958). Fungitoxic action of 8-hydroxyquinoline, pyridine-*N*-oxide-2-thiol and sodium dialkyldithiocarbamates, and their copper complexes. *Nature*, **182**, 1313–14.

Kaars Sijpesteijn, A. & Janssen, M. J. (1959). On the mode of action of dialkyldithiocarbamates on moulds and bacteria. *Antonie van Leeuwenhoek*, **25**, 422–38.

Kaars Sijpesteijn, A., Janssen, M. J. & Van der Kerk, G. J. M. (1956). The role of metals in the fungitoxic action of sodium dimethyldithiocarbamate. *Biochimica et biophysica acta*, **21**, 398.

Kaars Sijpesteijn, A., Janssen, M. J. & Van der Kerk, G. J. M. (1957). Investigations on organic fungicides. XI. The role of metals and chelating agents in the fungitoxic action of sodium dimethyldithiocarbamate (NaDDC). *Biochimica et biophysica acta*, **23**, 550–7.

Kaars Sijpesteijn, A., Luijten, J. G. A. & Van der Kerk, G. J. M. (1969). Organometallic fungicides. *Fungicides II*, ed. D. C. Torgeson, pp. 331–66. New York: Academic Press.

Kaars Sijpesteijn, A., Rijkens, F., Luijten, J. G. A. & Willemsens, L. C. (1962). On the antifungal and antibacterial activity of some trisubstituted organogermanium, organotin and organolead compounds. *Antonie van Leeuwenhoek*, **28**, 346–56.

Kaars Sijpesteijn, A. & Van der Kerk, G. J. M. (1952). Investigations on organic fungicides. VI. Histidine as an antagonist of tetramethylthiuram disulphide (TMTD) and related compounds. *Antonie van Leeuwenhoek*, **18**, 83–106.

Kaars Sijpesteijn, A. & Van der Kerk, G. J. M. (1954a). Investigations on organic fungicides. VIII. The biochemical mode of action of bisdithiocarbamates and diisothiocyanates. *Biochimica et biophysica acta*, **13**, 545–52.

Kaars Sijpesteijn, A. & Van der Kerk, G. J. M. (1954b). Investigations on organic fungicides. IX. The antagonistic action of certain imidazole derivatives and of α-keto acids on the fungitoxicity of dimethyldithiocarbamates. *Biochimica et biophysica acta*, **15**, 69–77.

Kaars Sijpesteijn, A. & Vonk, J. W. (1975). Decomposition of bisdithiocarbamates and metabolism by plants and microorganisms. *Environmental Quality and Safety*, supplement, vol. III, ed. F. Coulston & F. Korte, pp. 59–61. Stuttgart: Thieme Publishers.

Klöpping, H. L. (1951). Chemical constitution and antifungal action of sulphur compounds. PhD thesis, University of Utrecht, Netherlands.

Klöpping, H. L. & Van der Kerk, G. J. M. (1951). Investigations on organic fungicides. V. Chemical constitution and fungistatic activity of aliphatic bisdithiocarbamates and isothiocyanates. *Recueil des Travaux Chimiques des Pays Bas*, **70**, 949–61.

Lancashire, W. E. & Griffith, D. E. (1971). Biocide resistance in yeast: isolation and general properties of trialkyltin resistant mutants. *FEBS Letters*, **17**, 209–14.

Pluijgers, C. W., Vonk, J. W. & Thorn, G. D. (1971). Re-examination of the structure of ethylenethiuram monosulphide. *Tetrahedron Letters*, **18**, 1317–18.

Sisler, H. D. & Cox, C. E. (1954). Effects of tetramethylthiuram disulfide on metabolism of *Fusarium roseum*. *American Journal of Botany*, **41**, 338–45.

Sisler, H. D. & Cox, C. E. (1955). Effects of tetramethylthiuram disulfide on anaerobic breakdown of glucose by brewers' yeast. *American Journal of Botany*, **42**, 351–6.

Stockdale, M., Dawson, A. P. & Selwyn, M. J. (1970). Effects of trialkyltin and triphenyltin compounds on mitochondrial respiration. *European Journal of Biochemistry*, **15**, 342–51.

Thorn, G. D. & Ludwig, R. A. (1962). *The Dithiocarbamates and Related Compounds*. Amsterdam: Elsevier.

Upshall, A. & Johnson, P. E. (1981). Thiram-induced abnormal chromosome segregation in *Aspergillus nidulans*. *Mutation Research*, **89**, 297–301.

Van der Kerk, G. J. M. & Klöpping, H. L. (1952). Investigations on organic fungicides. VII. Further considerations regarding the relations between chemical structure and antifungal action of dithiocarbamate and bisdithiocarbamate derivatives. *Recueil des Travaux Chimiques des Pays-Bas*, **71**, 1179–97.

Vonk, J. W. (1971). Ethylenethiourea, a systemic decomposition product of nabam. *Mededelingen van de Faculteit Landbouwwetenschappen Gent Rijksuniversiteit*, **36**, 109–12.

Vonk, J. W. (1975). The fate of ethylenebisdithiocarbamates in plants and microorganisms. PhD thesis, University of Utrecht, Netherlands.

Vonk, J. W. & Kaars Sijpesteijn, A. (1976). Formation of ethylenethiourea from 5,6-dihydro-3H-imidazo[2,1-c]-1,2,4-dithiazole-3-thione by microorganisms and reducing agents. *Journal of Environmental Science*, **B11**, 33–47.

Watson, K. & Smith, J. E. (1967). Oxidative phosphorylation and respiratory control in mitochondria from *Aspergillus niger*. *Biochemical Journal*, **104**, 332–9.

Wren, A. & Massey, V. (1966). Lipoyl dehydrogenase from *Saccharomyces cerevisiae*. II. Kinetic and inhibitor studies. *Biochimica et biophysica acta*, **122**, 436–49.

Yoneyama, K., Sekido, S. & Misato, T. (1982). An inhibitory action of tetramethylthiuram disulfide on lipid synthesis of bacterial cells. *Journal of Pesticide Science*, **7**, 57–60.

Zdzienicka, M., Zielenska, M., Tudek, B. & Szymczyk, T. (1979). Mutagenic activity of thiram in Ames tester strains of *Salmonella typhimurium*. *Mutation Research*, **68**, 9–13.

8

Mode of action of carboxamides

P. J. KUHN

Shell Research Limited, Sittingbourne Research Centre, Sittingbourne, Kent, ME9 8AG, UK

Introduction

Since the discovery, in the mid-1960s, of the systemic antifungal properties of carboxin (structure I, Table 8.1) and oxycarboxin (structure II, Table 8.1; von Schmeling & Kulka, 1966), an extensive range of related materials have been synthesised; several of these have been introduced as agricultural fungicides (Edgington *et al.*, 1980). The 1,4-oxathiin ring in the two progenitors was replaced in later compounds by various planar and non-planar ring systems; these included benzene (structures III, IV), dihydropyran (V), pyridine (VI), furan (VII–XI), pyrazole (XII), thiophene (XIII), and thiazole (XIV); see Table 8.1. Most of the compounds, however, possess a common nucleus (Fig. 8.1) that was suggested by an early study of structure–activity relationships (SAR) to be vital for systemic fungicidal effects (ten Haken & Dunn, 1971). In view of the relationship of this substructure with crotonic acid (*trans*-2-butenoic acid), ten Haken & Dunn (1971) termed this class of fungicides the '*cis*-crotonanilides'.

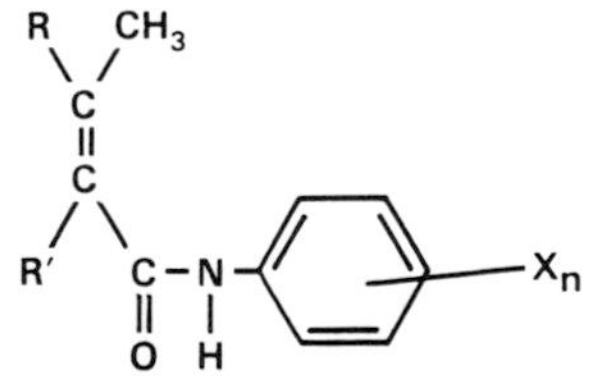

Fig. 8.1. The basic *cis*-crotonanilide structure suggested by ten Haken & Dunn (1971) to be a prerequisite for systemic fungicidal activity. $R = CH_3$, $R' = H$, or $R = R'$ plus a double bond forming a ring system. $X_n =$ one or more electron-donating groups or unsubstituted.

Table 8.1. *Representative carboxamide fungicides*

Common or chemical name	Structure	Other name	References
Carboxin	I	Vitavax	Edgington & Reinbergs (1966), Edgington, Walton & Miller (1966), Hardison (1966, 1967), von Schmeling & Kulka (1966) Tate & von Schmeling (1968)
Oxycarboxin	II	Plantvax	Edgington & Reinbergs (1966), Edgington *et al.* (1966), Hardison (1966, 1967), von Schmeling & Kulka (1966)
Mebenil	III	BAS 3050 F	Pommer (1968), Pommer & Kradel (1969), Pommer & Zwick (1974), Pommer *et al.* (1974)
Benodanil	IV	Calirus	Löcher, Hampel & Pommer (1974) Pommer *et al.* (1973, 1974), Pommer & Zwick (1974)
Pyracarbolid		Sicarol	Jank & Grossmann (1971)

2-Chloropyridyl
3-(3'-*tert*-butyl)-
carboxanilide

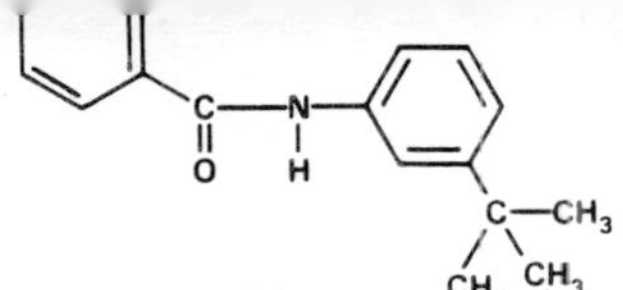

VI

Zeeh, Linhart & Pommer (1977)

Fenfuram

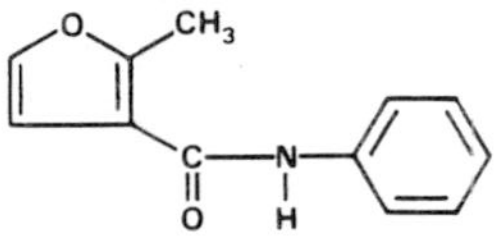

Panoram

VII

ten Haken & Dunn (1971)

Furcarbanil

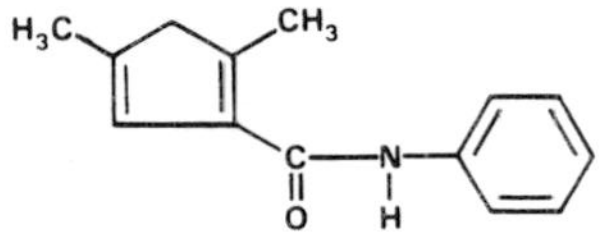

BAS 3191 F

VIII

Pommer (1972),
Pommer & Kradel (1970),
Pommer & Zwick (1974)

Methfuroxam

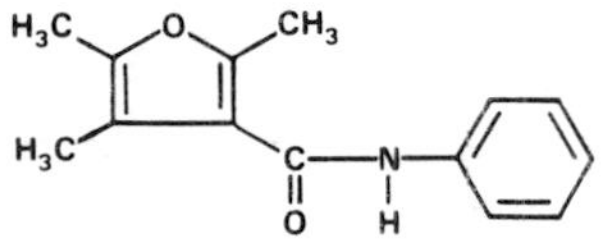

UBI H719

IX

Alcock (1978), Jackson (1979)

Cyclafuramid

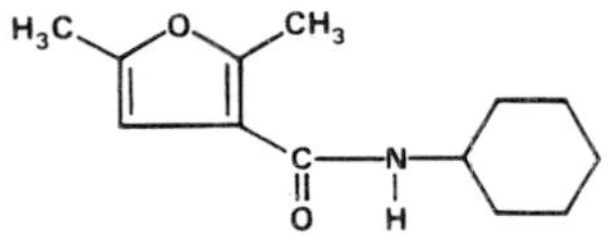

BAS 3270 F

X

Pommer, Hampel & Löcher (1971)

Table 8.1. (*contd.*)

Common or chemical name	Structure	Other name	References
Furmecyclox	XI	BAS 389 F	Papavizas, Lewis & O'Neill (1979), Pommer & Reuther (1978), Pommer & Zeeh (1977)
1,3,5-Trimethyl-pyrazole-4-carboxanilide	XII		Carter, Huppatz & Wain (1976)
3-Methylthiophene-2-carboxanilide	XIII		White & Thorn (1975, 1980)
2,4-Diemthylthiazole-5-carboxanilide	XIV	G696	Hardison (1971), Mathre (1971a), Snel, von Schmeling & Edgington (1970)

While the majority of the fungicides compiled in Table 8.1 are *cis*-crotonanilides in the strict sense, a few (for example, benodanil and cyclafuramid) do not conform exactly with this nomenclature. Despite this, however, the close affinity of these materials with the other compounds noted, in terms both of chemical structure and biological activity, suggests that their inclusion here is probably justified. So as to accommodate these slight variations in structure, the more general collective term 'carboxamides' is used here throughout.

In addition to the compounds listed in Table 8.1, most of which have been commercialised or have at least gone into development, many other carboxamides with fungicidal activity have been produced. Extensive SAR investigations have been carried out, inhibitory activity having been assessed *in vivo* against target diseases, and *in vitro* either towards fungal growth in culture, or versus selected cellular and subcellular biochemical systems. Interesting and important as these SAR studies undoubtedly are, limitations on space preclude any detailed consideration here. For information on this topic the reader is referred to papers by ten Haken & Dunn (1971), Mathre (1971*a*), Snel, von Schmeling & Edgington (1970), White & Thorn (1975, 1980), White, Thorn & Georgopoulos (1978), and to citations contained therein.

All of the carboxamide fungicides are more or less selective towards diseases caused by Basidiomycetes, their spectrum of activity including primarily smuts (Ustilaginales), rusts (Uredinales) and rots caused by *Rhizoctonia solani*. Generally the same pattern of specificity is also observed against fungal growth *in vitro* (Edgington, Walton & Miller, 1966). Although carboxamide fungicides are active first and foremost against Basidiomycetes, there are some reported instances where the spectrum of activity *in vitro* has been somewhat extended. A case in point is the compound 'F427', a derivative of carboxin substituted with a phenyl group in the 2′ position of the anilide ring. Activity against Basidiomycetes is retained, but in addition F427 is also inhibitory towards a number of Deuteromycetes (Edgington & Barron, 1967). The subject of selectivity, and in particular its possible biochemical/physiological basis, is discussed later in this chapter.

Investigations into the mode of action of carboxamides were initiated soon after the introduction of carboxin and oxycarboxin, and have continued since that time. Much of the work has been reviewed previously on several occasions in varying levels of detail (e.g. Lyr, 1977; Leroux, 1981; Kaars Sijpesteijn, 1982; Langcake, Kuhn & Wade, 1983), and early studies will be described here only in brief. Attention is instead focused on more recent investigations where the mechanism and site of action have been examined at the molecular level.

Early studies indicating succinate metabolism as the primary site of action

The excellent agreement between the activity of carboxamides *in vivo* and *in vitro* led Snel *et al.* (1970) to suggest that these fungicides probably had a direct rather than an indirect mode of action. In view of their predominant selectivity towards Basidiomycetes, much of the groundwork on the mechanism of action of carboxamides was undertaken using either sporidia from smuts such as *Ustilago maydis*, *Ustilago nuda* and *Ustilago hordei*, or mycelium of *Rhizoctonia solani*. These are technically accessible representatives since they can be cultured axenically with relative ease, in contrast to the rusts which are considerably more difficult to handle, despite advances in recent years. Preliminary investigations using intact cells concentrated on examining the effects of carboxin and oxycarboxin on the 'standard' vital processes of respiration, and biosynthesis of protein, lipid, DNA and RNA. Of these potential sites of action, the most consistent and profound effects of the fungicides were on respiration (Mathre, 1970; Ragsdale & Sisler, 1970). Particularly important was the observation by Mathre (1970) that in carboxin-treated teliospores of *U. nuda* and mycelium of *R. solani* incubated with D-[U-^{14}C]glucose or [^{14}C]acetate, there was a decreased incorporation of label into citrate, malate and fumarate, but an increased incorporation into succinate. This was the first indication of an effect on succinate metabolism and as such pointed the way for many subsequent investigations.

As studies progressed from whole cells to mitochondrial and submitochondrial fractions, inhibitory effects on respiration and on succinate oxidation in particular were confirmed (Mathre, 1971*b*; Ulrich & Mathre, 1972). Oxidation of succinate was strongly inhibited by carboxin whereas NADH oxidation was unaffected. For example, in mitochondria from *Ustilago maydis*, oxygen uptake in the presence of succinate was inhibited by carboxin with a K_i of 0.32 μM (Mathre, 1971*b*). By contrast, when NADH was supplied as substrate, no inhibitory effect was observed even at a carboxin concentration of 150 μM. Using mitochondria from the same organism and testing a range of carboxamides, White (1971) found a reasonably good correlation between inhibition of succinate oxidation and inhibition of fungal growth. Collectively these observations suggested that the site of action of carboxin and related compounds was in the Complex II (succinate–ubiquinone reductase) segment of the mitochondrial electron transport system (Fig. 8.2).

Resistant mutants are extremely useful probes to have available in any mode of action investigation and, in the case of carboxamides, the application of such fungicide-tolerant strains has proved invaluable. By comparing the growth responses of a wild-type representative of *Ustilago maydis* and an oxathiin-resistant strain, Georgopoulos, Alexandri & Chrysayi (1972) showed that the latter was, as expected, considerably less sensitive to carboxin (Fig. 8.3*a*). The dose–response curves for inhibition of succinoxidase activity (Fig. 8.3b) were in good agreement with those for growth, suggesting that the two were closely related. Resistance was apparently governed by a single gene mutation that modified the sensitivity of Complex II, not only towards carboxin but also towards thiazole carboxamides. The results and conclusions generated by this study are quite consistent with those discussed earlier and represent, therefore, strong supporting evidence for the putative site of action.

In addition to studies using fungi, carboxin has also been shown to inhibit succinate oxidation in bacterial (Tucker & Lillich, 1974), higher plant (Mathre, 1971*b*) and mammalian (Day, Arron & Laties, 1978) systems. Direct comparison of the potency of carboxin towards these systems is difficult, however, because of the varied nature of the subcellular preparations employed in assays. The problem is compounded by the use of different electron acceptors and varying assay conditions.

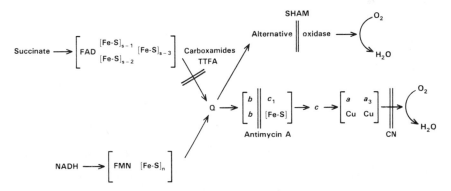

Fig. 8.2. Components of the mitochondrial electron transfer system including the cyanide-resistant alternative pathway. The sites of action of carboxamide fungicides and various other inhibitors are indicated (by double lines).

Taking these factors into account, however, published data certainly indicate that succinate-oxidising activities from Basidiomycete fungi, mammalian sources, and the bacterium *Micrococcus dentrificans* are considerably more sensitive than those from higher plants and the yeast *Saccharomyces cerevisiae* (Table 8.2).

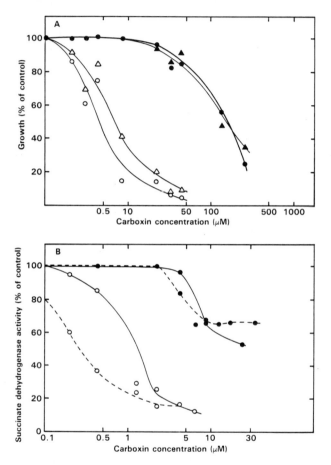

Fig. 8.3. (*a*) Carboxin dose–response curves of a wild-type and an oxathiin-resistant mutant of *U. maydis*: wild-type utilising (open circles) succinate or (open triangles) glucose; oxathiin-resistant mutant utilising (filled circles) succinate or (filled triangles) glucose. (*b*) Succinate dehydrogenase activity in *U. maydis* as a function of carboxin concentration: assays with crude extracts from (open circles) wild-type and (filled circles) oxathiin-resistant mutant sporidia; assays with mitochondrial preparations from (open circles, broken lines) wild-type and (filled circles, broken lines) oxathiin-resistant sporidia (from Georgopoulos *et al.*, 1972).

Table 8.2. *Inhibition of succinate oxidation by carboxin in preparations from fungal, mammalian, bacterial and higher plant systems*

Source	Preparation	Assay	Inhibitory activity of carboxin	Reference
Ustilago maydis	Mitochondria[b]	O_2 uptake	K_i 0.32 μM	Mathre (1971b)
Ustilago maydis	Mitochondria	Succinate–DCPIP[c] reductase	K_i 0.5 μM	White (1971)
Ustilago maydis	Mitochondria	Succinate–DCPIP reductase	I_{50}[d] 0.36 μM	White *et al.* (1978)
Ustilago maydis (carboxin-resistant mutant no. 92)[a]	Mitochondria	Succinate–DCPIP reductase	I_{50} 37.0 μM	
Rat liver	Mitochondria	O_2 uptake	K_i 6.0 μM	Day *et al.* (1978)
Beef heart	Electron transport particles – non-phosphorylating inner membrane preparations	Succinate–PMS[c] reductase	K_i 0.78 μM	
	Complex II	Succinate–DCPIP reductase	K_i 0.7–1.0 μM	Mowery *et al.* (1977)
		Succinate–DPB[c] reductase		
	Soluble	Succinate–PMS reductase	No inhibitory effect	
Micrococcus dentrificans	Membrane	O_2 uptake	K_i 16.0 μM	Tucker & Lillich (1974)
Saccharomyces cerevisiae	Mitochondria	O_2 uptake	K_i 270.0 μM	Mathre (1971b)
Phaseolus vulgaris	Mitochondria	Succinate–DCPIP reductase	26% inhibition at 100 μM	Ulrich & Mathre (1972)
Solanum tuberosum	Mitochondria	O_2 uptake	K_i 200.0 μM	Day *et al.* (1978)

From Langcake *et al.*, 1983.

[a] This mutant was highly resistant to carboxin, having an ED_{50} value of c. 37.0 μM.

[b] The term 'mitochondria' is used to indicate preparations containing intact organelles, as opposed to submitochondrial fractions. It does not imply the use of uncontaminated, homogeneous preparations.

[c] Abbreviations used: DCPIP, 2,6-dichlorophenolindophenol; PMS, phenazine methosulphate; and DPB, 2,3-dimethoxy-5-methyl-6-pentyl-1,4-benzoquinone.

[d] The I_{50} value is the concentration which inhibits the rate of DCPIP reduction by 50% as determined from a semilog plot of percentage inhibition against inhibitor concentration.

A digression on Complex II and succinate dehydrogenase

Before going on to discuss the site of action of carboxamide fungicides in detail, it is necessary to consider briefly the composition, structure and properties of Complex II and succinate dehydrogenase (SDH). The résumé that follows is based almost entirely on observations made using preparations from mammalian mitochondria since these have received by far the most attention. Although a consideration of mammalian systems in the present context may appear somewhat incongruous, it is in fact quite appropriate, since most if not all of the in-depth work on the mode of action of carboxamides has been undertaken using mitochondria from animal sources and, in particular, beef heart. The validity of this approach is open to discussion, but an important factor in its defence is the comparability of mitochondria from mammalian sources with those from susceptible fungi in terms of sensitivity to carboxin (Table 8.2).

The term Complex II is used to describe particulate preparations of that segment of the inner mitochondrial membrane that catalyse the transfer of reducing equivalents from succinate to coenzyme Q (ubiquinone). In addition to SDH itself, which constitutes approximately half of the total protein by weight (Hatefi & Stiggall, 1978), Complex II contains two other major polypeptides (Capaldi, Sweetland & Merli, 1977; Ackrell, Ball & Kearney, 1980). These were designated C_{II-3} and C_{II-4} by Capaldi *et al.* (1977) and assigned estimated molecular weights of 13 500 and 7000, respectively (cf. polypeptides 7 and 8 described by Hatefi & Galante, 1980). Preparations of Complex II also contain cytochrome c_1 and other peptide impurities from inactivated Complex III, and substantial quantities of lipid (Ziegler & Doeg, 1962; Hatefi & Stiggall, 1978).

SDH can be resolved from particulate preparations such as Complex II or electron transfer particles by a variety of treatments, the merits of which have been discussed by Ackrell, Kearney & Singer (1978). Characterisation of the purified soluble enzyme has shown it to be composed of two unequal subunits, a flavo-iron sulphur protein and an iron–sulphur protein, termed FP and IP, respectively (see for example Davis & Hatefi, 1971). The enzyme contains covalently bound FAD, nonhaem iron and acid-labile sulphur (Table 8.3). Together with cysteinyl residues, the nonhaem iron and acid-labile sulphides make up three iron–sulphur centres. These, along with the flavin, constitute the redox centres of the dehydrogenase which function in electron transfer. Various lines of evidence suggest that the FP subunit contains two

Table 8.3. *Composition of purified succinate dehydrogenase from beef heart*

Component	Molecular weight ($\times 10^{-3}$)	Covalently bound FAD (mol mol^{-1} of protein)	Nonhaem iron (mol mol^{-1} of protein)	Acid-labile sulphide (mol mol^{-1} of protein)
SDH	175 ± 10[a,b,e,h] 97 ± 3.88[c,f] 110 ± 10[a,g]	1	8	8
FP	68 ± 3[d,e] 70 ± 4.9[d,f] 68.5 ± 1[d,g]	1	4	4
IP	30 ± 1[d,e] 27 ± 1.35[d,f] 30 ± 0.5[d,g]	–	4	4

Molecular weights were estimated by: [a]density gradient ultracentrifugation; [b]gel column chromatography; [c]total flavin content; and [d]sodium dodecyl sulphate polyacrylamide gel electrophoresis; [e]Coles *et al.* (1972); [f]Davis & Hatefi (1971); [g]Righetti & Cerletti (1971); [h]The discrepancy between this and the other figures cited was rationalised with the enzyme being present in a monomer–dimer equilibrium, the degree of dissociation–association depending on temperature and protein concentration.

ferredoxin-type binuclear iron–sulphur clusters designated centres S-1 and S-2, while the smaller IP subunit contains a single tetranuclear high potential (HiPIP)-type cluster designated S-3 (Ohnishi & Salerno, 1982).

Soluble SDH, in contrast to Complex II, cannot transfer electrons to the physiological acceptor, ubiquinone; assays rely instead on succinate-reductase activities using artificial acceptors such as ferricyanide and phenazine methosulphate (PMS). Furthermore, SDH after removal from the membrane environment is no longer stabilised against oxygen. Depending on the method by which the SDH is prepared, however, it is possible to restore succinate–ubiquinone reductase and succinoxidase activities to a greater or lesser extent by reincorporation of the enzyme into alkali-treated submitochondrial particles or other membrane components (Bruni & Racker, 1968; Yamashita & Racker, 1969); preparations of SDH that exhibit this capacity are termed reconstitutively active (King, 1963).

Reconstitution of succinate–ubiquinone reductase activity can also be achieved by mixing soluble SDH with the peptides C_{II-3} and C_{II-4}, implying that these components may be critical to the catalytic properties of the enzyme in the membrane environment (Ackrell *et al.*, 1980). This was confirmed by treating preparations of the two peptides with chymotrypsin. Concomitant with the progressive disappearance of C_{II-3}, the ability to elicit ubiquinone reductase activity was lost. C_{II-4} was apparently unaffected by chymotryptic digestion as judged by electrophoresis; it was, however, unable to reconstitute ubiquinone reductase activity in the absence of intact C_{II-3}. Ackrell *et al.* (1980) speculated on the role of the two peptides, suggesting that C_{II-4} might bind SDH in the membrane, or bind and orient C_{II-3} to function in the reduction of ubiquinone. By analogy with the Q-binding protein characterised by Yu & Yu (1980, 1982) it is likely that one or other of the peptides serves to bind ubiquinone.

The way in which the various polypeptides of Complex II are arranged in the mitochondrial inner membrane has been investigated using chemical surface-labelling and immunological techniques (Merli *et al.*, 1979; Girdlestone, Bisson & Capaldi, 1981). The results obtained are consistent with the model illustrated in Fig. 8.4 (Girdlestone *et al.*, 1981). This shows the FP subunit to be held above the lipid bilayer and exposed on the matrix side of the membrane, with the smaller IP subunit also exposed on this surface but partially shielded from water by penetration into lipid. The peptides C_{II-3} and C_{II-4} are intercalated into the lipid bilayer. In a recent review (Ohnishi & Salerno, 1982), the scope of the model has been extended by indicating the likely positions of the prosthetic groups (Fig. 8.4).

I have attempted in this section to present a short overview on the structure and catalytic properties of Complex II and SDH. For a more detailed treatment of the subject several excellent reviews are available, including those by Hatefi & Stiggall (1976), Ohnishi (1979) and Singer, Gutman & Massey (1973).

Site of action of carboxamides within Complex II

While it has long been established that carboxamides inhibit succinate oxidation, the mechanism of inhibition and the site of action within Complex II have been more contentious issues. Particular attention has centred on whether or not they are identical with those of the inhibitor, thenoyltrifluoroacetone (TTFA). Based primarily on the

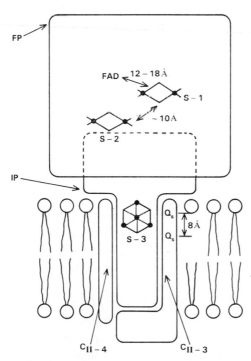

Fig. 8.4. Schematic representation of the topographical arrangement of Complex II including polypeptide subunits, prosthetic groups (FAD, centres S-1, S-2 and S-3), lipid molecules, and a pair of bound ubiquinone species (from Girdlestone *et al.*, 1981; Ohnishi & Salerno, 1982).

observations that succinate–PMS reductase activities of inner-membrane preparations and solubilised SDH from heart mitochondria were inhibited by TTFA but not by carboxin, Schewe et al. (1973) have suggested that the two compounds possess different sites of action. Specifically, it was proposed that TTFA acts between the FP and IP subunits, whereas carboxin reacts on the oxygen side of the enzyme at the junction with ubiquinone (Lyr & Schewe, 1975; Lyr, 1977). Mowery et al. (1977) were critical of some of the techniques used in the experiments upon which these conclusions were based, and their own investigations with beef-heart mitochondria failed to confirm any qualitative differences in the activities of the two inhibitors. Several different submitochondrial preparations and a variety of assays were used but, without exception, the characteristics of inhibition by carboxin and TTFA were identical, with respect to both kinetics and the maximum extents of inhibition that could be attained (Mowery et al., 1976, 1977). Carboxin and related carboxamides were more active, however, having K_i values in the range 0.002–3.0 μM as compared with 9.0–9.7 μM for TTFA. The lack of activity of TTFA against solubilised SDH (Ziegler, 1961) was confirmed as was the parallel inactivity of carboxin against this form of enzyme (Georgopoulos, Chrysayi & White, 1975; Ulrich & Mathre, 1972). As well as these similarities, further evidence consistent with a common site of action for TTFA and carboxin is provided by the fact that succinate oxidation in mitochondria from carboxin-resistant mutants of *Ustilago maydis* exhibits reduced sensitivity to TTFA (Georgopoulos & Ziogas, 1977).

One of the methods that has been used to elucidate the mechanism of action of carboxamides is electron paramagnetic resonance (epr) spectroscopy, a technique that can detect and characterise the iron–sulphur centres of Complex II that are involved in electron transfer. Ackrell et al. (1977) employed this technique in combination with the rapid freezing of enzyme-reaction mixtures to study the reduction and reoxidation phases associated with the activity of Complex II; in these experiments, succinate was used as reductant and the synthetic water-soluble Q_1 analogue, 2,3-dimethoxy-5-methyl-6-pentyl-1,4-benzoquinone (DPB), as oxidant. The carboxamide, 3'-methylcarboxin, had little or no effect on the rate of reduction of centre S-1 though the reduction of the HiPIP centre was somewhat affected. Much more striking, however, was the effect of the fungicide and of TTFA on the reoxidation of these iron–sulphur centres by DPB, both compounds proving highly inhibitory in this respect. This suggested that TTFA and carboxamides probably reacted at a site vicinal to the HiPIP centre, thereby inter-fering with the flow of electrons to ubiquinone. No evidence was obtained, however, to support earlier proposals that TTFA (Ziegler, 1961; Redfearn, Whittaker & Burgos, 1965) and carboxamides (Georgopoulos *et al.*, 1975) act by chelating one of the nonhaem iron components of the enzyme.

Although the epr studies discussed above were extremely valuable in indicating the point within Complex II at which electron transfer is probably interrupted, they provided no direct information on the actual locus of interaction between enzyme and fungicide. The nature of the receptor site has been investigated by examining the binding of car-boxamides to inner membranes, to unresolved and resolved Complex II, and to solubilised SDH, all prepared from beef-heart mitochondria (Coles *et al.*, 1978; Ramsay *et al.*, 1981). Two compounds have been employed, [G-^{3}H]2,4,5-trimethyl-3-carboxanilinofuran (TCF) (meth-furoxam, Structure IX, Table 8.1), and [G-^{3}H]3'-azido-5,6-dihydro-2-methyl-1,4-oxathiin-3-carboxanilide (azidocarboxin). The latter de-rivative, as well as being radioactive, also bears a photoaffinity label, exposure to ultraviolet (UV) light resulting in conversion into the nitrene.

Preliminary experiments, in which inner membrane preparations were exposed to TCF, revealed that inhibition of succinate oxidation could be reversed on dilution of assay mixtures with buffer. Even after repeated washings, however, a significant proportion of the labelled carboxamide remained bound to the particles, this representing compa-ratively tight non-specific binding at sites not responsible for inhibition of enzyme activity. The approach adopted by Coles *et al.* (1978) in order to characterise specific binding, relied on the fact that TCF and TTFA block succinate oxidation in an identical fashion, but while the car-boxamide exhibits considerable non-specific binding, TTFA binds only at sites involved directly in enzyme inhibition. A measure of specific binding could, therefore, be obtained by comparing the binding of different concentrations of TCF in the presence or absence of an excess of TTFA. The results of such an experiment, in this case using Complex II, are illustrated in Fig. 8.5. Scatchard plots of these data gave an estimated K_D for TCF of $1.0 \pm 0.1 \times 10^{-7}$ M at 4 °C. This compares with a K_i value of 1.1×10^{-7} M for inhibition of succinate–DCPIP reductase at 25 °C. Allowing for the difference in temperature, the excellent agree-ment between these figures represents strong evidence that the binding at low concentrations of TCF is responsible for inhibition of succinate oxidation. This is further supported by the observation that binding of low concentrations of TCF is reversed by TTFA in a competitive fashion, whereas at higher concentrations of the carboxamide binding is unaffected.

The lack of inhibitory activity of carboxamides against solubilised SDH has been mentioned previously. Consistent with this property, it was found that TCF did not bind to a reconstitutively active soluble preparation of the enzyme (Coles *et al.*, 1978).

In order to examine the binding site(s) in more detail Ramsay *et al.* (1981) reacted Complex II with azidocarboxin. After irradiation with UV light, assay mixtures were treated with sodium dodecylsulphate and subjected to polyacrylamide gel electrophoresis. Protein bands were located by staining with Coomassie blue, and radioactivity measured by liquid-scintillation counting after cutting gels into slices and dissolving

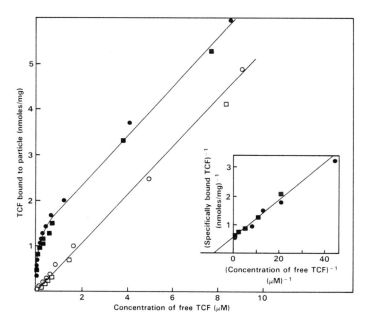

Fig. 8.5. Effect of TTFA on the binding of TCF to Complex II determined by equilibrium dialysis and by centrifugation: (filled squares) equilibrium dialysis without TTFA; (filled circles) centrifugation without TTFA; (open squares) equilibrium dialysis with TTFA; (open circles) centrifugation with TTFA. Inset, double-reciprocal plot showing the specifically bound component of TCF obtained by subtracting the amount of TCF bound in the presence of TTFA from that bound in its absence using (filled squares) equilibrium dialysis or (filled circles) centrifugation (from Coles *et al.*, 1978).

these directly in scintillation fluid. Using these procedures, it was possible to establish which components of Complex II formed adducts with the inhibitor.

Comparison of the peptide band pattern of the resolved membrane preparation with the distribution of radioactivity revealed that a substantial proportion of the bound inhibitor was associated with the low molecular weight polypeptides C_{II-3} and C_{II-4}. That a large part of this binding is specific was demonstrated by the marked reduction in the amount of azidocarboxin bound in the presence of TTFA. The non-linear relationship between azidocarboxin concentration and extent of incorporation into C_{II-3} and C_{II-4} (Fig. 8.6a) is additional evidence of specific binding. Based on the same criteria, binding to the phospholipid component by Complex II (Fig. 8.6b) was also judged to be partly specific. Compared to the polypeptides C_{II-3} and C_{II-4}, less inhibitor was found to be associated with the FP and IP subunits of SDH. Further-

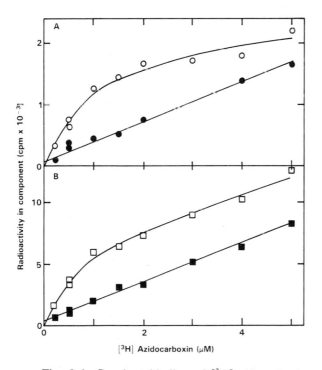

Fig. 8.6. Covalent binding of [³H]azidocarboxin to components of complex II: (filled squares) 70S subunit of SDH; (filled circles) 30S subunit of SDH; (open circles) polypeptides $C_{II-3} + C_{II-4}$; (open squares) phospholipids (from Ramsay *et al.*, 1981).

more, for both proteins, the interaction with azidocarboxin was apparently non-specific (Fig. 8.6a, b); this was despite the reduced amounts of azidocarboxin bound in the presence of TTFA. Notwithstanding the non-specific nature of the binding to FP and IP, these subunits are required for the normal interaction of azidocarboxin with the peptides of low molecular weight since, in resolved Complex II, specific binding to the latter components is destroyed.

Based on their results, Ramsay et al. (1981) concluded that the components of Complex II principally involved in the specific binding of azidocarboxin are the polypeptides C_{II-3} and C_{II-4}. Regarding the role of phospholipids, it was suggested that these also seem to form a part or to be adjacent to the binding site(s). This is probably all that can be said at this stage, and certainly it would be difficult to rationalise phospholipids themselves being the specific binding sites since they would appear to lack the degree of complexity necessary to confer such a property. An associated difficulty lies in selecting the most appropriate units by which to describe quantitatively the amount of carboxamide bound to phospholipid. Ramsay et al. (1981) expressed this in terms of nmol of inhibitor per 250 μg of protein (≈ 1 nmol of histidylflavin) or per nmol of phospholipid. If the carboxamide is bound to the phospholipids of Complex II indiscriminately, a situation which by definition is inconsistent with the idea of specific binding, then the amounts of inhibitor involved on a molar basis are relatively insignificant compared with those associated with C_{II-3} and C_{II-4}. Conversely, if 'specific' phospholipid molecules are involved, then the relative contribution of these to total specific binding becomes progressively more important the fewer of these molecules there are present. Specificity in this instance might be related to the proximity of the phospholipid to the receptor site rather than to chemical heterogeneity involving populations of distinct structural types. Obviously more work is required in order to clarify the position, but it is interesting to note that a hydrophobic region was central to the carboxin receptor model proposed by Schewe et al. (1979). This could reflect the participation of lipid, or of non-polar amino acids present in the polypeptides C_{II-3} and C_{II-4}.

Before concluding this discussion on binding studies it is important to raise an anomaly that was found in work with azidocarboxin. The amount of inhibitor specifically bound was equivalent to only 0.6 mol per mol of enzyme rather than the expected ratio of 1:1. The result would not appear to be one of chance since a quite comparable figure was found in the earlier investigation with TCF (Coles et al., 1978).

Three possible explanations for the discrepancy were put forward (Coles *et al.*, 1978):

(i) that only one in every two molecules of enzyme is capable of binding the inhibitor;

(ii) that the method employed to determine specific binding gives low titres because of the use of insufficient concentrations of TTFA;

(iii) that the enzyme exists in Complex II as a dimer.

The first explanation is at odds with the observation that enzyme activity can be completely inhibited by carboxamides, while the second may be ruled out since the same titres and K_D values were obtained at much higher TTFA concentrations (Coles *et al.*, 1978). As regards the last proposal, although SDH has been reported to form a dimer in solution (Coles *et al.*, 1972), it was emphasised by Coles *et al.* (1978) that there was no evidence to suggest association in the membrane environment. For the present, therefore, the basis of the substoichiometric relationship is unclear.

The cyanide-resistant alternative pathway of respiration – a second site of action?

Although by far the majority of work on the mode of action of carboxamides has centred on the inhibition of succinate oxidation, there have been several reports on the effects of these fungicides on the cyanide- and antimycin-A-insensitive, alternative pathway of respiration. The first such report was by Sherald & Sisler (1972) who suggested that carboxin selectively inhibited this pathway in *Ustilago maydis*. Lyr & Schewe (1975) extended the concept further and, based on studies using the yeast *Rhodotorula mucilaginosa*, proposed that an iron–sulphur component (the HiPIP centre S-3) of SDH was not only the branch point for the alternative pathway, but the alternative oxidase itself. This suggestion was obviously attractive since fungicide action at a single locus could explain inhibition of both succinate oxidation and the alternative pathway. Furthermore, the proposal did, at the time, appear consistent with the reported loss of alternative-pathway activity in a carboxin-resistant mutant of *U. maydis* (Georgopoulos & Sisler, 1970). A later study, however, failed to confirm differential effects on the alternative respiration of wild-type and carboxin-resistant mutants of this fungus; in both cases the fungicide lacked any specific inhibitory activity at this site (Ziogas & Georgopoulos, 1979).

As regards the suggestion that centre S-3 and the alternative oxidase are one and the same, subsequent investigations have shown this to be untenable. In the first place, the epr signal from centre 3 is unaffected by the classical alternative pathway inhibitor salicylhydroxamic acid (SHAM) (Rich & Bonner, 1978). Furthermore, mitochondrial particles from the arum lily *Sauromatum guttatum* that have been treated to remove SDH, nevertheless retain alternative oxidase activity with NADH (Rich & Bonner, 1977). Regarding the point of interaction between the cytochrome and alternative pathways, current thinking favours a site at the level of ubiquinone (Storey, 1976; Hanssens, Verachtert & von Jagow, 1978; Moore & Rich, 1980).

As well as investigations using fungal mitochondria, there has been at least one study on the effects of carboxamides on the alternative respiratory pathway of higher-plant mitochondria. Succinate oxidation in mitochondria from sweet potato (*Ipomoea batatas*) was sensitive to treatment with 3'-phenoxycarboxin in the presence of potassium cyanide or SHAM (Day *et al.*, 1978). By contrast, when malate was supplied as substrate, marked inhibition of oxygen consumption was observed only in mitochondria treated with potassium cyanide, i.e. with the cytochrome pathway blocked and respiration proceeding via the alternative oxidase. This suggests, therefore, that in sweet-potato mitochondria carboxamides inhibit both SDH and the alternative pathway.

With the limited and apparently conflicting data presently available, it is difficult to assess the significance of the alternative pathway as a second site of action for carboxamides. Clarification awaits not only further studies on carboxamide-treated-fungal and higher-plant systems, but also the acquisition of fundamental information on the alternative oxidase itself. The use of intact cells, an approach adopted by Sherald & Sisler (1972) and Lyr & Schewe (1975), can complicate the interpretation of results and, in general, studies with isolated mitochondria are to be preferred. This point might be emphasised in the light of the recent isolation from *Saccharomyces cerevisiae* of an extramitochondrial enzyme activity that exhibits characteristics typical of the alternative oxidase (Ainsworth, Ball & Tustanoff, 1980*a*,*b*) but was considered by Laties (1982) to be something quite distinct from it, possibly a mixed-function oxidase. A final point which should be borne in mind is that the alternative pathway in fungi and plants may not necessarily be identical. While there are undoubtedly similarities (e.g. sensitivity to SHAM and a reduced affinity for oxygen compared with that of cytochrome oxidase) it is conceivable that differences might also exist.

If it is established that the alternative pathway genuinely is a second site of action for carboxamides, then several intriguing questions are raised. For instance, do carboxamides inhibit the alternative oxidase by binding to a region similar in structure and conformation to the binding sites in Complex II? Taking this one step further, might the alternative oxidase contain binding peptides of low molecular weight analogous with C_{II-3} and C_{II-4}? What is the relationship between the sites of action of carboxamides and SHAM? Finally, and perhaps most important, is inhibition of the alternative pathway critical or indeed even relevant to disease control? If the alternative oxidase is itself of little or no significance to pathogen metabolism then, presumably, this second putative site of action for carboxamides is incidental in practical terms (albeit highly interesting from an academic standpoint).

Comments on selectivity

The phenomenon of selectivity or specificity, whereby organisms exhibit differential sensitivity towards a compound, is well known and has profound implications in that it may well determine the fate of a novel pesticide. It could be, for instance, that a particular fungicide is over-selective, the spectrum of diseases controlled being too limited to justify product development. Conversely, there may be insufficient selectivity such that the material is phytotoxic as well as fungitoxic. Similarly, selectivity may also affect a compound's toxicological status. These vital considerations apart, a thorough understanding of the basis of selectivity represents an intellectual challenge worthy of more attention than it has received to date. One could quite reasonably speculate that if more was known about the subject, it would then facilitate the design of compounds with an inbuilt and predetermined spectrum of activity.

The factors that are normally cited as determinants of selectivity are as follows:

(i) uptake and transport initially across the plasmalemma and from there to the site of action;

(ii) metabolism concomitant with either detoxification or activation;

(iii) differential sensitivity of the target site.

In the case of carboxamides, published information on the basis of selectivity towards different fungi is relatively limited and is scattered throughout the literature, but all three of the above explanations have been invoked or alluded to at one time or another. Mathre (1968) compared the uptake of carboxin and oxycarboxin by two sensitive (*Rhizoctonia solani* and *Ustilago maydis*) and two resistant (*Fusarium oxysporum* f.sp. *lycopersici* and *Saccharomyces cerevisiae*) fungi. Both the rate and extent of uptake were considerably greater in the two sensitive representatives, suggesting that selectivity reflected differential penetration of the fungicides. In another study, however, no differences were found in the overall accumulation of carboxin by sensitive and insensitive species (Lyr, Luthardt & Ritter, 1971). As regards metabolism as a possible basis for selectivity, information is confined to reports of the ability of some insensitive Phycomycete fungi to metabolise carboxin (Wallnöfer, 1968, 1969), furcarbanil (Wallnöfer *et al.*, 1972*a*), mebenil and 2-chlorobenzanilide (Wallnöfer, Safe & Hutzinger, 1972*b*). Finally, there is some evidence to suggest that selectivity involves differences in the susceptibility of the site of action to inhibition (Ulrich & Mathre, 1972).

A fuller understanding of the basis of selectivity of carboxamides awaits a comprehensive investigation. Such a study would of course be a major undertaking, not only because of the varied physiological and biochemical facets involved but also in view of the need to include representatives from all the major fungal taxonomic groups. For comparative purposes, it would be interesting to include compounds such as F427 and furmecyclox where the spectrum of activity is extended relative to that of carboxin and oxycarboxin. The fact that the former two compounds do have some activity outside the Basidiomycetes stresses that selectivity is relative and not absolute.

Should the selectivity of carboxamides towards fungi prove to be determined at the mitochondrial level, as shown in the case of *Ustilago maydis* and *Saccharomyces cerevisiae* (Mathre, 1971*b*; Ulrich & Mathre, 1972), then a detailed study of Complex II is warranted. In view of the work discussed earlier such a study should include a comparison of inhibitor binding. If differences in binding are revealed then, clearly, the next objective would be to compare structural features of the component parts of the receptor site. It is appreciated that these proposals are made at a time when there is still much to be learned about the binding proteins in Complex II from beef-heart mitochondria. Ultimately, however, it will be necessary to move towards more detailed studies with fungal mitochondria, perhaps to confirm the well-tried mammalian system as a valid model or, more importantly, to expose possible differences.

The resistance exhibited by fungicide-tolerant mutants can be considered an example of selectivity, and one which particularly lends itself to comparative study. Georgopoulos *et al.* (1972) concluded that the resistance of a carboxin-insensitive strain of *Ustilago maydis* was governed by an alteration in Complex II; in another study, using *Ustilago hordei*, tolerance was, however, considered to be only partly accounted for by this factor (Ben-Yephet, Dinoor & Henis, 1975). Irrespective of the full explanation, it is to be hoped that future studies will shed light on some interesting related issues. What, for instance, is the biochemical distinction between weakly and highly resistant strains? Similarly, what is the basis of negatively correlated cross-resistance? An example of this is provided by 4'-phenylcarboxin which is 13 times less active than carboxin against succinate oxidation in mitochondria from a wild-type strain of *U. maydis* but 19–59 times more toxic towards preparations from moderately resistant mutants (White *et al.*, 1978).

So far the discussion has been restricted to fungi, but carboxamides also show selectivity in the wider context of bacteria, plants and animals (Mathre, 1972). To some extent this would appear attributable to differences in the sensitivities of the succinate oxidising activities from the various sources (Table 8.2). Other considerations that have been expressed earlier with regard to fungi can probably be applied equally well in these cases.

Concluding remarks

It is apparent from this review that the mode of action of carboxamides is relatively well understood. To put this in the context of other systemic fungicides, understanding is probably on a par with that of benomyl and related inhibitors of cell division, and perhaps some-what ahead of that of inhibitors of sterol biosynthesis. Needless to say, this does not mean that the mode of action of carboxamides is a closed book. Many interesting areas for investigation remain, some of which have been highlighted in the text.

I would like to conclude by making three general points based on the studies described here, but probably with universal applicability to mode-of-action investigations. Firstly, the elucidation of the mode of action of a compound or class of compound is a long process, in the case of carboxamides having already spanned some 15 years. Secondly, progress is often made as a result of independent biochemical advances in related areas. Thirdly, a fuller understanding may be accelerated if the compound is adopted as a useful probe by recognised experts in the relevant field. The first point might be borne in mind by those disillusioned by slow progress or conflicting results, while the second and third emphasise that ultimate resolution relies on, and is synergised by, a cooperative multidisciplinary approach.

Acknowledgements. I am grateful to John R. Bowyer and Pieter ten Haken for useful discussions, and to Larry J. Mulheirn for helpful comments on the manuscript. Communications from Drs G. G. Laties, H. Lyr and E.-H. Pommer, are also acknowledged. Finally, I wish to thank staff of the Reprographics Department, Shell Research Ltd, Sittingbourne and Denise Taylor for their skilled preparation of the artwork and typescript, respectively.

References

Ackrell, B. A. C., Ball, M. B. & Kearney, E. B. (1980). Peptides from Complex II active in reconstitution of succinate-ubiquinone reductase. *Journal of Biological Chemistry*, **255**, 2761–9.

Ackrell, B. A. C., Kearney, E. B., Coles, C. J., Singer, T. P., Beinert, H., Wan, Y.-P. & Folkers, K. (1977). Kinetics of the reoxidation of succinate dehydrogenase. *Archives of Biochemistry and Biophysics*, **182**, 107–17.

Ackrell, B. A. C., Kearney, E. B. & Singer, T. P. (1978). Mammalian succinate dehydrogenase. In *Methods in Enzymology*, vol. LIII, *Biomembranes, Part D, Biological Oxidations: Mitochondrial and Microbial Systems*, ed. S. Fleischer & L. Packer, pp. 466–83. London: Academic Press.

Ainsworth, P. J., Ball, A. J. S. & Tustanoff, E. R. (1980*a*). Cyanide-resistant respiration in yeast. I. Isolation of a cyanide-insensitive NAD(P)H oxidoreductase. *Archives of Biochemistry and Biophysics*, **202**, 172–86.

Ainsworth, P. J., Ball, A. J. S. & Tustanoff, E. R. (1980*b*). Cyanide-resistant respiration in yeast. II. Characterization of a cyanide-insensitive NAD(P)H oxidoreductase. *Archives of Biochemistry and Biophyscs*, **202**, 187–200.

Alcock, K. T. (1978). Field evaluation of 2,4,5-trimethyl-*N*-phenyl-3-furancarboxamide (UBI-H719) against cereal smut diseases in Australia. *Plant Disease Reporter*, **62**, 854–8.

Ben-Yephet, Y., Dinoor, A. & Henis, Y. (1975). The physiological basis of carboxin sensitivity and tolerance in *Ustilago hordei*. *Phytopathology*, **65**, 936–42.

Bruni, A. & Racker, E. (1968). Resolution and reconstitution of the mitochondrial electron transport system. I. Reconstitution of the succinate-ubiquinone reductase. *Journal of Biological Chemistry*, **243**, 962–71.

Capaldi, R. A., Sweetland, J. & Merli, A. (1977). Polypeptides in the succinate-coenzyme Q reductase segment of the respiratory chain. *Biochemistry*, **16**, 5707–10.

Carter, G. A., Huppatz, J. L. & Wain, R. L. (1976). Investigations on fungicides. XIX. The fungitoxicity and systemic antifungal activity of certain pyrazole analogues of carboxin. *Annals of Applied Biology*, **84**, 333–42.

Coles, C. J., Singer, T. P., White, G. A. & Thorn, G. D. (1978). Studies on the binding of carboxin analogs to succinate dehydrogenase. *Journal of Biological Chemistry*, **253**, 5573–8.

Coles, C. J., Tisdale, H. D., Kenney, W. C. & Singer, T. P. (1972). Studies on succinate dehydrogenase. XXI. Quaternary structure of succinate dehydrogenase. *Physiological Chemistry and Physics*, **4**, 301–16.

Davis, K. A. & Hatefi, Y. (1971). Succinate dehydrogenase. I. Purification, molecular properties, and sub-structure. *Biochemistry*, **10**, 2509–16.

Day, D. A., Arron, G. P. & Laties, G. G. (1978). The effect of carboxins on higher plant mitochondria. *FEBS Letters*, **85**, 99–102.

Edgington, L. V. & Barron, G. L. (1967). Fungitoxic spectrum of oxathiin compounds. *Phytopathology*, **57**, 1256–7.

Edgington, L. V., Martin, R. A., Bruin, G. C. & Parsons, I. M. (1980). Systemic fungicides: a perspective after 10 years. *Plant Disease*, **64**, 19–23.

Edgington, L. V. & Reinbergs, E. (1966). Control of loose smut in barley with systemic fungicides. *Canadian Journal of Plant Science*, **46**, 336.

Edgington, L. V., Walton, G. S. & Miller, P. M. (1966). Fungicide selective for Basidiomycetes. *Science*, **153**, 307–8.

Georgopoulos, S. G., Alexandri, E. & Chrysayi, M. (1972). Genetic evidence for the action of oxathiin and thiazole derivatives on the succinic dehydrogenase system of *Ustilago maydis* mitochondria. *Journal of Bacteriology*, **110**, 809–17.

Georgopoulos, S. G., Chrysayi, M. & White, G. A. (1975). Carboxin resistance in the haploid, the heterozygous diploid, and the plant parasitic dicaryotic phase of *Ustilago maydis*. *Pesticide Biochemistry and Physiology*, **5**, 543–51.

Georgopoulos, S. G. & Sisler, H. D. (1970). Gene mutation eliminating antimycin A-tolerant electron transport in *Ustilago maydis*. *Journal of Bacteriology*, **103**, 745–50.

Georgopoulos, S. G. & Ziogas, B. N. (1977). A new class of carboxin-resistant mutants of *Ustilago maydis*. *Netherlands Journal of Plant Pathology*, **83** (Suppl 1), 235–42.

Girdlestone, J., Bisson, R. & Capaldi, R. A. (1981). Interaction of succinate-ubiquinone reductase (Complex II) with (arylazido)phospholipids. *Biochemistry*, **20**, 152–6.

Hanssens, L., Verachtert, H. & von Jagow, G. (1978). Participation of ubiquinone in the cyanide-insensitive respiration of *Moniliella tomentosa*. *Functions of Alternative Terminal Oxidases*, vol. 49. Proceedings of the 11th FEBS Meeting, Copenhagen, 1977, ed. H. Degn, D. Lloyd & G. C. Hill, pp. 47–53. Oxford: Pergamon.

Hardison, J. R. (1966). Chemotherapy of *Urocystis agropyri* in Merion Kentucky bluegrass (*Poa pratensis*) with two derivatives of 1,4-oxathiin. *Crop Science*, **6**, 384.

Hardison, J. R. (1967). Chemotherapeutic control of stripe smut (*Ustilago striiformis*) in grasses by two derivatives of 1,4-oxathiin. *Phytopathology*, **57**, 242–5.

Hardison, J. R. (1971). Chemotherapy of smut and rust pathogens in *Poa pratensis* by thiazole compounds. *Phytopathology*, **61**, 1396–9.

Hatefi, Y. & Galante, Y. M. (1980). Isolation of cytochrome b_{560} from Complex II (succinate-ubiquinone oxidoreductase) and its reconstitution with succinate dehydrogenase. *Journal of Biological Chemistry*, **255**, 5530–7.

Hatefi, Y. & Stiggall, D. L. (1976). Metal-containing flavoprotein dehydrogenases. In *The Enzymes*, vol. XIII, *Oxidation–Reduction, Part C, Dehydrogenases (II) Oxidases (II) Hydrogen Peroxide Cleavage*, ed. P. D. Boyer, pp. 175–297. London: Academic Press.

Hatefi, Y. & Stiggall, D. L. (1978). Preparation and properties of succinate : ubiquinone oxidoreductase (Complex II). In *Methods in Enzymology*, vol. LIII, *Biomembranes, Part D, Biological Oxidations: Mitochondrial and Microbial Systems*, ed. S. Fleischer & L. Packer, pp. 21–7. London: Academic Press.

Jackson, D. (1979). Methfuroxam, a new systemic fungicide for the control of

Basidiomycete pathogens. In *Proceedings of the 1979 British Crop Protection Conference – Pests and Diseases*, **2**, 549–555.

Jank, B. & Grossmann, F. (1971). 2-Methyl-5,6-dihydro-4-H-pyran-3-carboxylic acid anilide: a new systemic fungicide against smut diseases. *Pesticide Science*, **2**, 43–4.

Kaars Sijpesteijn, A. (1982). Mechanism of action of fungicides. In *Fungicide Resistance in Crop Protection*, ed. J. Dekker & S. G. Georgopoulos, pp. 32–45. Wageningen: Pudoc.

King, T. (1963). Reconstitution of respiratory chain enzyme systems XII. Some observations on the reconstitution of the succinate oxidase system from heart muscle. *Journal of Biological Chemistry*, **238**, 4037–51.

Langcake, P., Kuhn, P. J. & Wade, M. (1983). The mode of action of systemic fungicides. In *Progress in Pesticide Biochemistry and Toxicology,* vol. 3, ed. D. H. Hutson & T. R. Roberts, pp. 1–109. Chichester: Wiley.

Laties, G. G. (1982). The cyanide-resistant alternative path in higher plant respiration. *Annual Review of Plant Physiology*, **33**, 519–55.

Leroux, P. (1981). Les modes d'action des substances antifongiques à usages agricoles. *La Défense des Végétaux*, no. 207 (35 année), 59–83.

Löcher, F., Hampel, M. & Pommer, E.-H. (1974). Ergebnisse bei der Rostbekaempfung mit Benodil (BAS 317 00 F). *Mededelingen van de Faculteit Landbouwwetenschappen Rijksuniversiteit Gent*, **39**, 1079–89.

Lyr, H. (1977). Effect of fungicides on energy production and intermediary metabolism. In *Antifungal Compounds*, vol. 2, *Interactions in Biological and Ecological Systems*, ed. M. R. Siegel & H. D. Sisler, pp. 301–32. New York: Marcel Dekker.

Lyr, H., Luthardt, W. & Ritter, G. (1971). Wirkungsweise von Oxathiin-Derivaten auf die Physiologie sensitiver und insensitiver Hefearten. *Zeitschrift für Allgemeine Mikrobiologie*, **11**, 373–85.

Lyr, H. & Schewe, T. (1975). On the mechanism of the cyanide-insensitive alternative pathway of respiration in fungi and higher plants and the nature of the alternative terminal oxidase. *Acta Biologica et Medica Germanica*, **34**, 1631–41.

Mathre, D. E. (1968). Uptake and binding of oxathiin systemic fungicides by resistant and sensitive fungi. *Phytopathology*, **58**, 1464–9.

Mathre, D. E. (1970). Mode of action of oxathiin systemic fungicides. I. Effect of carboxin and oxycarboxin on the general metabolism of several Basidiomycetes. *Phytopathology*, **60**, 671–6.

Mathre, D. E. (1971*a*). Mode of action of oxathiin systemic fungicides. Structure–activity relationships. *Journal of Agricultural and Food Chemistry*, **19**, 872–4.

Mathre, D. E. (1971*b*). Mode of action of oxathiin systemic fungicides. III. Effect on mitochondrial activities. *Pesticide Biochemistry and Physiology*, **1**, 216–24.

Mathre, D. E. (1972). Effects of oxathiin systemic fungicides on various biological systems. *Bulletin of Environmental Contamination and Toxicology*, **8**, 311–16.

Merli, A., Capaldi, R. A., Ackrell, B. A. C. & Kearney, E. B. (1979). Arrangement of Complex II (succinate-ubiquinone reductase) in the mitochondrial inner membrane. *Biochemistry*, **18**, 1393–1400.

Moore, A. L. & Rich, P. R. (1980). The bioenergetics of plant mitochondria. *Trends in Biochemical Sciences*, **5**, 284–8.

Mowery, P. C., Ackrell, B. A. C., Singer, T. P., White, G. A. & Thorn, G. D. (1976). Carboxins: powerful selective inhibitors of succinate oxidation in animal tissues. *Biochemical and Biophysical Research Communications*, **71**, 354–61.

Mowery, P. C., Steenkamp, D. J., Ackrell, B. A. C., Singer, T. P. & White, G. A.

(1977). Inhibition of mammalian succinate dehydrogenase by carboxins. *Archives of Biochemistry and Biophysics*, **178**, 495–506.

Ohnishi, T. (1979). Mitochondrial iron–sulfur flavodehydrogenases. In *Membrane Proteins in Energy Transduction*, ed. R. A. Capaldi, pp. 1–80. New York: Marcel Dekker.

Ohnishi, T. & Salerno, J. C. (1982). Iron-sulfur clusters in the mitochondrial electron-transport chain. In *Iron-sulfur Proteins*, ed. T. P. Spiro, pp. 286–327. New York: Wiley.

Papavizas, G. C., Lewis, J. A. & O'Neill, N. R. (1979). BAS 389, a new fungicide for control of *Rhizoctonia solani* in cotton. *Plant Disease Reporter*, **63**, 567–73.

Pommer, E.-H. (1968). Substuierte Benzoesäureanilide als Fungizide mit selektiver Wirksamkeit. *Mededelingen van de Faculteit Landbouwwetenschappen Rijksuniversiteit Gent*, **33**, 1019–24.

Pommer, E.-H. (1972). The systemic activity of a new fungicide of the furane carbonic acid aniline group (BAS 3191 F). In *Proceedings of the 2nd International Congress of Pesticide Biochemistry (IUPAC)*, vol. V, ed. A. S. Tahori, pp. 397–404. London: Gordon & Breach Science Publishers.

Pommer, E.-H., Girgensohn, B., König, K.-H., Osieka, H. & Zeeh, B. (1974). Development of new systemic fungicides with carboxanilide structure. *Kemia Kemi*, **1**, 617–18.

Pommer, E.-H., Hampel, M. & Löcher, F. (1971). Results with 'BAS 3270 F' (76 943). In *Proceedings of the 6th British Insecticide and Fungicide Conference*, **2**, 587–96.

Pommer, E.-H., Jung, K., Hampel, M. & Löcher, F. (1973). BAS 3170 F (2-Jodbenzoesäureanilid), ein neues Fungizid zur Bekämpfung von Rostpilzen in Getreide.**39**. *Deutsche Pflanzenschutz-Tagung*, Stuttgart, Heft 151.

Pommer, E.-H. & Kradel, J. (1969). Mebenil (BAS 3050 F), a new compound with specific action against some Basidiomycetes. In *Proceedings of the 5th British Insecticide and Fungicide Conference*, **2**, 563–8.

Pommer, E.-H. & Kradel, J. (1970). 2,5-Dimethyl-furan-3-carbonsaeureanilid (BAS 3191 F), ein neuer wirkstoff zur bekaempfung samenbürtiger pilze an getreide. In *Abstracts of the 7th International Congress on Plant Protection*, Paris, 409.

Pommer, E.-H. & Reuther, W. (1978). Furmetamid, a new active ingredient for the control of wood-destroying Basidiomycetes. In *Proceedings of the 4th International Symposium on Biodeterioration*, Berlin, pp. 67–70.

Pommer, E.-H. & Zeeh, B. (1977). Substituted 3-furamides with specific activity against Basidiomycetes. *Pesticide Science*, **8**, 320–2.

Pommer, E.-H. & Zwick, W. (1974). Efficacy of systemic fungicides against Basidiomycetes. *Indian Phytopathology*, **27**, 53–8.

Ragsdale, N. N. & Sisler, H. D. (1970). Metabolic effects related to fungitoxicity of carboxin. *Phytopathology*, **60**, 1422–7.

Ramsay, R. R., Ackrell, B. A. C., Coles, C. J., Singer, T. P., White, G. A. & Thorn, G. D. (1981). Reaction site of carboxanilides and of thenoyltrifluoroacetone in Complex II. *Proceedings of the National Academy of Sciences, USA*, **78**, 825–8.

Redfearn, E. R., Whittaker, P. A. & Burgos, J. (1965). The interaction of electron carriers in the mitochondrial NADH$_2$ and succinate oxidase systems. In *Oxidases and Related Redox Systems*, ed. T. E. King, H. S. Mason & M. Morrison, pp. 943–59. New York: Wiley.

Rich, P. R. & Bonner, W. D. jr. (1977). Epr studies of higher plant mitochondria in relation to alternative respiratory oxidations. *Plant Physiology*, **59** *(Supplement)*, 59 (abstract).

Rich, P. R. & Bonner, W. D. jr. (1978). The nature and location of cyanide and antimycin resistant respiration in higher plants. In *Functions of Alternative Terminal Oxidases*, vol. 49, Proceedings of the 11th FEBS Meeting, Copenhagen, 1977, ed. H. Degn, D. Lloyd & G. C. Hill, pp. 149–58. Oxford: Pergamon.

Righetti, P. & Cerletti, P. (1971). Molecular parameters of the beef heart succinate dehydrogenase. *FEBS Letters*, **13**, 181–3.

Schewe, T., Müller, W., Lyr, H. & Zanke, D. (1979). Ein molekulares Rezeptormodell für Carboxin. In *Systemfungizide, V*, International Symposium, Reinhardsbrunn, 1977, ed. H. Lyr & C. Polter, pp. 241–51. Berlin: Akademie-Verlag.

Schewe, T., Rapoport, S., Böhme, G. & Kunz, W. (1973). Zum Angriffspunkt des Systemfungizids Carboxin in der Atmungskette. *Acta Biologica et Medica Germanica*, **31**, 73–86.

Sherald, J. L. & Sisler, H. D. (1972). Selective inhibition of antimycin A-insensitive respiration in *Ustilago maydis* and *Ceratocystis ulmi*. *Plant and Cell Physiology*, **13**, 1039–52.

Singer, T. P., Gutman, M. & Massey, V. (1973). Succinate dehydrogenase. In *Iron-sulfur Proteins*, vol. 1, ed. W. Lovenburg, pp. 227–54. New York: Academic Press.

Snel, M., von Schmeling, B. & Edgington, L. V. (1970). Fungitoxicity and structure–activity relationships of some oxathiin and thiazole derivatives. *Phytopathology*, **60**, 1164–9.

Storey, B. T. (1976). Respiratory chain of plant mitochondria. XVIII. Point of interaction of the alternative oxidase with the respiratory chain. *Plant Physiology*, **58**, 521–5.

Tate, H. D. & von Schmeling, B. (1968). Vitavax – ein neues systemisches Fungizid. *Pflanzenschutzberichte, Wien*, **37**, 193–6.

ten Haken, P. & Dunn, C. L. (1971). Structure-activity relationships in a group of carboxanilides systemically active against broad bean rust (*Uromyces fabae*) and wheat rust (*Puccinia recondita*). In *Proceedings of the 6th British Insecticide and Fungicide Conference*, **2**, 453–62.

Tucker, A. N. & Lillich, T. T. (1974). Effect of the systemic fungicide carboxin on electron transport function in membranes of *Micrococcus dentrificans*. *Antimicrobial Agents and Chemotherapy*, **6**, 572–8.

Ulrich, J. T. & Mathre, D. E. (1972). Mode of action of oxathiin systemic fungicides. V. Effect on electron transport of *Ustilago maydis* and *Saccharomyces cerevisiae*. *Journal of Bacteriology*, **110**, 628–32.

von Schmeling, B. & Kulka, M. (1966). Systemic fungicidal activity of 1,4-oxathiin derivatives. *Science*, **152**, 659–60.

Wallnöfer, P. (1968). Abbau von 1,4-Oxathiin-Derivaten durch den Schimmelpilz *Rhizopus japonicus*. *Naturwissenschaften*, **55**, 351.

Wallnöfer, P. (1969). Der mikrobielle Abbau des 1,4-Oxathiinderivats, 2,3-Dihydro-5-carboxanilido-6-methyl-1,4-oxathiin (DCMO). *Archiv für Mikrobiologie*, **64**, 319–26.

Wallnöfer, P. R., Königer, M., Safe, S. & Hutzinger, O. (1972*a*). Metabolism of the systemic fungicide 2,5-dimethyl-3-furancarboxylic acid anilide (BAS 3191) by *Rhizopus japonicus* and related fungi. *Journal of Agricultural and Food Chemistry*, **20**, 20–2.

Wallnöfer, P. R., Safe, S. & Hutzinger, O. (1972*b*). Metabolism of the systemic fungicides 2-methylbenzanilide and 2-chlorobenzanilide by *Rhizopus japonicus*. *Pesticide Biochemistry and Physiology*, **1**, 458–63.

White, G. A. (1971). A potent effect of 1,4-oxathiin systemic fungicides on succinate oxidation by a particulate preparation from *Ustilago maydis*. *Biochemical and Biophysical Research Communications*, **44**, 1212–19.

White, G. A. & Thorn, G. D. (1975). Structure–activity relationships of carboxamide fungicides and the succinic dehydrogenase complex of *Cryptococcus laurentii* and *Ustilago maydis*. *Pesticide Biochemistry and Physiology*, **5**, 380–95.

White, G. A. & Thorn, G. D. (1980). Thiophene carboxamide fungicides: structure–activity relationships with the succinate dehydrogenase complex from wild-type and carboxin-resistant mutant strains of *Ustilago maydis*. *Pesticide Biochemistry and Physiology*, **14**, 26–40.

White, G. A., Thorn, G. D. & Georgopoulos, S. G. (1978). Oxathiin carboxamides highly active against carboxin-resistant succinic dehydrogenase complexes from carboxin-selected mutants of *Ustilago maydis* and *Aspergillus nidulans*. *Pesticide Biochemistry and Physiology*, **9**, 165–82.

Yamashita, S. & Racker, E. (1969). Resolution and reconstitution of the mitochondrial electron transport system II. Reconstitution of succinoxidase from individual components. *Journal of Biological Chemistry*, **244**, 1220–7.

Yu, C.-A. & Yu, L. (1980). Isolation and properties of a mitochondrial protein that converts succinate dehydrogenase into succinate-ubiquinone oxidoreductase. *Biochemistry*, **19**, 3579–85.

Yu, C.-A. & Yu, L. (1982). Specific interaction between protein and ubiquinone in succinate-ubiquinone reductase. *Journal of Biological Chemistry*, **257**, 6127–31.

Zeeh, B., Linhart, F. & Pommer, E.-H. (1977). Nicotinic anilides. *German Offenlegungsschrift*, 26 11 601, 17 p.

Ziegler, D. M. (1961). The mechanism of coenzyme Q reduction in heart mitochondria. In *Biological Structure and Function*, ed. T. W. Goodwin & O. Lindberg, pp. 253–64. New York: Academic Press.

Ziegler, D. M. & Doeg, K. A. (1962). Studies on the electron transport system. XLIII. The isolation of a succinic-coenzyme Q reductase from beef heart mitochondria. *Archives of Biochemistry and Biophysics*, **97**, 41–50.

Ziogas, B. N. & Georgopoulos, S. G. (1979). The effect of carboxin and of thenoyl trifluoroacetone on cyanide-sensitive and cyanide-resistant respiration of *Ustilago maydis* mitochondria. *Pesticide Biochemistry and Physiology*, **11**, 208–17.

9

Antifungal activity of substituted 2-aminopyrimidines

D. W. HOLLOMON

Department of Insecticides and Fungicides, Rothamsted Experimental Station, Harpenden, Herts AL5 2JQ

Historical

The pyrimidine nucleus is often associated with biological activity. Many pyrimidines are antibiotics and some show significant antifungal activity (Rader, Monroe & Whetstone, 1952; Dekker, 1962). Screening of phosphorylated pyrimidines related to the insecticide diazinon by workers at Jealott's Hill Research Station, revealed a series of O,O-diethyl-O(2-dimethylamino-4-methylpyrimid-6-yl) phosphorothionates with protectant fungicidal activity against certain powdery mildews. Additional structure–activity studies showed that removing the phosphorothionate moeity did not destroy activity, and the resulting 2-amino-4-hydroxypyrimidines were also systemic in plants. Three members of this 2-aminopyrimidine group of fungicides have subsequently been exploited commercially. Dimethirimol (Elias *et al.*, 1968) and ethirimol (Bebbington *et al.*, 1969) are used to control powdery mildew on cucurbits (*Sphaerotheca fuliginea, Erysiphe cichor-*

Diazinon

Dimethirimol

Ethirimol

Bupirimate

acearum) and cereals (*Erysiphe graminis*) respectively, whilst bupirimate (Finney, Farrel & Bent, 1975) is used against mildews on some fruit (*Podosphaera leucotricha*), and ornamental crops (*Sphaerotheca pannosa*).

Spectrum of activity

Several workers have examined the effects of 2-aminopyrimidine fungicides on fungi. Frequently, formulated chemicals were incorporated into agar at commercially recommended rates, and significant inhibitory effects on fungi other than powdery mildews were seldom observed. This specificity towards mildews has been confirmed with pure compounds, to avoid the complications which may arise due to the presence of formulating agents. Of 15 fungi tested (Hollomon, 1979*a*; Hollomon & Chamberlain, 1981), only the five powdery mildews were inhibited by ethirimol, although even these differed somewhat in their sensitivity. Dimethirimol failed to inhibit spore germination in *Botrytis fabae*, *Puccinia recondita*, or *Venturia inaequalis*, and only marginally inhibited germination of *Phytophthora infestans* zoosporangia (Bent, 1970). 2-aminopyrimidines had little effect on the growth of *B. cinerea* (Grindle, 1981) or on aphid pathogens used with integrated pest control systems (Wilding & Probyn, 1980; Hall, 1981). Bupirimate, dimethirimol and ethirimol are only used commercially to control powdery mildews. Phytotoxicity problems have not, so far, been reported, although high dose rates may reduce root length and chromosome volume in barley (Bennett, 1971).

Uptake, translocation, metabolism and degradation

Although 2-aminopyrimidine fungicides are frequently used as foliar sprays, dimethirimol and ethirimol are also applied as a soil drench and seed dressing respectively. Uptake from soil depends on its organic-matter content and acidity as well as other factors but, once adsorbed onto soil particles, 2-aminopyrimidines are released slowly; this provides a continuous supply of fungicide to the growing crop (Graham-Bryce & Coutts, 1971). Movement within plants is apoplastic and both dimethirimol and ethirimol accumulate at leaf margins. Ethirimol, or derivatives, may even be released onto the leaf surface when applied as a seed treatment. Bupirimate does not readily enter leaves (Teal & Cavell, 1975) and, when applied through roots, remains confined within veins. It is, however, sufficiently volatile to have a pronounced vapour-phase activity and to move across leaf surfaces (Finney *et al.*, 1975).

Metabolism of 2-aminopyrimidines is similar in both barley and cucumber plants (Cavell, Hemmingway & Teal, 1971). Degradation is rapid, and ethirimol fed to barley through its roots has a half-life of no more than 4 d (Calderbank, 1971), although some of the degradation products may be fungitoxic. Degradation primarily involves *N*-dealkylation (Fig. 9.1) but other changes resulting in the formation of more water-soluble glucosides (and possibly phosphates) occur, as well as hydroxylation of the 5*n*-butyl group. Bupirimate is readily hydrolysed in acid solutions on leaf surfaces to form ethirimol. These changes have been examined in whole leaves, but metabolism in different cell layers, especially the epidermis which is the only layer entered directly by mildew, may be somewhat different (Hollomon, 1977). In short-term experiments, ethirimol was not substantially degraded in either extracts

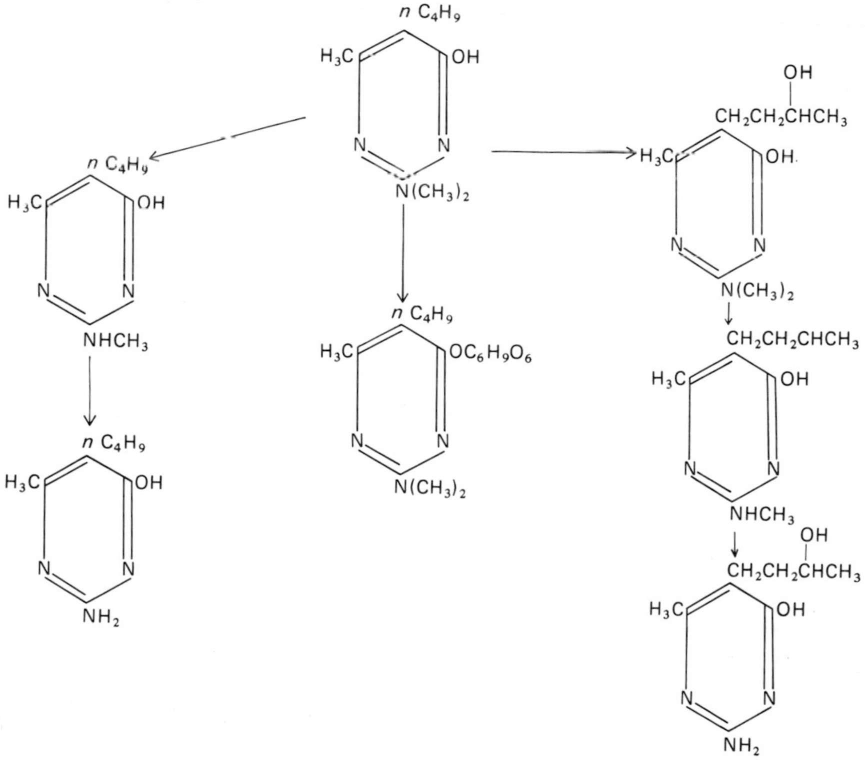

Fig. 9.1. Degradation products of dimethirimol in cucumber leaves.

from *Erysiphe graminis* conidia (Hollomon & Chamberlain, 1981) or haustoria of *Erysiphe pisi* (Manners & Gay, 1980).

Haustoria play an important role in absorption of nutrients by powdery mildews (Bushnell & Gay, 1978), and may provide a pathway for entry of systemic fungicides. Haustorial complexes isolated from peas infected with *Erysiphe pisi* rapidly accumulated ethirimol against a concentration gradient to at least 60 fold (2.4 mM) the ambient concentration in 50 mM phosphate buffer, pH 6.8 (Manners & Gay, 1982). Protoplasts from mesophyll cells did not accumulate ethirimol in this way. Manners & Gay suggest that accumulation of ethirimol analogues is significantly correlated with their partition coefficient between olive oil/water, and that lipophilicity plays some role in their uptake. However, uptake kinetics were similar at pH 4.2 (Manners & Gay, 1980), when ethirimol largely exists in a protonated form and is, therefore, less lipid soluble. Significant influx of cations into haustorial complexes also occurred (Manners & Gay, 1982), which may explain why ethirimol still accumulated when ionised. Concentrations of ethirimol achieved within haustoria are in excess of its aqueous solubility, suggesting that ethirimol is either attached to particulate matter, or dissolved in lipid droplets. How the fungicide is made available to inhibit soluble enzymes within the cytoplasm is not clear. Indeed, accumulation of 2-aminopyrimidines within haustorial complexes is not directly related to their fungitoxicity (Table 9.1).

In laboratory studies, isomeric photodimers of 2-aminopyrimidines are formed in degassed solutions exposed to ultraviolet (UV) light, by self-addition across the 5–6 double bond of the parent pyrimidine (Wells, Pollard & Sen, 1979). Similar products have not been detected in soil, water or plants treated with 2-aminopyrimidines and exposed to sunlight. This may be because atmospheric oxygen competes with excited molecules of the pyrimidinol, or that concentrations found in soil, water, or plants are considerably lower than those used in laboratory irradiation studies (0.4 mM). Other photochemical degradation products of ethirimol have been identified (Cavell, 1979), including desethylethirimol which may be formed through reactions catalysed by riboflavin and pteridines as photoreceptors.

Mechanism of action
Time of action

2-aminopyrimidines act at several stages during mildew development. Germination is inhibited (Bent, 1970), but probably at

Table 9.1. *Accumulation of 2-aminopyrimidine analogues by isolated haustorial complexes*

Compound	R_1	R_2	R_3	Accumulation with time[a]			Partition coefficient (olive oil/water)[a] (log_{10})	Fungicidal activity against	
				1 min	5 min	60 min		Barley mildew[b]	cucumber mildew[c]
R31671	CH_3	CH_3	nC_3H_7	64	83	122	0.60	—[d]	100
Dimethirimol	CH_3	CH_3	nC_4H_9	41	58	—	0.70	1.0	92
R31681	H	H	nC_4H_9	28	52	73	0.30	0.003	NA[e]
Ethirimol	C_2H_5	H	nC_4H_9	27	54	62	0.33	1.0	100
R33452	H	CH_3	nC_4H_9	31	42	58	0.18	0.22	94
R31805	CH_3	CH_3	CH_3	0	1	5	0.52	NA	—

[a]Data from Manners & Gay (1982). Accumulation is expressed as the ratio of the concentration within haustorial complexes to the concentration in the external solution (40 μM).
[b]Data from Hollomon & Chamberlain (1981). Fungicidal activity expressed relative to that of ethirimol; small numbers indicate lower activity.
[c]Unpublished data of B. K. Snell (quoted by Woodcock, 1972). Fungicidal activity is expressed as % leaf area free from mildew following systemic treatment.
[d]—: Data not available.
[e]NA: not active.

concentrations above those normally found in treated plants. Appressoria formation, an essential step in primary infection, is also inhibited (Fig. 9.2). However, the critical events occur well before appressoria appear (Hollomon, 1977) for, if ethirimol is applied to leaves 8 h after inoculation, yet some 6–12 h before appressoria actually form, development continues normally until haustoria are present. Effects occurring before the pathogen has penetrated the host may result from fungicide present on the leaf surface, although exchange of compounds, including fungicides, between host and pathogen may occur as early as 2 h after inoculation. Cytological evidence suggests limited penetration of the cuticle by protuberances observed beneath germ-tubes within 6 h after inoculation (Kunoh, Tsuzuki & Ishizaki, 1978). Reciprocal translocation of ions and fluorescent dye also occurs within this time (Kunoh & Ishizaki, 1981; Kunoh, Yamamori & Ishizaki, 1982). Some unidentified stimulus stops movement of host cytoplasm in epidermal cells, and this cytoplasm aggregates beneath germ-tubes long before penetration of the cell wall occurs (Aist, 1976). 2-aminopyrimidines also affect later stages of mildew development and may show antisporulant activity.

Reversal experiments

Powdery mildews are obligate parasites and, if complications of host metabolism are to be avoided, only conidia are available for biochemical studies of the mode of action of the fungicide. Conse-

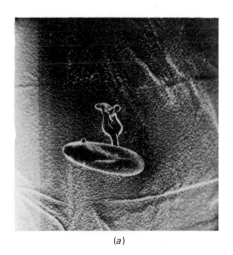 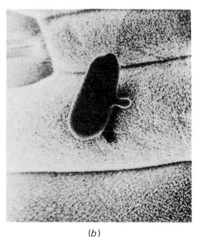

(a)   (b)

Fig. 9.2. Inhibition of appressoria formation by ethirimol. (a) Without ethirimol; (b) with ethirimol.

Table 9.2. *Effects of metabolites on the fungitoxicity of ethirimol on cucumber powdery mildew*[a]

Reversing agent	Ability to antagonise the effect of ethirimol (0 lowest, 10 highest)
Riboflavin	10
Adenine	8
Guanine	7
Folic acid	6
Pyridoxal-5-phosphate	6
Adenosine	6
5-aminolevulinic acid	4
Glycine	3
Thymidine	3
Orotic acid	2
Cytosine	2
Serine	2
Methionine	1
Betaine	1
Valine	1
Leucine	1
Formic acid	1
Succinic acid	1
Citric acid	0.5
Tryptophan	0.5
Formimino-glutamate	0.5
Glutamic acid	0

From Slate *et al.*, 1972.
[a]Cotyledon-stage cucumber plants in water-culture were supplied with ethirimol (10^{-5}–10^{-6} M) over 3–4 d. Reversing agents were maintained in the culture solution at 10^{-4} M (except riboflavin whose concentration was 10^{-6} M). Plants were infected with mildew conidia at the time of adding reversing agents. 'Reversal' was scored as the ability to promote mildew growth earlier than on plants treated with ethirimol only.

quently, indirect approaches have been adopted to locate possible sites of action of 2-aminopyrimidines. Reduction in toxicity of both ethirimol and dimethirimol through addition of various metabolites to fungicidal solutions was examined using cucumber, wheat and barley mildew systems (Sampson, 1969; Bent, 1970; Hollomon, 1979*b*). Effects on germination, primary infection and disease levels were recorded, and Table 9.2 shows results for one series of experiments in which 22 metabolites were tested, often at concentrations well above physiological levels (Slade *et al.*, 1972). Many antagonists which reduced the toxicity of ethirimol are involved in C-1 metabolism, either as cofactors, C-1 donors or C-1 products, suggesting that 2-aminopyrimidine fungicides might interfere with *de novo* purine biosynthesis, possibly through inhibition of pyridoxal-catalysed supply of C-1 units from serine (Sampson, 1969). But incorporation of (^{14}C) formate, a C-1 source, into purines was not inhibited by dimethirimol, nor were several enzymes involved in *de novo* purine biosynthesis, including serine hydroxymethylase. Furthermore, glycine was incorporated by barley mildew into protein but not ribonucleic acid, indicating that *de novo* purine biosynthesis probably does not operate during primary infection, at least in this mildew (Hollomon, 1979*b*).

Reversal by riboflavin was clearly due to direct photo-inactivation of the fungicide (Bent, 1970) and similar degradation of the fungicide may occur in the presence of folate. Reversal of the antifungal effect by pyridoxal-5-phosphate was not confirmed with wheat or barley mildew but, in all three systems, adenine or adenosine appear to somewhat reduce the toxicity of the fungicide. Pyrimidines were not particularly effective reversal agents.

Inhibitors

Similarities in the way some purine analogues and ethirimol affect mildew development provides further evidence that 2-aminopyrimidines interfere with purine metabolism, albeit after the purine ring system has formed. Kinetin (6-furfuryl adenine), shown earlier by Dekker (1963) to restrict mildew development, inhibits appressoria formation more effectively than subsequent colony development (Hollomon, 1979*b*). Ethirimol resistant strains of *Erysiphe graminis* were also cross-resistant to kinetin (Table 9.3). Although somewhat less active, isopentenyl adenine, 6-methylpurine, 6-methylaminoadenine and some other purine analogues, showed similar patterns of activity. Pyrimidines were either inactive or, like 6-azauracil and 2-thiouracil, affected colony development but not appressoria formation. Compounds thought to block different steps in *de novo* purine biosynthesis were not fungicidal. Cycloheximide and griseofulvin, which inhibit protein synthesis and microtubule formation respectively, inhibit mildew development, but their pattern of activity differs from that of ethirimol, and it is unlikely that 2-aminopyrimidines directly inhibit these metabolic events. Similar experiments using *Sphaerotheca fuliginea*, whilst not in complete agreement, confirmed that purine analogues can affect mildew development (Gorter & Nel, 1974; Nel & Gorter, 1974).

Table 9.3. *Cross-resistance between inhibitors of mildew development*

| | Bioassay | | | | | |
| | Appressoria formation (ED$_{50}$ × 10^{-6} M) | | | Colony length^a (ED$_{50}$ × 10^{-5} M) | | |
Compound	R^b	S^b	R/S	R	S	R/S
Ethirimol	12	1.8	6.7	3.6	3.0	1.2 NS
Dimethirimol	4.6	1.2	3.8	52.0	54.0	0.9 NS
6-Furfuryl adenine	14	1.7	8.2	+^{c}	+^{c}	—
6-Methyl purine	>7400^d	410	>18^d	386	340	1.1 NS
6-Azauracil	—	—	—^{e}	48	79	0.6 NS
Cycloheximide	35	38	0.9 NS	8.5	7	1.2 NS
Griseofulvin	78	310	0.26 NS	—	—^{e}	—

^{a}Bioassay based on measurement of length of mildew colonies, at varying fungicide concentrations, 48 h after inoculation. For fuller experimental details see Hollomon, 1977.

^{b}R = Resistant (strain 32B5); S = Sensitive (strain 23D5), see Hollomon, 1975 for details.

^{c}Some effect, but ED$_{50}$ value above maximum solubility in water.

^{d}6-Methylpurine did not inhibit the resistant strain at the highest concentration possible; ED$_{50}$ values and R/S were not calculated.

^{e}No effect at maximum solubility in water.

Radioisotope experiments

Slade *et al.* (1972) reported that [^{14}C] formate, sprayed onto mildewed cucumber leaves, was not incorporated into protein or RNA if dimethirimol was supplied via the roots. RNA synthesis was inhibited by ethirimol during primary infection of barley mildew, an effect less pronounced with ethirimol-resistant strains than with sensitive ones (Hollomon, 1979*b*). Metabolism of adenine and adenosine was examined by ion-exchange high pressure liquid chromatography of acid-soluble extracts from mildew, incubated for 24 h on barley leaves to which [2-^{3}H] adenine or [2-^{3}H] adenosine was applied before inoculation. Adenine and its nucleoside were both metabolised to other purines (Table 9.4), and many steps involved in purine salvage appear to operate during appressoria formation. Ethirimol did not limit uptake of either radiolabelled compound, but incorporation into most other metabolites was inhibited, suggesting that ethirimol interferes with adenine-salvage reactions (Fig. 9.3).

Purine   salvage   reactions

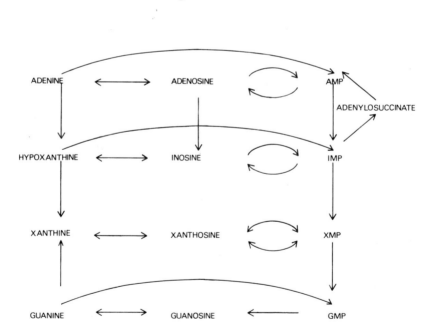

Fig. 9.3. Possible pathways of purine salvage. Inclusion of a pathway does not necessarily signify that it is important in powdery mildews.

Table 9.4. *Effect of ethirimol on [2-³H] adenine and [2-³H] adenosine metabolism during appressoria formation in Erysiphe graminis*

Metabolite	[2-³H] adenine[a]			[2-³H] Adenosine[a]		
	No ethirimol	Ethirimol[b]	% control	No ethirimol	Ethirimol	% control
Adenine	21.74	22.63	104	0.61	0.84	137
Adenosine	6.80	5.54	81	5.54	8.50	153
Inosine	18.14	5.09	28	9.41	2.82	30
Hypoxanthine	3.28	2.90	88	2.74	2.14	78
Guanosine	—[c]	—	—	2.93	1.76	60
Xanthosine	—	—	—	2.14	1.07	50
Adenine nucleotides	6.90	0.70	10	7.54	1.13	15
Inosine nucleotides	1.49	0.28	19	3.39	1.52	45
Guanosine nucleotides	—	—	—	1.45	0.49	34
Xanthosine nucleotides	—	—	—	1.53	0.40	26
Total (excluding unmetabolised precursor)	36.62	14.52	40	31.77	12.17	38

[a]Picomoles incorporated by 3×10^5 appressoria in 24h. The specific activity of the [2-³H] adenine was $17.8\,\mathrm{Ci\,mM^{-1}}$; the specific activity of the [2-³H] adenosine was $21.0\,\mathrm{Ci\,mM^{-1}}$. For further experimental details see Hollomon & Chamberlain, 1981.
[b]Erithmol concentration was $9.6 \times 10^{-5}\,\mathrm{M}$.
[c]Not measured.

Inhibition of adenosine deaminase

Several enzymes involved in the purine salvage pathway (Fig. 9.3) were detected in extracts from *Erysiphe graminis* conidia and mildewed barley. Of these only adenosine deaminase (EC 3.5.4.4; adenosine aminohydrolase, ADA-ase), which catalyses the hydrolytic deamination of adenosine to form inosine, was inhibited by ethirimol (Hollomon & Chamberlain, 1981). This enzyme was not detected in healthy peas or barley (Table 9.5), and is generally absent from plants (Zielke & Suelter, 1971). All fungi examined showed some ADA-ase activity, but only in powdery mildews was this inhibited by ethirimol. This specificity correlates well with the disease-control spectrum of 2-aminopyrimidines, and may arise through some feature which is unique to mildew ADA-ase.

Kinetic parameters of ADA-ase may differ significantly between tissues in the same organism (Henderson & Patterson, 1973), but many properties of ADA-ase from barley mildew conidia and infected plants are broadly similar to those for the same enzyme from other sources. Mildew ADA-ase is a soluble enzyme which binds deoxyadenosine more tightly than adenosine, but deaminates adenosine faster. Adenine and adenylate are not substrates, and activity is strongly inhibited by coformycin, a potent inhibitor of ADA-ase from many sources (Agarwal, Sagar & Parks, 1975). Gel filtration on Sephacryl S-300 reveals only a single form (MW 300 000) of the enzyme in infected barley, whereas multiple forms commonly exist in other organisms (Kelley, Daddona & Van der Weyden, 1977). The mildew enzyme remains a single protein after further purification on DEAE Sephacel, chromatofocusing (P_I4.4–4.6), and agarose gel electrophoresis. The largest form of the enzyme in human kidney (MW 300 000) comprises a single catalytic subunit (MW 36 000) together with a large accessory protein of unknown function (Daddona & Kelley, 1978). It is not known whether the mildew enzyme is similar but interestingly, ADA-ase in *Erysiphe graminis* conidia is somewhat larger than that in infected plants.

Inhibition by ethirimol is pH dependent and non-competitive ($K_i = 2.32 \times 10^{-5}$ M with deoxyadenosine as substrate; 9.5×10^{-6} M with adenosine as substrate). Equilibrium dialysis and gel filtration of the enzyme plus [^{14}C] ethirimol showed that the fungicide is not a tight-binding inhibitor.

Table 9.5. *Effect of ethirimol on ADA-ase from different sources*

	ADA-ase activity[a]		Inhibition (%)
	No ethirimol	Ethirimol (200 µg ml⁻¹)	
Powdery mildews			
Erysiphe graminis conidia	48.0	0.8	98
Erysiphe pisi conidia and mycelium	6.7	0.4	94
Erysiphe polygoni conidia	60.7	38.1	37
Sphaerotheca fuliginea conidia	8.9	4.5	42
Unidentified mildew from *Senecio vulgaris*	20.5	4.3	21
Other fungi			
Penicillium sp. mycelium	76.9	72.2	6
Cladosporium cucumerinum spores and mycelium	11.4	14.8	None
Fusarium sp. mycelium	34.9	33.5	4
Uromyces appendiculatus germlings	4.4	4.4	None
Uromyces fabae germlings	8.5	7.5	12
Puccinia obtegens germlings	4.3	4.7	None
Rhynchosporium secalis spores and mycelium	2.3	2.5	None
Neurospora crassa mycelium	8.6	8.2	9
Saccharomyces cerevisiae	0.8	0.7	12
Streptomyces scabies	7.9	8.5	None
Plants			
Healthy barley	*[b]	*	*
Mildew infected	4.8	2.8	42
Healthy peas	*	*	*
Mildew infected	3.2	0.8	25
Others			
Calf intestine (purified)	54,100	48,300	11
Myzus persicae	*	*	*
Musca domestica	33.2	30.4	9

[a] Picomole substrate converted per mg protein per 10 min at 25 °C. Adenosine concentration in all assays was 3.6×10^{-4} M. For other experimental details see Hollomon, 1979a.
[b] ADA-ase activity not detected.

Structure–activity studies

Hollomon & Chamberlain (1981) examined how structural variation in a series of 2-dimethylaminopyrimidines affects mildew development and inhibition of ADA-ase. Alterations at R_1, R_2 and R_3 (Table 9.6) generally produced less effective ADA-ase inhibitors and poorer fungicides. Since substitution of the 5n-butyl group (R_2) results in loss of both enzyme inhibition and fungicidal action, this group may not only ensure adequate lipophilicity for membrane permeability, but is essential for correct binding of the fungicide to its active site. Binding probably involves a hydrophobic part of the enzyme, as has been demonstrated for a series of 9-alkyl adenosine inhibitors of ADA-ase (Baker, 1967), and suggests that the n-butyl group is buried within the enzyme whilst the 2-substituted amino group is on its surface. 2-aminopyrimidines exist in solution as the tautomeric 1,4-dihydro-4-oxopyrimidines, and the oxygen function at position 4 (R_1) is important for activity. When this oxygen function is changed, analogues are inactive as enzyme inhibitors but some still show fungicidal activity. Analogues (Table 9.6) such as 4-chloro (11), and 4-benzoyloxymethoxy (13), may well owe their activity to conversion by intact organisms to hydroxypyrimidines (Woodcock, 1972). Many 2-substituted analogues were good ADA-ase inhibitors, but act as fungicides only after infection is established (Table 9.7). Differences in partition coefficients suggest that mobility through membranes might vary amongst 2-substituted analogues, but this does not fully explain their lack of activity. However, factors other than lipid solubility may govern entry of pyrimidines into germinating spores (Cummins & Mitchison, 1967). Despite anomalies, these structure–activity studies provide some additional evidence that ADA-ase is a site of action of 2-aminopyrimidine fungicides.

Role of adenosine metabolism in mildew infection

2-aminopyrimidine fungicides may inhibit ADA-ase, but why should this affect mildew development? Mildews do not synthesise purines *de novo*, and ADA-ase may be important in salvaging purines from the host for use by the pathogen to synthesise adenine and guanine nucleotides. However, inosine does not reverse the toxicity of ethirimol, and enzymes required to convert inosine to adenylate and guanylate appear absent from mildew.

As in most resting cells, the energy charge

$$[ATP] + 0.5[ADP]/[ATP] + [ADP] + [AMP]$$

Table 9.6. *Effect of 2-dimethylaminopyrimidines on ADA-ase and mildew growth*

Compound	R_1	R_2	R_3	ADA-ase inhibition	Bioassay[a,b] Appressoria formation	Colony length
1	OH	nC_4H_9	CH_3(dimethirimol)	1a	1a	1a
2	OH	H	CH_3	NA[c]	NA	NA
3	OH	CH_3	CH_3	NA	NA	NA
4	OH	$CH(CH_3)_2$	CH_3	NA	NA	NA
5	OH	C_6H_{13}	CH_3	0.05b	0.048b	0.16a
6	OH	nC_4H_9	H	0.06b	0.020d	NA
7	OH	H	H	NA	NA	NA
8	OH	nC_4H_9	C_2H_5	0.83a	0.769a	10.0b
9	OH	nC_4H_9	CH_3NH	NA	0.200c	0.55a
10	OH	nC_4H_9	CH_3S	NA	NA	10.0b
11	Cl	nC_4H_9	CH_3	NA	0.100c	10.0b
12	OCH_3	nC_4H_9	CH_3	NA	0.001e	1.6a
13	$OCH_2OCOC_6H_5$	nC_4H_9	CH_3	NA	0.800a	10.0b
14	CH_3	nC_4H_9	CH_3	NA	0.002c	1.6a

[a]Activities are expressed relative to dimethirimol and shown as reciprocals. Compounds with lower activities than dimethirimol are, therefore, poorer fungicides. Values for dimethirimol are: k_1 for ADA-ase = 5.4×10^{-4} M; ED_{50} appressoria formation = 1.2×10^{-6} M: colony length = 5.4×10^{-4} M.
[b]Values followed by the same letter are not significantly different ($P = 0.01$).
[c]NA, no activity.

Table 9.7. *Effect of 4-hydroxy-5-n-butyl-6-methyl pyrimidines on ADA-ase and mildew growth*

Compound	R_4	Partition coefficient[a]	ADA-ase inhibition	Bioassay[b,c] Appressoria	Colony length
15	H	1.38	50.00a	0.006a	0.77ab
16	NH_2	1.50	2.00b	0.003a	0.38ab
17	OH	1.26	0.20e	0.002a	NA
18	SH	1.81	0.05	NA[d]	NA
19	$NHCH_3$	1.64	1.67c	0.22b	0.15b
20	$NH(C_2H_5)$ (Ethirimol)	2.21	50.00a	1.00c	16.67c
1	$N(CH_3)_2$ (Dimethirimol)	1.93	1.00d	1.00c	1.00a
21	$NH \cdot NH$-phenyl	—	13.14g	0.06g	4.09a
22	pyrrole	—	2.16b	0.19b	3.75a

[a]Octanol: water $\log_{10}$, see 20.
[b]Activities are expressed as in Table 9.6, relative to dimethirimol.
[c]Values followed by the same letter are not significantly different ($P = 0.01$).
[d]NA: no activity.

(Atkinson, 1968) in mildew conidia is low, due primarily to a large AMP pool (D. W. Hollomon, unpublished observation). ADA-ase, a non-equilibrium enzyme, may act as a sink removing adenosine which in turn stimulates AMP breakdown by action of 5′ nucleotidase. By lowering AMP levels, energy charge would increase and stimulate metabolic activity required for germination and development.

Adenosine levels probably regulate the activity of many enzymes involved in purine salvage (Arch & Newsholme, 1978). In healthy barley, adenosine is converted to adenine by nucleosidase/phosphorylase, or to adenylate by adenosine kinase. After infection, both ADA-ase and S-adenosyl homocysteine hydrolase also compete for adenosine. Fig. 9.4 shows that the activity of some of these enzymes changes considerably during infection (D. W. Hollomon & J. A. Butters, unpublished results) suggesting that adenosine levels change also. If high concentrations of adenosine inhibit adenosine kinase (as seems likely) by reducing adenosine levels, ADA-ase may ensure increased kinase activity, and a greater supply of adenylate for transfer to mildew. Indeed, Table 9.4 shows that ethirimol inhibits adenylate synthesis, yet it does not appear to inhibit any of the enzymes involved in its synthesis. Until more is known about the regulation of purine metabolism during mildew development, the role of ADA-ase will remain uncertain.

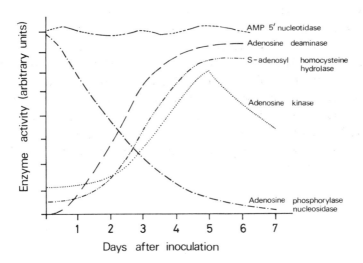

Fig. 9.4. Changes in activity of enzymes involved in adenosine metabolism during mildew infection.

Resistance

Strains of barley powdery mildew, resistant to levels of ethirimol which otherwise inhibit wild type strains, have been isolated from field crops (Hollomon, 1975). Levels of resistance vary but, in assays using ethirimol-treated seedlings, may be 10–20 times more resistant than the wild-type strain. Ethirimol-resistant strains are cross-resistant to other 2-aminopyrimidine fungicides but, so far, the biochemical mechanism of this resistance is not known. Sensitivity of the target enzyme is not reduced, nor its amount increased in resistant strains (Hollomon, 1979a). Recent experiments suggest that increased uptake from barley of adenine, which will influence ethirimol toxicity, may be associated with ethirimol resistance (D. W. Hollomon, unpublished results). It is likely, however, that several different biochemical mechanisms contribute to ethirimol resistance, since more than one heritable factor seems to be involved (Hollomon, 1981). These genetic factors are primarily additive in their effects. All attempts to induce ethirimol-resistant mutants with UV irradiation, ethyl methane sulphonate, or N-methyl-N'nitro-N-nitrosoguanidine, according to the procedures of Gabriel, Lisker & Ellingboe (1982) have so far been unsuccessful (J. A. Butters, personal communication).

Ethirimol-resistant strains occurred more frequently in mildew populations from barley cultivars with the Mlas (=Mla_{12}) host-plant-resistance gene (Wolfe & Dinoor, 1973), but genetic analysis failed to show any clear association between ethirimol resistance and the corresponding pathogen-virulence gene (Hollomon, 1981). An association might exist involving some factor controlling disease development in Mlas cultivars, but which was not detected in phenotypes characterised by largely qualitative virulence tests (Moseman, 1968). If they occur, such links would certainly influence how fungicides and genetic host-plant resistance might be integrated to control powdery mildew. Depending on the nature of any linkage, fungicide resistance might provide useful markers in genetic manipulation approaches to identification of virulence-gene products.

References

Agarwal, R. P., Sagar, S. M. & Parks, R. E. Jr. (1975). Adenosine deaminase from human erythrocytes purification and effects of adenosine analogs. *Biochemical Pharmacology*, **24**, 693–701.

Aist, J. R. (1976). Papillae and related wound plugs of plant cells. *Annual Review of Phytopathology*, **14**, 145–63.

Atkinson, D. E. (1968). The energy charge of the adenylate pool as a regulatory parameter: interaction with feedback modifiers. *Biochemistry*, **7**, 4030–4.

Arch, J. R. S. & Newsholme, E. A. (1978). Activities and some properties of 5' nucleotidase, adenosine kinase and adenosine deaminase in tissues from vertebrates and invertebrates in relation to the control of the concentration and physiological role of adenosine. *Biochemical Journal*, **174**, 965–77.

Baker, B. R. (ed.) (1967). *Design of Active-site-directed Irreversible Enzyme Inhibitors*, pp. 285–300. New York: Wiley.

Bebbington, R. M., Brooks, D. H., Geoghegan, M. J. & Snell, B. K. (1969). Ethirimol: a new systemic fungicide for the control of cereal powdery mildews. *Chemistry and Industry*, 1512.

Bennett, M. D. (1971). Effects of ethirimol on cytological characters in barley. *Nature*, **230**, 406.

Bent, K. J. (1970). Fungitoxic action of dimethirimol and ethirimol. *Annals of Applied Biology*, **66**, 103–13.

Bushnell, W. R. & Gay, J. (1978). Accumulation of solutes in relation to the structure and function of haustoria in powdery mildews. In *The Powdery Mildews*, ed. D. M. Spencer, pp. 183–235. London: Academic Press.

Calderbank, A. (1971). Metabolism and mode of action of dimethirimol and ethirimol. *Acta Phytopathologica Academiae Scientiarum Hungaricae*, **6**, 355–63.

Cavell, B. D. (1979). Methods used in the study of the photochemical degradation of pesticides. *Pesticide Science*, **10**, 177–80.

Cavell, B. D., Hemmingway, R. J. & Teal, G. (1971). Some aspects of the metabolism and translocation of the pyrimidine fungicides. *Proceedings 6th British Insecticide and Fungicide Conference*, **2**, 431–7.

Cummins, J. E. & Mitchison, J. M. (1967). Adenine uptake and pool formation in the fission yeast *Schizosaccharomyces pombe*. *Biochemica et biophysica acta*, **136**, 108–20.

Daddona, P. E. & Kelley, W. N. (1978). Human adenosine deaminase binding protein. *Journal Biological Chemistry*, **253**, 4617–23.

Dekker, J. (1962). Systemic control of powdery mildew by 6-azauracil and some other purine and pyrimidine derivatives. *Mededelingen Landbouwwhogeschool Opzoekingsstations Gent*, **27**, 1214–21.

Dekker, J. (1963). Effect of kinetin on powdery mildew. *Nature*, **197**, 1027–8.

Elias, R. S., Shephard, M. C., Snell, B. K. & Stubbs, J. (1968). 5-*n*-butyl-2-dimethyamino-4-hydroxy-6-methylpyrimidine: a systemic fungicide. *Nature*, **219**, 1160.

Finney, J. R., Farrell, G. M. & Bent, K. J. (1975). Bupirimate – a new fungicide for the control of powdery mildews on apples and other crops. In *Proceedings 8th British Insecticide and Fungicide Conference*, 667–73.

Gabriel, D. W., Lisker, N. & Ellingboe, A. H. (1982). The induction and analysis of two classes of mutations affecting pathogenicity in an obligate parasite. *Phytopathology*, **72**, 1026–8.

Gorter, G. J. M. A. & Nel, D. D. (1974). Systemic fungicidal effects on powdery mildews of metabolic inhibitors and related compounds in laboratory tests. I. Pyrimidine and purine analogues. *Phytophylactia*, **6**, 209–12.

Graham-Bryce, I. J. & Coutts, J. (1971). Interactions of pyrimidine fungicides with soil and their influence on uptake by plants. *Proceedings 6th British Insecticide and Fungicide Conference*, 419–26.

Grindle, M. (1981). Variations among field isolates of *Botrytis cinerea* in their sensitivity to antifungal compounds. *Pesticide Science*, **12**, 305–12.

Hall, R. A. (1981). Laboratory studies on the effects of fungicides, acaricides and insecticides on the entomopathogenic fungus, *Verticillium lecanii*. *Entomologia experimentalis et applicata*, **29**, 39–48.

Henderson, J. F. & Paterson, A. R. P. (1973). *Nucleotide Metabolism: an Introduction*. New York: Academic Press.

Hollomon, D. W. (1975). Behaviour of a barley powdery mildew strain tolerant to ethirimol. *Proceedings 8th British Insecticide and Fungicide Conference*, 51–8.

Hollomon, D. W. (1977). Laboratory evaluation of ethirimol activity. In *Crop Protection Agents*, ed. N. R. McFarlane, pp. 505–15. London: Academic Press.

Hollomon, D. W. (1979a). Specificity of ethirimol in relation to inhibition of the enzyme adenosine deaminase. *Proceedings British Crop Protection Conference; Pests and Diseases*, 251–6.

Hollomon, D. W. (1979b). Evidence that ethirimol may interfere with adenine metabolism during primary infection of barley powdery mildew. *Pesticide Biochemistry and Physiology*, **10**, 181–9.

Hollomon, D. W. (1981). Genetic control of ethirimol resistance in a natural population of *Erysiphe graminis* f.sp. *hordei*. *Phytopathology*, **71**, 536–40.

Hollomon, D. W. & Chamberlain, K. (1981). Hydroxypyrimidine fungicides inhibit adenosine deaminase in barley powdery mildew. *Pesticide Biochemistry and Physiology*, **16**, 158–9.

Kelley, W. N., Daddona, P. E. & Van der Weyden, M. B. (1977). Characterization of human adenosine deaminase. In *Purine and Pyrimidine Metabolism*, CIBA Foundation Symposium (N.S. **48**), 277–93.

Kunoh, H. & Ishizaki, T. (1981). Cytological studies of early stages of powdery mildew in barley and wheat. VII. Reciprocal translocation of a fluorescent dye between barley coleoptile cells and conidia. *Physiological Plant Pathology*, **18**, 207–11.

Kunoh, H., Tsuzuki, T. & Ishizaki, H. (1978). Cytological studies of early stages of powdery mildew in barley and wheat. IV. Direct ingress from superficial primary germ-tubes and appressoria of *Erysiphe graminis hordei* on barley leaves. *Physiological Plant Pathology*, **13**, 327–83.

Kunoh, H., Yamamori, K. & Ishizaki, H. (1982). Cytological studies of early stages of powdery mildew in barley and wheat. VIII. Autoflorescence at penetration sites of *Erysiphe graminis hordei* on living barley coleoptiles. *Physiological Plant Pathology*, **21**, 373–9.

Manners, J. M. & Gay, J. (1980). Fluxes and accumulation of ethirimol in haustoria of *Erysiphe pisi* and protoplasts of *Pisum sativum*. *Annals of Applied Biology*, **96**, 283–93.

Manners, J. M. & Gay, J. (1982). Accumulation of systemic fungicides and other solutes by haustorial complexes isolated from *Pisum sativum* infected with *Erysiphe pisi*. *Pesticide Science*, **13**, 195–203.

Moseman, J. G. (1968). Reactions of barley to *Erysiphe graminis* f.sp. *hordei* from North America, England, Ireland and Japan. *Plant Disease Reporter*, **52**, 463–7.

Nel, D. D. & Gorter, G. J. M. A. (1974). Systemic fungicidal effects on powdery mildews of metabolic inhibitors and related compounds in laboratory tests. II. Antibiotics. *Phytophylactica*, **6**, 213–16.

Rader, W. E., Monroe, C. M. & Whetstone, R. R. (1952). Tetrahydropyrimidine derivatives as potential foliage fungicides. *Science*, **115**, 124–5.

Sampson, M. J. (1969). The mode of action of a new group of species specific fungicides. *Proceedings 5th British Insecticide and Fungicide Conference*, 483–7.

Slade, P. Cavell, B. D., Hemmingway, R. J. & Sampson, M. J. (1972). Metabolism and mode of action of dimethirimol and ethirimol. *Pesticide Chemistry*, **V**, 295–303.

Teal, G. & Cavell, B. D. (1975). Degradation of bupirimate fungicide on apples and in water. *Proceedings 8th British Insecticide and Fungicide Conference*, 25–30.

Wells, C. H., Pollard, S. J. & Sen, D. (1979). Photochemistry of some systemic pyrimidine fungicides. *Pesticide Science*, **10**, 171–6.

Wilding, N. & Probyn, P. J. (1980). Effect of fungicides on development of *Entomophthora aphidis*. *Transactions British Mycological Society*, **75**, 297–302.

Wolfe, M. S. & Dinoor, A. (1973). The problems of fungicide tolerance in the field. *Proceedings 7th British Insecticide and Fungicide Conference*, 11–19.

Woodcock, D. (1972). Structure–activity relationships. In *Systemic Fungicides*, ed. R. W. Marsh, pp. 34–85. London: Longmans.

Zielke, C. L. & Suelter, C. H. (1971). Purine, purine nucleoside and purine nucleotide aminohydrolases. In *The Enzymes*, vol. 4, ed. P. D. Boyer, pp. 47–78. New York: Academic Press.

10

Antifungal activity of dicarboximides and aromatic hydrocarbons and resistance to these fungicides

P. LEROUX and R. FRITZ

Institut National de la Recherche Agronomique, Laboratoire de Phytopharmacie, Centre National de Recherches Agronomiques, Etoile de Choisy, 78000 Versailles, France

Introduction

Iprodione, procymidone and vinclozolin are generally classified as 'dicarboximides' or 'cyclic-imides'; their common structure is a 3,5-dichlorophenyl-*N*-cyclic-imide (Fig. 10.1). The presence of chlorines in the benzene ring and their position are important for good antifungal activity (Takayama & Fujinami, 1979). Other compounds in this group are illustrated in Fig. 10.1. Captafol, captan, ditalimfos and folpet, which are also dicarboximides, do not belong to this particular group since they lack a benzene ring.

Although the dicarboximides are mainly used against Helotiaceae, (*Botrytis*, *Monilinia* and *Sclerotinia*) they are also toxic to many other fungi. For example, they can inhibit fungi in the Basidiomycetes (*Corticium*, *Rhizoctonia*, *Tilletia*, *Typhula*, *Ustilago* and some rusts), the Zygomycetes (*Mucor*, *Rhizopus* and some Entomophthorales), and the Ascomycetes (*Aspergillus*, *Cochliobolus*, *Diaporthe*, *Glomerella*, *Leptosphaeria*, *Penicillium*, *Rosellinia*, *Venturia* and some oidia). Among the Adelomycetes, *Sclerotium cepivorum* and species of *Alternaria*, *Cercospora*, *Colletotrichum*, *Fusarium*, *Helminthosporium*, *Phoma*, *Thielaviopsis* and *Verticillium* are sensitive to dicarboximides. However, these fungicides show a low toxicity towards Oomycetes, yeasts, and fungi like *Fusarium oxysporum* and *Pseudocercosporella herpotrichoides* (Menager *et al.*, 1971; Lacroix *et al.*, 1974; Buchenauer, 1976; Fritz, 1977; Lartaud *et al.*, 1977; Pommer & Zeeh, 1982). The registered uses of iprodione, procymidone and vinclozolin in France are shown in Table 10.1.

1_ oxazolidines – diones :

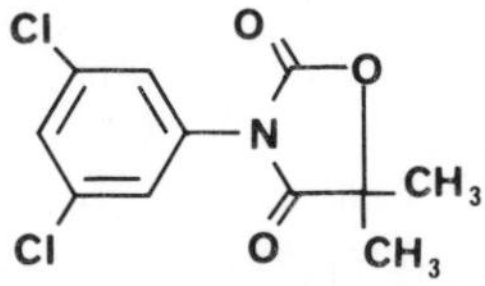

dichlozolin (DDOD)

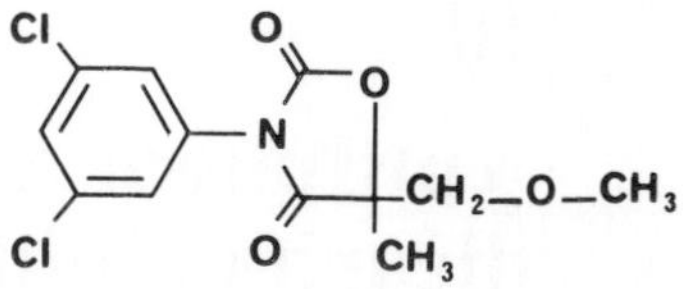

myclozolin

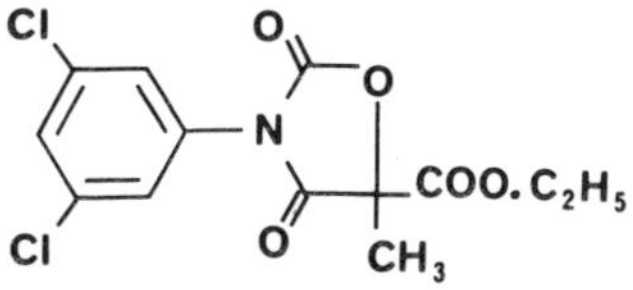

M 8164 (Serinal)

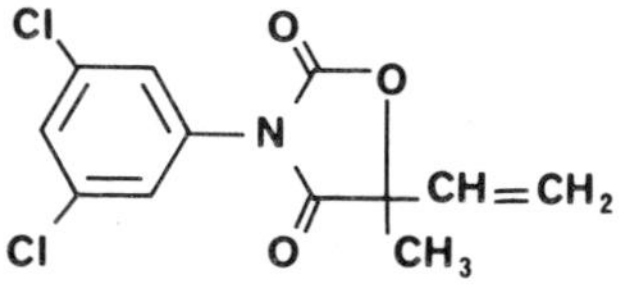

vinclozolin

2_ succinimides :

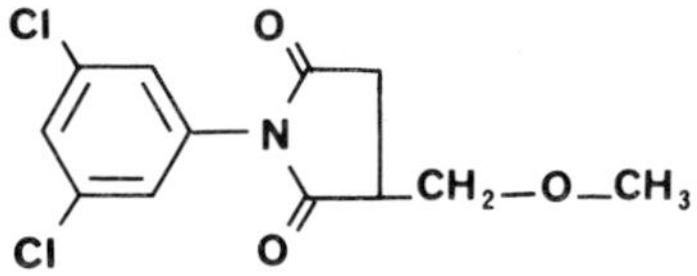

Co 6054

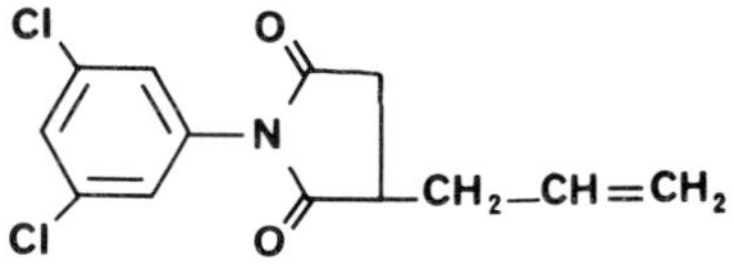

Co 4462

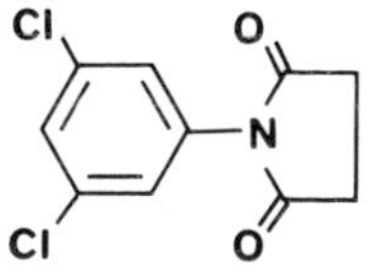

dimethachlon (DSI)

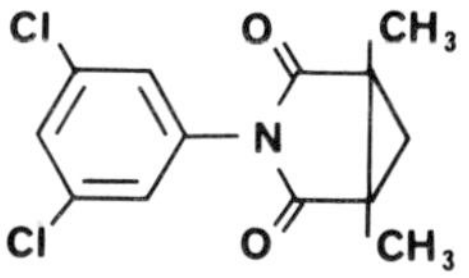

procymidone

3_ hydantoïnes :

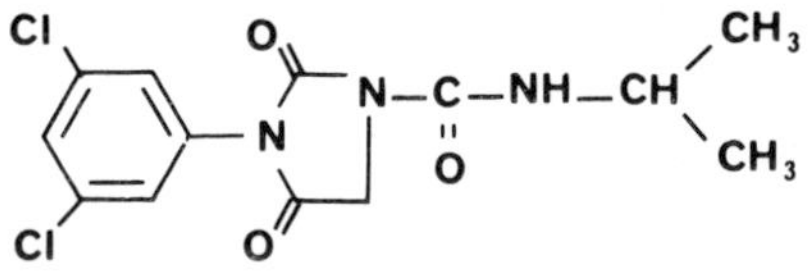

iprodione

Fig. 10.1. Structure of some dicarboximide fungicides.

Table 10.1. *Registered uses of dicarboximides in France*

Pathogens	Plants	Fungicides	Rates of applications[a]
Botrytis cinerea	Vine	Iprodione Procymidone Vinclozolin	75 g hl^{-1} 75 g hl^{-1} 75 g hl^{-1}
Botrytis cinerea	Strawberry French bean Tomato	Iprodione Procymidone Vinclozolin	100 g hl^{-1} 75 g hl^{-1} 100 g hl^{-1}
Sclerotinia sp.	Lettuce	Iprodione	75 g hl^{-1}
Botrytis cinerea	Chicory	Procymidone Vinclozolin	75 g hl^{-1} or 30 kg ha^{-1} (soil treatment for chicory only) 75 g hl^{-1} or 1500 g ha^{-1} (soil treatment)
Sclerotium cepivorum	Allium	Iprodione Vinclozolin	150 g per q seeds 150 g per q seeds
Monilinia sp.	Stone fruits	Iprodione Vinclozolin	75 g hl^{-1} 50 g hl^{-1}
Alternaria brassicae	Rape	Iprodione Procymidone	750 g ha^{-1} 750 g ha^{-1}
Sclerotinia sclerotiorum	Rape	Vinclozolin	750 g ha^{-1}
Phoma betae	Beet	Iprodione	150 g per q seeds
Rhizoctonia solani	Potato	Iprodione	400 g hl^{-1} (soaking of tubers) or 10 g per q (pulverisation on tubers)
Corticium fusiforme	Turf-grass	Iprodione	6000 g ha^{-1}

[a]Units used are hl: hectolitre; ha: hectare; q: quintal.

These dicarboximides can be considered as new fungicides since they have novel structures. However, their antifungal activity is probably similar to that of fungicides belonging to the 'aromatic hydrocarbon group' (Georgopoulos & Zaracovitis, 1967), such as biphenyl, dicloran, hexachlorobenzene, O-phenylphenol, quintozene and tecnazene (Fig. 10.2). The fungicides chloroneb and tolclofos-methyl and the herbicides dichlobenil and chlorthiamid (Fig. 10.2) have probably the same mode of action.

Most of these chemicals (the dicarboximides and the aromatic hydrocarbons) are good protective fungicides, but some are systemic

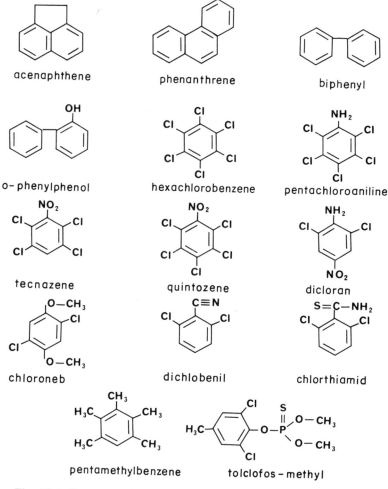

Fig. 10.2. Structure of some aromatic fungicides.

fungicides; translocation of dichlozolin (Menager *et al.*, 1971), iprodione (Cayley & Hide, 1980), myclozolin (Pommer & Zeeh, 1982), procymidone (Hisada, Kato & Kawase, 1977; Cooke *et al.*, 1979), and chloroneb (Thapliyal & Sinclair, 1970; Thorn, 1973) has been observed in plants.

This chapter describes the mode of action of dicarboximides and aromatic hydrocarbons and discusses resistance to these fungicides.

Resistance to dicarboximides and aromatic hydrocarbons

Dicarboximides have been reviewed by Beever & Byrde (1982) and Pommer & Lorenz (1982) and aromatic hydrocarbons have been reviewed by Dekker (1976), Georgopoulos (1977) and Georgopoulos & Zaracovitis (1967).

Resistance in the laboratory

It is very easy to isolate strains of fungi resistant to iprodione, procymidone or vinclozolin in the laboratory. Resistant strains have been obtained for *Alternaria alternata* (McPhee, 1980), *Alternaria kikuchiana* (Kato, Hisada & Kawase, 1979), *Aspergillus nidulans* (Leroux, Fritz & Gredt, 1982*a*; Beever, 1983), *Botrytis cinerea* (see references in Beever & Byrde, 1982 and Pommer & Lorenz, 1982), *Botrytis squamosa* (Presly *et al.*, 1980), *Botrytis tulipae* (Chastagner & Vassey, 1979), *Cochliobolus miyabeanus* (Kato *et al.*, 1979), *Monilinia fructicola* (Sztejnberg & Jones, 1978; Ritchie, 1983), *Monilinia laxa* (Katan & Shabi, 1981), *Penicillium expansum* (Leroux, Gredt & Fritz, 1978; Rosenberg & Meyer, 1981; Beever, 1983), *Rhizoctonia solani* (Kato *et al.*, 1979); *Rhizopus nigricans* (Leroux, Gredt & Fritz, 1981), *Sclerotinia minor* (Brenneman & Stipes, 1983), *Sclerotinia sclerotiorum* (Kato *et al.*, 1979) and *Ustilago maydis* (Leroux *et al.*, 1978).

In the case of aromatic hydrocarbons, studies have been made with:
- (i) acenaphthene on *Neurospora sitophila* and *Penicillium chrysogenum* (see references in Georgopoulos & Zaracovitis, 1967);
- (ii) biphenyl on *Diplodia natalensis, Nectria haematococca, Neurospora crassa, Penicillium digitatum, Penicillium italicum* and *Sclerotium rolfsii* (see references in Georgopoulos & Zaracovitis, 1967);
- (iii) chlorinated nitrobenzenes on *Aspergillus nidulans* (Threlfall, 1968), *Botrytis allii* (Priest & Wood, 1961), *Botrytis cinerea* (Esuruoso & Wood, 1971), *Fusarium caeruleum* (McKee, 1951), *Nectria haematococca, Sclerotium rolfsii* (Georgopoulos

& Zaracovitis, 1967) and *Thanatephorus cucumeris* (Meyer & Parmenter, 1968);

(iv) dicloran on *Botrytis cinerea* (Webster, Ogawa & Bose, 1970; Esuruoso & Wood, 1971; Leroux, Fritz & Gredt, 1977; Chastagner & Ogawa, 1979), *Gilbertella persicaria* (Ogawa, Ramsey & Moore, 1963), *Rhizopus stolonifer* (Webster, Ogawa & Moore, 1968) and *Ustilago maydis* (Leroux *et al.*, 1978);

(v) chloroneb on *Aspergillus nidulans, Aspergillus niger, Penicillium expansum* (Van Tuyl, 1977), *Mucor mucedo* (Lyr & Casperson, 1982b; Lyr & Werner, 1982) and *Ustilago maydis* (Tillman & Sisler, 1973; Van Tuyl, 1977; Leroux *et al.*, 1978).

Strains resistant to dicarboximides can be obtained from spores (with or without mutagenic treatment) at frequencies between 1×10^{-8} and 1×10^{-4} (see references in Beever & Byrde, 1982 and Pommer & Lorenz, 1982). Strains of *Ustilago maydis* resistant to chloroneb have been obtained from sporidia at a frequency of 2×10^{-4} (Van Tuyl, 1977). With filamentous fungi, sectors resistant to dicarboximides or aromatic hydrocarbons are easily obtained from colonies grown on fungicide-containing media (see references in Pommer & Lorenz, 1982).

Resistance in the field

Strains of several pathogens 'resistant' to dicarboximides or aromatic hydrocarbons have been isolated from infected plants (Table 10.2) but a reduction in fungicide efficacy was not always observed in practice. However, in the case of *Penicillium* sp. on citrus treated with biphenyl and sodium-*o*-phenylphenate (Eckert, 1982), *Tilletia foetida* on wheat after seed treatments with hexachlorobenzene (Kuiper, 1965; Skorda, 1977) and *Botrytis cinerea* on plants treated with dicarboximides (see references in Beever & Byrde, 1982 and Pommer & Lorenz, 1982), failures of chemical control have been reported. In France, the main problem concerns grey mould (*B. cinerea*) on grapevines and surveys have revealed that the frequency of isolation of strains of this fungus resistant to dicarboximides is related to the annual number of fungicide treatments given to the crop (Leroux & Gredt, 1983). Three situations can be recognised:

(i) in the south of France, where only one (or less) fungicide application is given each year, the dicarboximides have always provided good control of grey mould;

(ii) in Alsace, Beaujolais, Bourgogne and Val de Loire, where at

Table 10.2. *Fungi isolated from plants which are resistant to aromatic hydrocarbons and dicarboximides*

Pathogens	Plants	Fungicides	Countries	References
Penicillium digitatum, Penicillium italicum	Citrus	Biphenyl and sodium-*o*-phenylphenate	USA	Harding, 1962; Eckert, 1982
Rhizoctonia solani	Cotton	Quintozene	USA	Shalta & Sinclair, 1963
Rhizopus arrhizus	Stone-fruits	Dicloran	USA	Weber & Ogawa, 1965
Sclerotium cepivorum	Allium	Dicloran	USA	Locke, 1969
Tilletia foetida	Wheat	Hexachlorobenzene	Australia	Kuiper, 1965
			Greece	Skorda, 1977
Botrytis cinerea	Vine	Dicarboximides	Germany, France, Switzerland	References mentioned by
Botrytis cinerea	Strawberry	Dicarboximides	Belgium, UK, Israel	Beever & Byrde, 1982 or
Botrytis cinerea	Cultures under greenhouses	Dicarboximides	Greece, Israel, Japan, UK	Pommer & Lorenz, 1982
Fusarium nivale	Turf-grass	Iprodione	USA	Chastagner & Vassey, 1982
Monilinia fructicola	Stone-fruits	Dicarboximides	USA	Sztejnberg & Jones, 1978
Sclerotinia homeocarpa	Turf-grass	Iprodione	USA	Detweiter, Vargas & Danneberg, 1983; Pennuci & Jackson, 1983

least two treatments are given each year, dicarboximides are more or less effective in controlling grey mould;

(iii) in Champagne, where the vine growers apply four fungicide treatments per year, the population of *Botrytis cinerea* is made up mainly of strains which are resistant to dicarboximides and benzimidazoles; these fungicides are thus totally ineffective in controlling grey mould in this area.

Patterns of cross-resistance

In the case of *Botrytis cinerea* (using strains selected in the laboratory or isolated from the field), cross-resistance always occurs between iprodione, procymidone and vinclozolin (Table 10.3). This cross-resistance (resistance to two or more fungicides, mediated by the same genetic factor) also extends to other dicarboximides (dichlozolin, myclozolin, M 8164, Co 4462 and Co 6054; Table 10.3; Katan, 1982), to dicloran and to quintozene (Leroux *et al.*, 1977; Gullino & Garibaldi, 1979; Maraite *et al.*, 1981). This pattern of cross-resistance is also observed with fungi like *Aspergillus nidulans*, *Penicillium expansum* and *Ustilago maydis* (Leroux *et al.*, 1978, 1982a). Cross-resistance is also observed to other aromatic compounds. These include:

(i) polynuclear aromatics; acenaphthene, fluorene, naphthalene, phenanthrene, biphenyl and amino-, bromo-, chloro-, hydroxy- or nitro-derivatives of biphenyl (Georgopoulos & Vomvoyanni, 1965; Threlfall, 1972; Tillman & Sisler, 1973; Leroux *et al.*, 1982a);

(ii) chlorinated benzenes; di-, tri-, tetra-, penta- and hexachloro-benzenes (Tillman & Sisler, 1973; Skorda, 1977; Leroux *et al.*, 1982a); Priest & Wood (1961) tested some bromo- and iododerivatives;

(iii) chlorinated anilines; di- (in particular the 3,5-isomer), tri-, tetra- and pentachloroanilines (Leroux *et al.*, 1982a).

(iv) chlorinated nitrobenzenes; mono-, di-, tri-, tetra- (in particular tecnazene) chloronitrobenzenes and quintozene (Brook, 1952; Priest & Wood, 1961; Georgopoulos, 1963b; Threlfall, 1968; Leroux *et al.*, 1982a);

(v) chlorinated nitroanilines; 2,4-dichloro-6-nitroaniline, 2,3,5,6-tetrachloronitroaniline (Priest & Wood, 1961; Leroux *et al.*, 1982a) and dicloran;

(vi) various aromatic hydrocarbons; chloroneb (Tillman & Sisler, 1973; Van Tuyl, 1977; Leroux *et al.*, 1978; Lyr & Casperson,

Table 10.3. *Effects of various chemicals on mycelial growth and spore germination of Botrytis cinerea*

Chemicals	EC_{50}^{a} (μg ml^{-1}) Spore germination (sensitive strain: S)[b]	Mycelial growth (sensitive strain: S)[a]	Mycelial growth (moderately resistant strain R^{+})	Mycelial growth (highly resistant strain: RR)[b]
Dichlozolin	0.5	0.15	2	>300
Iprodione	2	0.3	3.5	300
Myclozolin	0.4	0.15	2.5	>300
Procymidone	1.5	0.2	2.5	>300
Vinclozolin	0.5	0.15	2	>300
M 8164	3	1	12.5	>300
Acenaphthene	8	3	10	>300
Biphenyl	25	10	—[d]	>300
Chloroneb	100	2	6	>300
Chlorthiamid	—[d]	8	20	90
Dichlobenil	15	3	20	>300
Dicloran	5	0.8	6	>300
Pentachloroaniline	—[d]	5	50	>300
Pentachlorobenzene	30	2.5	5	>300
Pentamethylbenzene	—[d]	75	—[d]	>300
Phenanthrene	>100	2	—[d]	>300
O-phenylphenol	—[d]	0.5	2.5	30
Quintozene	>100	1.5	—[d]	>300
Tecnazene	—[d]	0.5	—[d]	>300
Tolclofos-methyl	60	0.5	2.5	>300
Etridiazol	—[d]	10	30	90
Methanol[c]	—[d]	12	20	80

[a]EC_{50} represents the fungicide concentration which inhibited mycelial growth rate or spore germination to 50% of the control value.
[b]S and R strains were isolated from French vineyards, strain RR was selected *in vitro*.
[c]For methanol, the EC_{50} is in μl ml^{-1}
[d]Not tested.

1982*b*; Lyr & Werner, 1982) chlorthiamid, dichlobenil, pentamethylbenzene and tolclofos-methyl (Table 10.3; Leroux & Gredt, 1980; Leroux *et al.*, 1982*a*).

Priest & Wood (1961) also mentioned unsubstituted benzene. The fungicidal activity obtained with oxadiazon (a herbicide; Leroux & Gredt, 1980), 3-phenylindole (Van Tuyl, 1977), diphenylether compounds (Leroux & Gredt, 1980; Lyr *et al.*, 1983), methylene blue and brilliant cresyl green (Threlfall, 1972) have yet to be confirmed.

Results obtained with some alcohols, e.g. methanol (Leroux & Gredt, 1982*a*), etridiazol (Table 10.3; Lyr & Casperson, 1982*b*; Lyr & Werner, 1982) and carbon dioxide (Meyer & Parmenter, 1968) indicate that cross-resistance can occur between dicarboximides and non-aromatic compounds.

Strains resistant to dicarboximides remain sensitive to dithiocarbamates (e.g. thiram), sulphenimides (e.g. captafol, captan, dichlofluanid and folpet), carboxamides (e.g. carboxin and benodanil), phenol derivatives (e.g. DNOC, (2,6-dinitro-*o*-cresol), ioxynil and pentachlorophenol), chlorothalonil, polyoxin B, edifenphos and sterol inhibitors (Leroux *et al.*, 1977; P. Leroux, unpublished results). Sometimes resistant strains selected on media containing sterol inhibitors (fenarimol and triadimefon) are also resistant to dicarboximides (Fuchs & De Waard, 1982; P. Leroux, unpublished results). Finally, cross-resistance is never observed between benzimidazoles (and thiophanates) and dicarboximides. However, it is possible to obtain strains which are resistant to both these fungicides; examples of multiple resistance (resistance to two fungicides, mediated by different genetic factors) are common in *Botrytis cinerea* (Leroux *et al.*, 1977; Chastagner & Ogawa, 1979; Leroux, Lafon & Gredt, 1982*b*).

Levels of resistance

To estimate the degree (or level) of resistance towards a fungicide, it is necessary to compare its effect on sensitive and resistant strains. Tests done on mycelial growth or spore germination allow a calculation to be made of the fungicide concentration which gives a 50% inhibition (EC_{50}) of mycelial (colony) growth or spore germination. The ratio between the EC_{50} values of the resistant and sensitive strains gives a measure of the level of resistance of the former strain. However, in many publications it is not possible to calculate this value because the authors have only tested one fungicide concentration.

Several strains of *Botrytis cinerea* resistant to dicarboximides have been isolated (Leroux *et al.*, 1977; Leroux & Gredt, 1982*a, b*; Beever,

1983). Under laboratory conditions it is easy to select strains with high levels of resistance to dicarboximides (greater than 1000) whereas such resistant isolates are rarely isolated from the field. Strains of *B. cinerea* isolated from various plants in different countries had levels of resistance towards iprodione, procymidone and vinclozolin between 8 and 30 when tests were made on mycelial growth (Table 10.3; Maraite *et al.*, 1981; Katan, 1982; Leroux & Gredt, 1982*a*, *b*; Pappas, 1982). With spore-germination tests, similar values were obtained with iprodione; whereas higher resistance degrees were observed with procymidone and vinclozolin (Leroux & Gredt, 1982*a*; Pappas, 1982). This variation in resistance levels obtained with dicarboximides is probably correlated with their solubilities in water (iprodione is 5–10 times more soluble in water than procymidone or vinclozolin).

The studies conducted with other fungi (*Aspergillus nidulans, Penicillium expansum, Ustilago maydis, Sclerotinia minor*, etc.) indicate that strains with high levels of resistance are easily induced *in vitro* (Leroux *et al.*, 1982*a*; Beever, 1983; Brenneman & Stipes, 1983). Strains of *Fusarium nivale* and *Sclerotinia homeocarpa* isolated from turf-grass had a degree of resistance to iprodione of 10 and 100 respectively (Chastagner & Vassey, 1982; Detweiter; Vargas & Dannebers, 1983). Harding (1962) observed that isolates of *Penicillium digitatum* obtained from packing houses were 4 or 5 times more resistant to sodium-O-phenylphenate than were sensitive strains.

We have mentioned above that cross-resistance occurs between dicarboximides and many other compounds toxic to fungi. However, for a particular isolate, the level of resistance varies between these chemicals (Table 10.3). For example, a strain of *Penicillium expansum* selected as being resistant to iprodione was highly resistant (resistance level greater than 50) to dicarboximides, acenaphthene, dicloran and pentachlorobenzene, had a moderate degree of resistance (between 5 and 50) to biphenyl, *o*-phenylphenol and pentamethylbenzene, and was relatively sensitive (resistance level below 5) to chlorthiamid and *m*-, *p*-phenylphenol (Leroux *et al.*, 1982*a*). The differences in the behaviour (fungicidal activity) of phenylphenol isomers are similar to those observed previously with isomers of chlorinated nitrobenzenes (Brook, 1952; Priest & Wood, 1961; Georgopoulos, 1963*b*).

Negatively correlated cross-resistance
The terms 'negatively correlated cross-resistance' and 'collateral sensitivity' (Georgopoulos, 1977) have been used to describe a situation where a particular mutation increases resistance to one

fungicide but increases sensitivity to a second fungicide. This type of phenomenon has been observed between benzimidazole fungicides and some carbamate herbicides (Leroux & Gredt, 1982*a*).

Garibaldi & Gullino (1979) showed that dicarboximide-resistant strains of *Botrytis cinerea* were more sensitive to copper oxychloride than wild strains, but we have been unable to isolate strains with these characteristics (we also tested cupric chloride and sulphate; P. Leroux, unpublished results). On the other hand, a negatively correlated cross-resistance was observed in many dicarboximide-resistant strains to mineral salts like potassium chloride, sodium chloride and calcium chloride (Fig. 10.3). This phenomenon was observed in *Botrytis cinerea* and various other fungi (*Aspergillus nidulans, Penicillium expansum* and *Ustilago maydis*). The same type of negatively correlated cross-resistance was also observed between dicarboximides and sugars (glucose, lactose, mannose, sucrose and xylose), glycerol, mannitol and sorbitol (Fig. 10.3) (Leroux & Gredt, 1982*a*; Beever, 1983), and these results indicate that there is a correlation between dicarboximide resistance and osmotic sensitivity. In our experiments we always found a direct relationship between the level of fungicide resistance and the degree of osmotic sensitivity. However, this correlation is not always observed (Beever, 1983).

We also tested the sensitivity of various strains to 2-deoxy-D-glucose and L-sorbose; these hexose analogues are known to have fungitoxic effects (Moore, 1981). Using *Botrytis cinerea*, we did not observe cross-resistance or collateral sensitivity between dicarboximides and these sugars (P. Leroux, unpublished results). However, Georgopoulos & Zaracovitis (1967) observed that mutants of *Nectria haematococca* and *Neurospora crassa* resistant to biphenyl were less affected by L-sorbose than their respective wild types.

Genetics of resistance

Information about the genetics of dicarboximide resistance is lacking. However, the patterns of cross-resistance described above suggest that the same genes probably confer resistance to dicarboximides and aromatic hydrocarbons; a single gene locus may confer resistance to fungicides belonging to both groups (Georgopoulos, 1977).

In *Nectria haematococca*, five loci for resistance to chlorinated nitrobenzenes were identified and they all gave similar levels of resistance (Vomvoyanni & Georgopoulos, 1966; Georgopoulos,

1977). However, the mutation frequency is different for each locus (Georgopoulos, 1977).

Threlfall (1968) found that resistance of *Aspergillus nidulans* to quintozene was conferred by two recessive genes belonging to linkage group III. Van Tuyl (1977) studied a locus for chloroneb resistance which was an allele of *pcnb A* previously described by Threlfall (1968). Georgopoulos & Zaracovitis (1967) showed that a single gene was responsible for resistance to aromatic hydrocarbons in *Neurospora crassa* and *Neurospora sitophila*. Beever & Byrde (1982) noted that certain osmotic mutants of *N. crassa* (*os-1* or *os-4* in linkage Group I) were also resistant to dicarboximides. In addition they selected some

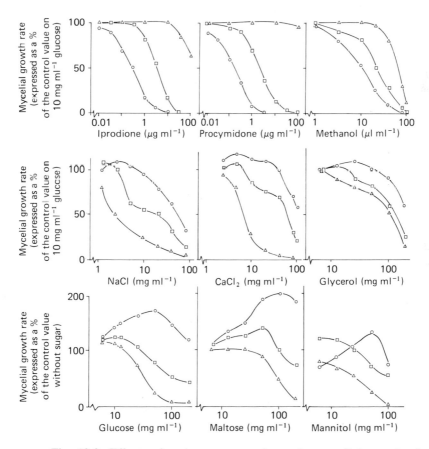

Fig. 10.3. Effects of various compounds on the mycelial growth of *Botrytis cinerea* (see also Leroux & Gredt, 1982*a*). Circles: sensitive strain (S-type); squares: resistant strain (R-type); triangles: highly resistant strain (RR-type).

strains resistant to these fungicides which were osmotically sensitive but they did not study the genetic mechanisms involved.

The genetics of sodium-*o*-phenylphenate-resistance in *Penicillium expansum* and *Penicillium italicum* was investigated by Beraha & Garber (1966, 1980); three unlinked recessive genes and one dominant gene were involved respectively. In *Ustilago maydis* one gene confers resistance to chloroneb (Tillman & Sisler, 1973; Van Tuyl, 1977).

When resistance to dicarboximides or aromatic hydrocarbons is studied in multinucleate fungi (*Botrytis cinerea, Mucor* sp., or *Rhizopus* sp.), it is found that heterokaryosis affects the stability of the resistance. When grown in the absence of fungicide, strains of *Botrytis cinerea* resistant to iprodione, procymidone, vinclozolin, dicloran or quintozene either produce resistant spores or a mixture of resistant and sensitive spores (see references in Georgopoulos, 1977; Beever & Byrde, 1982; Pommer & Lorenz, 1982). In the first case, resistance is probably homokaryotic but the second response indicates the heterokaryotic nature of the resistant strains. Emergence in *Rhizopus stolonifer* of resistance to dicloran and in *Thanatephorus cucumeris* of resistance to quintozene or biphenyl was also attributed to the breakdown of heterokaryons (Meyer & Parmenter, 1968; Webster *et al.*, 1968). Dekker (1976) suggests also that 'in multinucleate fungi, heterokaryosis may serve as the mechanism in maintaining low levels of fungicide tolerance'. This may be true for strains of *Botrytis cinerea* isolated from nature. However, most low-level resistant strains of *B. cinerea* isolated from French vineyards, maintained their resistance after subculture on fungicide-free media (P. Leroux, unpublished results).

Mechanisms of resistance

Various experiments have indicated that resistance to dicarboximides or aromatic hydrocarbons may be due to decreased membrane permeability to the fungicide, or an increased detoxification of the fungicide or a decreased affinity between the fungicide and the target site.

Uptake of dicarboximides by sensitive and resistant strains of *Botrytis cinerea, Penicillium expansum* or *Ustilago maydis* is generally similar (Leroux *et al.*, 1978; Pappas *et al.*, 1979; Leroux & Gredt, 1982*a*; Lorenz & Eichhorn, 1982). Procymidone or iprodione uptake by mycelia of *B. cinerea* is a rapid and reversible process, the rate of which depends on the external concentration of the dicarboximide and is not influenced by temperature or the metabolic activity of the fungus (Hisada & Kawase, 1980; Leroux & Gredt, 1982*b*; P. Leroux, unpublished results). Aromatic hydrocarbon fungicides like dicloran, chloroneb, biphenyl and quintozene do not reduce the uptake of iprodione or procymidone (P. Leroux, unpublished results). According to Hisada & Kawase (1980) 'the lack of binding (of procymidone) to the cell wall and the preferential binding to the protoplast lead to the conclusion that the target site of this fungicide may be on the cytoplasmic membrane'. Finally it seems that there is little metabolism of iprodione by sensitive or resistant strains of *B. cinerea* (in liquid cultures for an incubation period of 24 h; P. Leroux, unpublished results).

The results obtained when *Mucor mucedo* (Lyr & Casperson, 1982*b*; Lyr & Werner, 1982) or *Ustilago maydis* (Tillman & Sisler, 1973) were treated with chloroneb, were similar to those mentioned above for the dicarboximides, i.e. a variation in uptake or metabolism of the fungicide was not responsible for resistance. Lyr & Werner (1982) suggest that resistance in *M. mucedo* could be due to reduced binding of chloroneb to mitochondrial membranes and this might be related to the low tyrosine content of the latter.

Studies conducted with chlorinated nitrobenzenes (quintozene and tecnazene) indicate that resistance can be due to reduced uptake or increased detoxification of these fungicides. Nakanishi & Oku (1970) suggest that resistance of *Botrytis cinerea* to quintozene was mainly due to increased metabolism of the fungicide. They showed that quintozene was rapidly converted to pentachloroaniline and pentachlorothioanisole by a resistant strain but, in a sensitive strain, the latter metabolite was not detected and only a low concentration of pentachloroaniline was observed. Threlfall (1968) found that uptake of tecnazene by *Aspergillus nidulans* was five times greater in sensitive than in resistant strains. Threlfall (1968, 1972) also observed that sensitive strains utilised glucose at a slower rate than resistant strains and that hyphae of resistant strains had a higher hexosamine content than hyphae of sensitive strains.

According to Georgopoulos, Zafiratos & Georgiadis (1967), biphenyl affects membrane-transport processes and resistant strains of *Nectria haematococca* seem to possess transport systems which have a reduced affinity to biphenyl.

Biological properties of resistant strains

Most experiments with dicarboximides have been made with *Botrytis cinerea* and, from the reviews of Beever & Byrde (1982) and Pommer & Lorenz (1982), it appears that mycelial growth rate, sporulation, sclerotial production and pathogenicity of resistant strains of this fungus are usually lower than those of sensitive strains. However, in both groups, the range of these characteristics is broad and can overlap. We have shown that moderately resistant strains (R-type) are often similar to sensitive strains, whereas highly resistant strains (RR-type) normally show reduced vigour. Strains of *Sclerotinia homeocarpa*, resistant to iprodione, are less virulent than sensitive strains (Detweiter *et al.*, 1983), but in *Fusarium nivale* resistance does not affect virulence although it causes a decrease in mycelial-growth rate (Chastagner & Vassey, 1982); resistant strains of *Monilinia fructicola* sporulate less and are less virulent than sensitive strains (Ritchie, 1983). Resistance to aromatic hydrocarbons is generally, but not always, accompanied by a decrease in vigour and pathogenicity (Georgopoulos, 1963*a*; Chastagner & Ogawa, 1979; Eckert, 1982). For example, Georgopoulos (1963*a*) showed that mutations at a gene locus in *Nectria haematococca* responsible for resistance to chlorinated nitrobenzenes gave both highly and weakly pathogenic strains. A relationship was observed between the gene locus responsible for reduced pathogenicity and the locus responsible for resistance to quintozene.

Most competition tests and field observations suggest that strains resistant to dicarboximides are less fit than sensitive strains. The results obtained with *Botrytis cinerea* have been discussed by Beever & Byrde (1982) and Pommer & Lorenz (1982). In our experiments we have established that moderately resistant strains survive the winter in vineyards and, in the absence of fungicide treatment, their frequency decreases during the summer and at the beginning of autumn (before the harvest). This phenomenon was pronounced following serious infections of *B. cinerea*. The reduced fitness of the resistant strains which are also osmotically sensitive was probably due to the high concentration of sugars in the grapes; the same factor could explain the absence of strains with high levels of fungicide resistance (Leroux & Gredt, 1982*a*). Few experiments have been made with other fungi, although it has been reported that dicarboximide-resistant strains of *Monilinia fructicola* have reduced parasitic fitness (Ritchie, 1983) and strains of *Sclerotinia homeocarpa*, resistant to iprodione, disappear rapidly in the absence of fungicide treatment (J. Paviot, personal communication).

Mode of action

In the above account it has been shown that fungicides can show cross-resistance. This observation suggests that the compounds concerned have the same mode of action. This hypothesis will be discussed below.

Morphological and cytological studies

Effect on conidia, mycelium and flagella. Dicarboximides, like aromatic hydrocarbons, inhibit mycelial growth more than spore germination (Table 10.3; Sharples, 1961; Lankov, 1971; Buchenauer, 1976; Fritz, Leroux & Gredt, 1977; Pappas & Fisher, 1979; Davis & Dennis, 1981; Reilly & Lamoureux, 1981; Craig & Peberdy, 1983). Kato (1983) observed the inhibitory effects of aromatic hydrocarbons on the motility of *Phytophthora capsici* zoospores. Biphenyl, chloroneb, dicloran, quintozene and tolclofos-methyl $(8\,\mu g\,ml^{-1})$ inhibited flagellar movement but chlorothalonil, which does not show cross-resistance with these fungicides, did not. It is surprising that procymidone does not affect zoospore motility; only a slight inhibition of the motility of zoospores of *Plasmopara viticola* was observed at $40\,\mu g\,ml^{-1}$ (Kato, 1983).

Morphological abnormalities. Many authors have reported that dicarboximide and aromatic hydrocarbons cause distortion, swelling and bursting of cells of *Botrytis cinerea* (Sharples, 1961; Esuruoso, Price & Wood, 1968; Lankov, 1971; Buchenauer, 1976; Hisada & Kawase, 1977; Eichhorn & Lorenz, 1978; Albert, 1979, 1981; Davis & Dennis, 1981; Kato, 1983), *Ustilago avenae* (Buchenauer, 1976), *Ustilago maydis* (Kato, 1983), *Aspergillus nidulans* (Craig & Peberdy, 1983), *Nectria haematococca* (Georgopoulos *et al.*, 1967), *Phytophthora capsici, Pseudoperonospora cubensis* (Kato, 1983) and *Rhizopus arrhizus* (Weber & Ogawa, 1965). These fungicides had similar effects on plants and plant cells (Lewis, Weber & Venketeswaran, 1969; Verloop, 1972; Umetsu, Satoh & Matsuda, 1976).

Only growing mycelia are affected by these fungicides (Hisada & Kawase, 1977; Eichhorn & Lorenz, 1978; Albert, 1979). They induce hyphal branching (Esuruoso *et al.*, 1968; Threlfall, 1968; Buchenauer, 1976; Lyr & Casperson, 1982*a*), and result in the formation of wide hyphae with short 'cells'. They also sometimes cause conidia and hyphal 'cells' to burst. Hyphal bursting is influenced by pH (Eichhorn & Lorenz, 1978; Albert, 1979) and the composition (Sharples, 1961) and consistency of the culture medium (Weber & Ogawa, 1965; Georgopoulos *et al.*, 1967). Conidia burst during germination; subapical as well as apical hyphal 'cells' also sometimes burst (Weber &

Ogawa, 1965; Hisada & Kawase, 1977; Eichhorn & Lorenz, 1978). However, Albert (1979) found that apical 'cells' of fungal hyphae were the first to burst. This bursting phenomenon was not observed with a resistant strain (Weber & Ogawa, 1965).

Lewis *et al.* (1969) found that germination and development of seedlings of *Nicotiana tabacum* was inhibited by $10 \mu g \, ml^{-1}$ of dicloran; plants grown in media containing dicloran formed small distorted roots. When plants were treated with lower concentrations (1 or $5 \mu g \, ml^{-1}$) of dicloran, root growth was stimulated. Verloop (1972) reported that dichlobenil at $0.03 \, mg \, ml^{-1}$ inhibited root hair growth and caused swelling of the elongation zone. Umetsu *et al.* (1976) found that dichlobenil caused soybean cells to swell and burst. These authors observed similar results with coumarin but not with colchicine.

Effect on the cell wall. The effects of dicarboximides and aromatic hydrocarbons on cell walls have been observed by light and electron microscopy (Threlfall, 1968; Werner, Lyr & Casperson, 1978; Casperson & Lyr, 1982; Lyr & Casperson, 1982*a, b*) and by using protoplasts and sphaeroplasts (Umetsu *et al.,* 1976; Hisada & Kawase, 1977, Meyer & Herth, 1978; Albert, 1979, 1981; Galbraith & Shields, 1982).

Biphenyl, chloroneb, dicloran, drazoxolon, etridiazol, quintozene, a nitrogen atmosphere and high concentrations of glucose, caused thickening of the walls of hyphae of *Mucor mucedo* (Lyr & Casperson, 1982*a*). This thickening increased as a function of fungicide concentration and was not inhibited by cycloheximide. Uncoupling agents such as 2,4-dinitrophenol, 2,4,5- or 2,4,6-trichlorophenol and pentachlorophenol did not induce anomalous cell wall synthesis.

Hisada & Kawase (1977) reported that, in the presence of procymidone, protoplasts obtained from the hyphae of *Botrytis cinerea* were stable and regenerated cell walls. However, this wall seemed imperfect since normal hyphae were not formed. Dichlobenil, and to a lesser extent coumarin, inhibited cell wall formation by protoplasts of *Nicotiana tabacum*. The removal of dichlobenil resulted in an enhanced rate of wall synthesis (Meyer & Herth, 1978; Galbraith & Shields, 1982).

Albert (1981) reported that conidia of *Botrytis cinerea* formed sphaeroplasts when incubated in isotonic nutritive solutions in the presence of vinclozolin ($100 \mu g \, ml^{-1}$); these protoplast-like structures grew in a chain-like manner and burst when placed in water. He also obtained sphaeroplasts from hyphae which had abnormally thickened walls and did not burst in water. Hisada & Kawase (1977) did not obtain sphaeroplasts from hyphae treated with procymidone, even under conditions of high osmotic pressure. When cultured with dichlobenil or coumarin, soybean cells became swollen and spherical; at high concentrations they burst (Umetsu *et al.*, 1976).

Lyr & Casperson (1982*a*) and Werner *et al.* (1978) compared abnormal wall synthesis in the presence of aromatic hydrocarbons with the process involved in the conversion from mycelial to yeast phase growth in dimorphic fungi. Fisher (1977) showed that dicloran and 2,4,-dichloro-6-nitroaniline induced the formation of yeast-like cells in *Actinomucor elegans* and *Mucor pusillus*, two species which are not dimorphic under normal growth conditions. This kind of response was also obtained with actinomycin D, 6-azauracil, cycloheximide, griseofulvin, 2-phenylethanol, sodium azide and thiabendazole; these compounds have very different modes of action. Fisher (1977) suggested that the formation of yeast-like cells by fungi is correlated with a shift in metabolism towards glycolysis. Protoplasts of *Botrytis cinerea* grew in a yeast-like manner in presence of procymidone or vinclozolin (Hisada & Kawase, 1977; Albert, 1981).

Effect on cellular division and hereditary process. Dicarboximides and aromatic hydrocarbons (vinclozolin, tolclofos-methyl and chloroneb) inhibit cell division in *Ustilago* sp. without having an immediate effect on growth (increase in dry weight) or nuclear division (Tillman & Sisler, 1973; Buchenauer, 1976; Kato, 1983). A similar result was obtained with tobacco protoplasts and dichlobenil (Meyer & Herth, 1978). However, Hisada *et al.* (1978) observed that procymidone inhibited cell and nuclear division in *Botrytis cinerea* without affecting its dry weight. In contrast, Kato *et al.* (1979) observed that procymidone increased the number of nuclei in *B. cinerea*.

Therefore, nuclear division may not be affected by these fungicides. However, certain genetic effects have been reported. These were shown by genetic analysis (mutagenic effects, appearance of sectors, haploidisation and mitotic recombination) and cytological observations (chromosomal aberrations). Acenaphthene (Ark, 1946) and quintozene (Clarke, 1971) have mutagenic properties. Aromatic hydrocarbons induced the formation of mutant sectors in *Fusarium caeruleum* (McKee, 1951) and *Aspergillus nidulans* (Threlfall, 1968; Georgopoulos, Kappas & Hastie, 1976; Azevedo, Santana & Bonatelli, 1977). Georgopoulos *et al.* (1976) suggest that an ability to induce somatic segregation is the only property common to all aromatic hydrocarbons and that, therefore, the primary effect of this group of fungicides is on nuclei. Azevedo *et al.* (1977) reported that chloroneb increased haploidisation and increased the frequency of non-disjunctions. Kappas (1978) found that at least three different mechanisms (non-disjunction, mitotic crossing-over and chromosome breakage-deletions) were responsible for the recombination activity of actinomycin D, chloroneb, dicloran and benomyl on *Aspergillus nidulans*. He suggested that chromosome breakage-deletion was responsible for chloroneb and dicloran causing haploidisation. Georgopoulos, Sarris & Ziogas (1979) studied the effect of dicarboximides on mitotic instability. They concluded that these compounds, like aromatic hydrocarbons, increased the frequency of non-disjunction or caused chromosome breakage-deletion. Production of polyploidy has been reported in *Tradescantia* with acenaphthene (Nebel, 1938), in *Allium* with paradichlorobenzene (Carey & McDonough, 1943) and in meiotic cells of barley with dicloran (Wuu & Grant, 1967).

Effect on ultrastructure. Chloroneb, some of its degradation products and quintozene caused lysis of the inner membrane of mitochondria of *Mucor mucedo*, an enlargement of the perinuclear space and an increased formation of vacuoles (Werner *et al.*, 1978; Casperson & Lyr, 1982; Lyr & Werner, 1982). The cause of this mitochondrial degradation is not yet known. Vinclozolin did not act like chloroneb and quintozene because it caused enlargement of the perinuclear space but no mitochondrial degradation (Lyr & Casperson, 1982*b*).

Dichlobenil did not disrupt microtubules in epidermal cells of *Vigna angularis* (Hogetsu, Shibaoka & Shimokoriyama, 1974), but inhibited the formation of Golgi vesicles in tobacco protoplasts (Meyer & Herth, 1978).

Biochemical studies

Effect on respiration. Dicarboximides and most aromatic hydrocarbons have no effect on respiration (Weber & Ogawa, 1965; Threlfall, 1968; Hock & Sisler, 1969; Tillman & Sisler, 1973; Buchenauer, 1976; Fisher, 1977; Fritz *et al.*, 1977; Blakemore & Carey, 1978; Hisada *et al.*, 1978; Pappas & Fisher, 1979; Reilly & Lamoureux, 1981). However, Rehm (1969) showed that sodium *o*-phenylphenate inhibited isocitrate-dehydrogenase and Lyr & Werner (1982) found that chloroneb decreased oxygen consumption of mitochondria isolated from a sensitive strain of *Mucor mucedo*. H. Lyr (personal communication) suggests that chloroneb may interfere with NADPH-cytochrome c reductase.

Effect on cell membrane integrity. Dicarboximides did not affect the efflux of electrolytes from *Botrytis cinerea*, *Sclerotinia sclerotiorum* or *Ustilago avenae* (Buchenauer, 1976; Pappas & Fisher, 1979) or the uptake of ^{14}C-glucose (Hisada *et al.*, 1978; Reilly & Lamoureux, 1981). Hisada & Kawase (1977) found that procymidone increased the loss of ^{14}C glucose from mycelium of *B. cinerea* but this release was caused by cell bursting.

Contradictory results have been obtained with aromatic hydrocarbons. Georgopoulos *et al.* (1967) showed that biphenyl did not cause general membrane damage to *Nectria haematococca* but inhibited uptake of ^{32}P and stimulated uptake of ^{42}K. Craig & Peberdy (1983) observed that dicloran increased the conductivity of culture filtrates of *Aspergillus nidulans*, suggesting that it causes an increase in the membrane permeability. Sharom & Mellors (1980) reported that polychlorinated biphenyls inhibited such membrane enzymes as 5'-nucleotidase and ATPase.

Effect on nucleic acid and protein synthesis. Information about this subject is presented in Table 10.4. Dicarboximides and aromatic hydrocarbons have little or no effect on protein synthesis but may inhibit or stimulate nucleic acid synthesis. An exception was reported by Weber & Ogawa (1965) who observed that, in *Rhizopus arrhizus*, dicloran only affected protein synthesis. Nucleoside phosphorylation was never affected by these chemicals.

Effect on lipid synthesis. Lipid metabolism is generally little affected by dicarboximides or dicloran (Buchenauer, 1976; Fritz *et al.*, 1977; Hisada *et al.*, 1978; Pappas & Fisher, 1979; Reilly & Lamoureux, 1981; Craig & Peberdy, 1983). However, some lipid modifications have been reported. A decrease in the incorporation of ^{14}C-acetate into lipids was observed with dicloran (Craig & Peberdy, 1983) and procymidone (Hisada *et al.*, 1978). Dicarboximides can alter the composition of fungal lipids by increasing the amount of free fatty acids present (Buchenauer, 1976; Fritz *et al.*, 1977), or by inhibiting triglyceride synthesis (Buchenauer, 1976; Pappas & Fisher, 1979). With the exception of the results of Pappas & Fisher (1979), who report that iprodione induced accumulation of 4,4-dimethylsterols in *Botrytis cinerea*, it seems

Table 10.4. *Effects of dicarboximides and aromatic hydrocarbons on protein and nucleic acid synthesis*

Chemicals	Organisms	Effect on protein and nucleic acids synthesis	References
Polychlorobiphenyls	bacteria	Inhibition of nucleic acids synthesis; mainly DNA	Blakemore, 1978
Vinclozolin	*Ustilago avenae*	Low inhibition of protein and nucleic acids synthesis (DNA synthesis was the most affected)	Buchenauer, 1976
Dicloran	*Aspergillus nidulans*	Stimulation of RNA synthesis	Craig & Peberdy, 1983
Dicloran, iprodione, vinclozolin	*Botrytis cinerea*	Inhibition of nucleic acids synthesis (at low concentrations dicloran stimulated RNA synthesis)	Fritz *et al.*, 1977
Dichlobenil	*Nicotiana tabacum*	No effect on DNA and protein synthesis (RNA synthesis was not tested)	Galbraith & Shields, 1982
Procymidone	*Botrytis cinerea*	Inhibition of protein and nucleic acids synthesis (no effect on nucleoside phosphorylation)	Hisada *et al.*, 1978
Chloroneb	*Rhizoctonia solani*	Inhibition of nucleic acids synthesis; mainly DNA (no effect on nucleoside phosphorylation)	Hock & Sisler, 1969

Dicloran	*Nicotiana tabacum*	Stimulation of nucleic acids synthesis at low concentrations and inhibition at high concentrations	Lewis et al., 1965
Iprodione, procymidone, vinclozolin	*Botrytis cinerea*	Iprodione inhibited protein and DNA synthesis; the other dicarboximides did not	Pappas & Fisher, 1979
Diphenyl, sodium-*O*-phenylphenate	enzymes	Inhibition of adenosine deaminase	Rehm, 1969
Iprodione	*Sclerotinia sclerotiorum*	Inhibition of nucleic acids synthesis; mainly DNA	Reilly & Lamoureux, 1981
Dicloran	*Botrytis cinerea*	Stimulation of nucleic acids synthesis; mainly RNA (protein synthesis was not tested)	Sharples, 1961
Quintozene, tecnazene	*Aspergillus nidulans*	Stimulation of DNA synthesis (protein and RNA synthesis were not tested)	Threlfall, 1968
Chloroneb	*Ustilago maydis*	Inhibition of nucleic acids synthesis; mainly DNA	Tillman & Sisler, 1973
Dicloran	*Rhizopus arrhizus*	Inhibition of protein synthesis (no effect on nucleic acids synthesis)	Weber & Ogawa, 1965

unlikely that dicarboximides or aromatic hydrocarbons inhibit sterol synthesis (Buchenauer, 1976; P. Leroux, unpublished results).

Effect on polysaccharide synthesis. Dicarboximides and aromatic hydrocarbons may affect wall composition. Macris & Georgopoulos (1969) reported that quintozene reduced the hexosamine content of the cell wall of *Neurospora crassa*, but Threlfall (1972) observed the contrary in *Aspergillus nidulans*. Tillman & Sisler (1973) report that chloroneb does not affect the amount of hexose or hexosamine in walls of *Ustilago maydis*. Lyr & Casperson (1982*a*) observed that quintozene induced thickening of the cell wall of *Mucor mucedo*. Craig & Peberdy (1983) showed that a concentration of dicloran which caused a 50% inhibition of growth of *A. nidulans*, increased chitin synthase activity. Dichlobenil has been reported to act as a specific inhibitor of cellulose synthesis (Hogetsu *et al.*, 1974; Montezinos & Delmer, 1980; Galbraith & Shields, 1982).

The results obtained with dicarboximides are also difficult to explain. Hisada *et al.* (1978) showed that procymidone increased the dry-weight of the wall of *Botrytis cinerea* without changing its composition. They also found that procymidone increased the incorporation of ^{14}C-acetate, ^{14}C-glucose and ^{14}C-glucosamine into the wall. However, Pappas & Fisher (1979) found that iprodione, procymidone and vinclozolin slightly reduced the incorporation of ^{14}C-glucosamine and ^{14}C-*N*-acetyl-glucosamine into the wall of *B. cinerea*. Reilly & Lamoureux (1981) reported that iprodione increased the incorporation of ^{14}C-acetate into *Sclerotinia sclerotiorum*, but slightly decreased the incorporation of ^{14}C-glucose and ^{14}C-glucosamine into the wall.

Conclusions and discussion

The data reported above suggest that dicarboximides and aromatic hydrocarbons have the same modes of action. In addition to their similar spectrum of activity, they also have identical effects on fungal morphology and metabolism and always show cross-resistance. Except for chlorinated nitrobenzenes, resistance to these fungicides is not due to decreased permeability or increased metabolism. It is probable that the difference between sensitive and resistant strains concerns the site of action of these fungicides.

One possibility is that the primary effect of these fungicides is on nuclei and this would explain the genetic effects caused by dicarbox-imides and aromatic hydrocarbons. These inhibitory effects may involve nucleic acid synthesis. Georgopoulos *et al.* (1979) stated that 'it is certain that the effect of dicarboximides on the chromosomes and, possibly on mitotic spindle, is a main reason for the fungitoxicity', but Beever & Byrde (1982) suggest that it is premature to come to this conclusion.

The second possibility is that these fungicides affect the cytoplasmic membrane and/or the cell wall. Disturbances of membrane permeability and cell wall synthesis, morphological abnormalities and the osmotic sensitivity of resistant strains could be attributed to changes in membrane integrity. These fungicides may inhibit some enzyme of the plasma membrane, either directly or by having an affect on lipids. They may have a non-specific, physical toxicity based on their lipid solubility, or they may inhibit the synthesis of certain lipids. Support for the hypothesis that they affect cell wall synthesis include the following:

(i) for Galbraith & Shields (1982) it is possible 'that the site of dichlobenil inhibition resides not at the level of enzymes of polysaccharide synthesis but perhaps at the level of membrane fusion events responsible for continued expression of synthesis';

(ii) for Tillman & Sisler (1973) 'interference of chloroneb with an enzymatic plasticising of the cell wall might account for the failure of the treated sporidia (of *Ustilago maydis*) to bud. The toxicant might act directly on the enzyme or accumulate at the membrane–cell wall interface and prevent contact of membrane bound enzyme with the cell wall. The consequence of such interferences could be the cessation of growth because of the encasement of the protoplast in a rigid cell wall capsule';

(iii) for Lyr & Casperson (1982*a*), cell wall abnormalities induced by the aromatic hydrocarbons could be the consequence of a primary effect on mitochondria (with modification in intracellular ATP levels).

Other possibilities have been suggested and further studies need to be made to understand the mode of action of dicarboximides and aromatic hydrocarbons and to discover the mechanism of resistance to these chemicals.

References

Albert, G. (1979). Wirkungsmechanismus und wirksamkelt von vinchlozolin bei *Botrytis cinerea* (Pers.). Inaugural dissertation, Universität zu Bonn, F.R.G.

Albert, G. (1981). Sphäroplastenbildung bei *Botrytis cinerea*, hervorgerufen durch vinclozolin. *Zeitschrift für Pflanzenkrankheiten und Pflanzenschutz*, **88**, 337–42.

Ark, P. A. (1946). Mutation in certain phytopathogenic bacteria induced by acenapthene. *Journal of Bacteriology*, **51**, 699–701.

Azevedo, J. L., Santana, E. P. & Bonatelli, R. (1977). Resistance and mitotic instability to chloroneb and 1,4-oxathiin in *Aspergillus nidulans*. *Mutation Research*, **48**, 163–72.

Beever, R. E. (1983). Osmotic sensitivity of fungal variants resistant to dicarboximide fungicides. *Transactions of the British Mycological Society*, **80**, 327–31.

Beever, R. E. & Byrde, R. J. W. (1982). Resistance to the dicarboximide fungicides. In *Fungicide Resistance in Crop Protection*, 1982, ed. J. Dekker & S. G. Georgopoulos, pp. 101–17, Wageningen: Pudoc.

Beraha, L. & Garber, E. D. (1966). Genetics of phytopathogenic fungi. XV. A genetic study of resistance to sodium ortho-phenylphenate and soldium dehydroacetate in *Penicillium expansum*. *American Journal of Botany*, **53**, 1041–7.

Beraha, L. & Garber, E. D. (1980). A genetic study of resistance to thiabendazole and sodium *O*-phenylphenate in *Penicillium italicum* by parasexual cycle. *Botanical Gazette*, **141**, 204–9.

Blakemore, R. P. (1978). Effects of polychlorinated biphenyls on macromolecular synthesis by a heterotrophic marine bacterium. *Applied and Environmental Microbiology*, **35**, 329–36.

Blakemore, R. P. & Carey, A. E. (1978). Effects of polychlorinated biphenyls on growth and respiration of heterotrophic marine bacteria. *Applied and Environmental Microbiology*, **35**, 323–8.

Brenneman, T. B. & Stipes, R. J. (1983). Sensitivity of *Sclerotinia minor* from peanut to dicloran, iprodione and vinclozolin. *Phytopathology*, **73**, 964.

Brook, M. (1952). Differences in the biological activity of 2,3,5,6-tetrachloronitrobenzene and its isomers. *Nature*, **170**, 1022.

Buchenauer, H. (1976). Preliminary studies on the mode of action of vinclozolin. *Mededelingen van de Faculteit Landbouwwetenschappen, Rijksuniversiteit, Gent*, **41**, 1509–19.

Carey, M. A. & McDonough, E. S. (1943). On the production of polyploidy in *Allium* with paradichlorobenzene. *The Journal of Heredity*, **34**, 238–40.

Casperson, G. & Lyr, H. (1982). Wirkung von Pentachlornitrobenzol (PCNB) auf die Ultrastruktur von *Mucor mucedo* und *Phytophthora cactorum*. *Zeitschrift für Allgemeine Mikrobiologie*, **22**, 219–26.

Cayley, G. R. & Hide, G. A. (1980). Uptake of iprodione and control of diseases on potato stems. *Pesticide Science*, **11**, 15–19.

Chastagner, G. A. & Ogawa, J. M. (1979). DCNA-benomyl multiple resistance in strains of *Botrytis cinerea*. *Phytopathology*, **69**, 699–702.

Chastagner, G. A. & Vassey, W. E. (1979). Tolerance of *Botrytis tulipae* to glycophen and vinclozolin. *Phytopathology*, **69**, 914.

Chastagner, G. A. & Vassey, W. E. (1982). Occurrence of iprodione tolerant *Fusarium nivale* under field conditions. *Plant Disease*, **66**, 112–14.

Clarke, C. H. (1971). The mutagenic specificities of pentachloronitrobenzene and captan, two environmental mutagens. *Mutation Research*, **11**, 247–8.

Cooke, B. K., Pappas, A. C., Jordan, V. W. L. & Wertern, N. M. (1979). Translocation of benomyl, prochloraz and procymidone in relation to control of *Botrytis cinerea* in strawberries. *Pesticide Science*, **10**, 467–72.

Craig, G. D. & Peberdy, J. F. (1983). The mode of action of s-benzyl-*O*,*O*-di-isopropyl phosphorothioate and of dicloran on *Aspergillus nidulans*. *Pesticide Science*, **14**, 17–24.

Davis, R. P. & Dennis, C. (1981). Properties of dicarboximide-resistant strains of *Botrytis cinerea*. *Pesticide Science*, **12**, 521–35.

Dekker, J. (1976). Acquired resistance to fungicides. *Annual Review of Phytopathology*, **14**, 405–28.

Detweiter, A. R., Vargas, J. M. & Danneberg, T. K. (1983). Resistance of *Sclerotinia homeocarpa* to iprodione and benomyl. *Plant Disease*, **67**, 617–30.

Eckert, J. W. (1982). Case study. 5. *Penicillium* decay of citrus fruits. In *Fungicide Resistance in Crop Protection*, 1982, ed. J. Dekker & S. G. Georgopoulos, pp. 231–50, Wageningen: Pudoc.

Eichhorn, K. W. & Lorenz, D. H. (1978). Untersuchungen über die Wirkung von Vinclozolin gegenüber *Botrytis cinerea in vitro*. *Zeitschrift für Pflanzenkrankheiten und Pflanzenschutz*, **85**, 449–60.

Esuruoso, O. F., Price, T. V. & Wood, R. K. S. (1968). Germination of *Botrytis cinerea* conidia in the presence of quintozene, tecnazene and dichloran. *Transactions of the British Mycological Society*, **51**, 405–10.

Esuruoso, O. F. & Wood, R. K. S. (1971). The resistance of spores of resistant strains of *Botrytis cinerea* to quintozene, tecnazene and dicloran. *Annals of Applied Biology*, **68**, 271–9.

Fisher, D. J. (1977). Induction of yeast-like growth in mucorales by systemic fungicide and other compounds. *Transactions of the British Mycological Society*, **68**, 397–402.

Fritz, R (1977). Action de quelques fongicides sur des entomophthorales pathogènes de pucerons. *Phytiatrie-Phytophurmacie*, **26**, 193–200.

Fritz, R., Leroux, P. & Gredt, M. (1977). Mécanisme de l'action fongitoxique de la promidione (26019 RP ou glycophène), de la vinchlozoline et du dicloran sur *Botrytis cinerea* Pers. *Phytopathologische Zeitschrift*, **90**, 152–63.

Fuchs, A & De Waard, M. A. (1982). Resistance to ergosterol-biosynthesis inhibitors. I. Chemistry and phenomenoligal aspects. In *Fungicide Resistance in Crop Protection*, 1982, ed. J. Dekker & S. G. Georgopoulos, pp. 71–86 Wageningen: Pudoc.

Galbraith, D. W. & Shields, B. A. (1982). The effects of inhibitors on tobacco protoplast development. *Physiologia Plantarum*, **55**, 25–30.

Garibaldi, A. & Gullino, M. L. (1979). Ulteriori osservazioni sulla resistenza di *Botrytis cinerea* a vinclozolin. *Phytopathologica Mediterranea*, **19**, 147–51.

Georgopoulos, S. G. (1963a). Pathogenicity of chlorinated-nitrobenzenes tolerant strains of *Hypomyces solani* f. *cucurbitae* race 1. *Phytopathology*, **53**, 1081–5.

Georgopoulos, S. G. (1963b). Tolerance to chlorinated-nitrobenzenes in *Hypomyces solani* f. *cucurbitae* and its mode of inheritance. *Phytopathology*, **53**, 1086–93.

Georgopoulos, S. G. (1977). Development of fungal resistance to fungicides. In *Antifungal compounds*, vol. 2, 1977, ed. M. R. Siegel & H. D. Sisler, pp. 439–95. New York: Marcel Dekker Inc.

Georgopoulos, S. G., Kappas, A. & Hastie, A. C. (1976). Induced sectoring in diploid *Aspergillus nidulans* as a criterion of fungitoxicity by interference with hereditary processes. *Phytopathology*, **66**, 217–20.

Georgopoulos, S. G., Sarris, M. & Ziogas, B. N. (1979). Mitotic instability in *Aspergillus nidulans* caused by the fungicides iprodione, procymidone and vinclozolin. *Pesticide Science*, **10**, 389–92.

Georgopoulos, S. G. & Vomvoyanni, V. E. (1965). Differential sensitivity of diphenyl-sensitive and diphenyl-resistant strains of fungi to chlorinated nitrobenzenes and to some diphenyl derivatives. *Canadian Journal of Botany*, **43**, 765–75.

Georgopoulos, S. G., Zafiratos, C. & Georgiadis, E. (1967). Membrane functions and tolerance to aromatic hydrocarbon fungitoxicants in conidia of *Fusarium solani*. *Physiologia Plantarum*, **20**, 373–81.

Georgopoulos, S. G. & Zaracovitis, C. (1967). Tolerance of fungi to organic fungicides. *Annual Review of Phytopathology*, **5**, 109–30.

Gullino, M. L. & Garibaldi, A. (1979). Osservazioni sperimentali sulla resistenza di isolamenti Italiani di *Botrytis cinerea* a vinclozolin. *La Difesa delle Piante*, **6**, 341–50.

Harding, P. R. jr. (1962). Differential sensitivity to sodium orthophenyl phenate by biphenyl-sensitive and biphenyl-resistant strains of *Penicillium digitatum*. *Plant Disease Reporter*, **46**, 100–4.

Hisada, Y., Kato, T. & Kawase, Y. (1977). Systemic movement in cucumber plants and control of cucumber grey mould by a new fungicide, S 7131. *Netherlands Journal of Plant Pathology*, **83** (sup. 1), 71–8.

Hisada, Y., Kato, T. & Kawase, Y. (1978). Mechanism of antifungal action of procymidone in *Botrytis cinerea*. *Annals of the Phytopathological Society of Japan*, **44**, 509–18.

Hisada, Y. & Kawase, Y. (1977). Morphological studies of antifungal action of *N*-(3′,5′-dichlorophenyl)-1,2-dimethyl-cyclopropane-1,2-dicarboximide on *Botrytis cinerea*. *Annals of the Phytopathological Society of Japan*, **43**, 151–8.

Hisada, Y. & Kawase, Y. (1980). Reverse binding of procymidone to a sensitive fungus *Botrytis cinerea*. *Journal of Pesticide Science*, **5**, 559–64.

Hock, W. K. & Sisler, H. D. (1969). Specificity and mechanism of antifungal action of chloroneb. *Phytopathology*, **59**, 627–32.

Hogetsu, T., Shibaoka, H. & Shimokoriyama, M. (1974). Involvement of cellulose synthesis in actions of gibberellin and kinetin on cell expansion. 2,6-dichlorobenzonitrile as a new cellulose-synthesis inhibitor. *Plant and Cell Physiology*, **15**, 389–93.

Kappas, A. (1978). On the mechanisms of induced somatic recombination by certain fungicides in *Aspergillus nidulans*. *Mutation Research*, **51**, 189–97.

Katan, T. (1982). Resistance to 3,5-dichlorophenyl-*N* cyclic imide (dicarboximides) fungicides in the grey mould pathogen *Botrytis cinerea* on protected crop. *Plant Pathology*, **31**, 133–41.

Katan, T. & Shabi, E. (1981). Resistance to dicarboximide fungicides in laboratory isolates of *Monilinia laxa*. *Netherland Journal of Plant Pathology*, **87**, 242.

Kato, T. (1983). Mode of antifungal action of a new fungicide, tolclofos-methyl. *Pesticide Chemistry: Human Welfare and the Environment*, 5th International Congress of Pesticide Chemistry, vol. 3, ed. J. Miyamoto & P. C. Kearney, pp. 153–7. Oxford: Pergamon Press.

Kato, T., Hisada, Y. & Kawase, Y. (1979). Nature of procymidone-tolerant *Botrytis cinerea* strains obtained *in vitro*. In *Proceedings of Seminar on Pest resistance to Pesticides, Palm Springs, U.S.A.* in press.

Kuiper, J. (1965). Failure of hexachlorobenzene to control common bunt of wheat. *Nature*, **206**, 1219–20.

Lacroix, L., Bic, G., Burgaud, L., Guillot, M., Leblanc, R., Riottot, R. & Sauli, M. (1974). Etude des propriétés antifongiques d'une nouvelle famille de dérivés de l'hydantoine et en particulier du 26019 RP. *Phytiatrie-Phytopharmacie*, **23**, 165–74.

Lankov, R. K. (1971). Growth responses of strains of *Botrytis cinerea* tolerant and susceptible to 2,6-dichloro-4-nitroaniline. *Phytopathologie*, **61**, 900.

Lartaud, G., Marchand, D., Duchondoris, J., Martin, N., Benoist, M. & Pommer, E. H. (1977). La vinchlozoline. *Phytiatrie-Phytopharmacie*, **26**, 239–50.

Leroux, P., Fritz, R. & Gredt, M. (1977). Etudes en laboratoire de souches de *Botrytis cinerea* résistantes à la dichlozoline, au dicloran, au quintozène, à la vinchlozoline et au 26019 RP (glycophène). *Phytopathologische Zeitschrift*, **89**, 347–58.

Leroux, P., Fritz, R. & Gredt, M. (1982*a*). Cross-resistance between 3,5-dichlorophenyl cyclic imide fungicides (dichlozolin, iprodione, procymidone, vinchlozolin) and various aromatic compounds. In *Systemische Fungizide und Antifungale Verbindungen*, VI Internationale symposium, Mai 1980, ed. H. Lyr & C. Polter, pp. 79–88. Berlin: Akademie-Verlag.

Leroux, P. & Gredt, M. (1980). Effets de divers herbicides sur la croissance mycélienne de souches de *Botrytis cinerea* et de *Penicillium expansum* sensibles ou résistantes à certains fongicides. *Weed Research*, **20**, 249–54.

Leroux, P. & Gredt, M. (1982*a*). Effets d'alcools primaires, de polyols, de sels minéraux et de sucres sur des souches de *Botrytis cinerea* sensibles ou résistantes à l'iprodione et à la procymidone. *Comptes rendus de l'Académie des Sciences, Paris, série III*, **294**, 53–6.

Leroux, P. & Gredt, M. (1982*b*). Phénomènes de résistance de *Botrytis cinerea* aux fongicides. *La Défense des Végétaux*, **213**, 3–17.

Leroux, P. & Gredt, M. (1983). Resistance of *Botrytis cinerea* to fungicides. In *Systemische Fungizide und Antifungale Verbindungen*, VII Internationale symposium, Mai 1983, ed. H. Lyr & C. Polter. Berlin: Akademie Verlag.

Leroux, P., Gredt, M. & Friz, R. (1978). Etudes en laboratoire de souches de quelques champignons phytopathogènes résistantes à la dichlozoline, à la dicyclidine, à l'iprodione, à la vinchlozoline et à divers fongicides aromatiques. *Mededelingen van de Faculteit Landbouwwetenschappen Rijkuniversiteit, Gent*, **43**, 881–9.

Leroux, P., Gredt, M. & Fritz, R. (1981). Resistance to 3,5-dichlorophenyl-*N*-cyclic imide fungicides. *Netherland Journal of Plant Pathology*, **87**, 244–5.

Leroux, P., Lafon, R. & Gredt, M. (1982*b*). La résistance de *Botrytis cinerea* aux benzimidazoles et aux imides-cycliques: situation dans les vignobles alsaciens, bordelais et champenois. *EPPO Bulletin*, **12**, 137–43.

Lewis, J., Weber, D. J. & Venketeswaran, S. (1969). Mode of action of 2,6-dichloro-4-nitroaniline in plant tissue culture. *Phytopathology*, **59**, 93–7.

Locke, S. B. (1969). Botran tolerance of *Sclerotinia cepivorum* isolates from fields with different Botran-treatment histories. *Phytopathology*, **59**, 13.

Lorenz, D. H. & Eichhorn, K. W. (1982). *Botrytis cinerea* and its resistance to dicarboximide fungicides. *EPPO Bulletin*, **12**, 125–9.

Lyr, H. & Casperson, G. (1982*a*). Anomalous cell wall synthesis in *Mucor mucedo* (L.) Fres. induced by some fungicides and other compounds related to the problem of dimorphism. *Zeitschrift für Allgemeine Mikrobiologie*, **22**, 245–54.

Lyr, H. & Casperson, G. (1982*b*). On the mechanism of action and the phenomenon of cross resistance of aromatic hydrocarbon fungicides and dicarboximide fungicides. *Acta Phytopathologica Academiae Scientiarum Hungaricae*, **17**, 317–26.

Lyr, H., Kempter, G., Kluge, E., Müller, H. M. & Zollfrank, G. (1983). On the fungicidal activity of some new diphenylether compounds. *Systemische Fungizide und Antifungale Verbindungen*, VII Internationale symposium, Mai 1983, ed. H. Lyr & C. Polter. Berlin: Akademie Verlag.

Lyr, H. & Werner, P. (1982). On the mechanism of action of the fungicide chloroneb. *Pesticide Biochemistry and Physiology*, **18**, 69–76.

McKee, R. K. (1951). Mutations appearing in *Fusarium coeruleum* cultures treated with tetrachloronitrobenzene. *Nature*, **167**, 611.

McPhee, W. J. (1980). Some characteristics of *Alternaria alternata* strains resistant to iprodione. *Plant Disease*, **64**, 847–9.

Macris, B. & Georgopoulos, S. G. (1969). Reduced hexosamine content of fungal cell wall due to the fungicide pentachloronitrobenzene. *Phytopathology*, **59**, 879–80.

Maraite, H., Gilles, G., Meunier, S., Weyns, J. & Bal, E. (1981). Resistance of *Botrytis cinerea* to dicarboximide fungicides in strawberry fields. *Parasitica*, **36**, 90–101.

Menager, M. L., Tissier, Comelli & Alluis (1971). La dichlozoline. *Phytiatrie-Phytopharmacie*, **20**, 169–82.

Meyer, Y. & Herth, W., (1978). Chemical inhibition of cell wall formation and cytokinesis, but not of nuclear division in protoplasts of *Nicotiana tabacum* L. cultivated *in vitro*. *Planta*, **142**, 253–62.

Meyer, R. W. & Parmenter, J. R. (1968). Changes in chemical tolerance associated with heterokaryosis in *Thanatephorus cucumeris*. *Phytopathology*, **58**, 472–5.

Montezinos, D. & Delmer, D. P. (1980). Characterization of inhibitors of cellulose synthesis in cotton fibers. *Planta*, **143**, 305–11.

Moore, D. (1981). Effects of hexose analogues on fungi, mechanisms of inhibition and of resistance. *New Phytologist*, **84**, 487–515.

Nakanishi, T. & Oku, H. (1970). Mechanism of selective toxicity of fungicide: absorption, metabolism and accumulation of pentachloro-nitrobenzene by phytopathogenic fungi. *Annals of the Phytopathological Society of Japan*, **36**, 67–73.

Nebel, B. R. (1938). Colchicine and acenaphthene as polyploidizing agents. *Nature*, **141**, 257.

Ogawa, J. M., Ramsey, R. H. & Moore, C. I. (1963). Behaviour of variants of *Gilbertella persicaria* arising in medium containing 2,6-dichloro-4-nitroaniline. *Phytopathology*, **53**, 97–100.

Pappas, A. C. (1982). Inadequate control of grey mould on cyclamen by dicarboximide fungicides in Greece. *Journal of Plant Disease and Protection*, **89**, 52–8.

Pappas, A. C., Cooke, B. K. & Jordan, V. W. L. (1979). Insensitivity of *Botrytis cinerea* to iprodione, procymidone and vinclozolin and their uptake by the fungus. *Plant Pathology*, **28**, 71–6.

Pappas, A. C. & Fisher, D. (1979). A comparison of the mechanisms of action of vinclozolin, procymidone, iprodine and prochloraz against *Botrytis cinerea*. *Pesticide Science*, **10**, 239–46.

Pennucci, A. & Jackson, N. (1983). Tolerance of *Sclerotinia homeocarpa* to iprodione and chlorothalonil. *Phytopathology*, **73**, 372.

Pommer, E. H. & Lorenz, G. (1982). Resistance of *Botrytis cinerea* to dicarboximide fungicides – a literature review. *Crop Protection*, **1**, 221–30.

Pommer, E. H. & Zeeh, B. (1982). Miclozolin, ein neuer Wirkstoff aus der Klasse der Dicarboximide. *Mededelingen van de Faculteit Landbouwwetenschappen Rijkuniversiteit Gent*, **47**, 935–42.

Presly, A. H., Maude, R. B., Miller, J. M. & Large, A. (1980). Tolerance in *Botrytis squamosa* to iprodine. *National Vegetable Research Station (U.K.), Annual Report 1979*, 62–3.

Priest, D. & Wood, R. K. S. (1961). Strains of *Botrytis allii* resistant to chlorinated nitrobenzenes. *Annals of Applied Biology*, **49**, 445–60.

Rehm, H. J. (1969). Inhibiting action of sodium-*o*-phenylphenate (SOPP) and biphenyl on specific reactions of the metabolism of microorganisms. *Proceedings of the First International Citrus Symposium*, **3**, ed. H. D. Chapman, pp. 1325–31. Riverside, California: University of California.

Reilly, C. C. & Lamoureux, L. (1981). The effects of the fungicide, iprodione, on the mycelium of *Sclerotinia sclerotiorum*. *Phytopathology*, **71**, 722–7.

Ritchie, D. F. (1983). Mycelial growth, peach fruit-rotting capability and sporulation of strains of *Monilinia fructicola* resistant to dichloran, iprodione, procymidone and vinclozolin. *Phytopathology*, **73**, 44–7.

Rosenberg, D. A. & Meyer, F. W. (1981). Post harvest fungicides for apples: development of resistance to benomyl, vinchlozoline and iprodione. *Plant Disease*, **65**, 1010–13.

Sharom, F. J. & Mellors, A. (1980). Effects of polychlorinated biphenyls on biological membranes. Physical toxicity and molar volume relationships. *Biochemical Pharmacology*, **29**, 3311–17.

Shalta, M. N. & Sinclair, J. B. (1963). Tolerance to pentachloronitrobenzene among cotton isolates of *Rhizoctonia solani*. *Phytopathology*, **53**, 1407–11.

Sharples, R. O. (1961). The fungitoxic effects of dicloran on *Botrytis cinerea*. *Proceedings of the British Insecticide and Fungicide Conference 1961*, pp. 327–36. London: Association of British Manufacturers of Agricultural Chemicals.

Skorda, E. A. (1977). Insensitivity of wheat bunt to hexachlorobenzene and quintozene in Greece. *Proceedings of the 1977 British Crop Protection Conference, Brighton*, pp. 67–71. London: British Crop Protection Council.

Sztejnberg, A. & Jones, A. L. (1978). Tolerance of the brown rot fungus *Monilinia fructicola* to iprodione, procymidone and vinclozolin fungicides. *Phytopathology News*, **12**, 187–8.

Takayama, C. & Fujinami, A. (1979). Quantitative structure-activity relationships of antifungal *N*-phenyl-succinimides and *N*-phenyl-1,2-dimethylcyclopropane-dicarboximides. *Pesticide Biochemistry and Physiology*, **12**, 163–71.

Thapliyal, P. N. & Sinclair, J. B. (1970). Uptake of three systemic fungicides by germinating soybean seed. *Phytopathology*, **70**, 1373–5.

Thorn, G. D. (1973). Uptake and metabolism of chloroneb by *Phaseolus vulgaris*. *Pesticide Bichemistry and Physiology*, **3**, 137–40.

Threlfall, R. J. (1968). The genetics and biochemistry of mutant of *Aspergillus nidulans* resistant to chlorinated nitrobenzenes. *Journal of General Microbiology*, **52**, 35–44.

Threlfall, R. J. (1972). Effect of pentachloronitrobenzene (PCNB) and other chemicals on sensitive and PCNB-resistant strains of *Aspergillus nidulans*. *Journal of General Microbiology*, **71**, 173–80.

Tillman, R. W. & Sisler, H. D. (1973). Effect of chloroneb on the growth and metabolism of *Ustilago maydis*. *Phytopathology*, **63**, 219–25.

Umetsu, N., Satoh, S. & Matsuda, K. (1976). Effects of 2,6-dichlorobenzonitrile on suspension-cultured soybean cells. *Plant and Cell Physiology*, **17**, 1071–5.

Van Tuyl, J. M. (1977). Genetics of fungal resistance to systemic fungicides. *Mededelingen Landbouwhogeschool Wageningen*, **77–2**, 1–136.

Verloop, A. (1972). Fate of the herbicide dichlobenil in plants and soil in relation to its biological activity. *Residue Reviews*, **43**, ed. F. A. Gunther & J. D. Gunther, pp. 55–103. Springer-Verlag: Berlin.

Vomvoyanni, V. E. & Georgopoulos, S. G. (1966). Dosage–response relationships in *Hypomyces*. *Phytopathology*, **56**, 1330–1.

Weber, J. R. & Ogawa, J. M. (1965). The mode of action of 2,6-dichloro-4-nitroaniline in *Rhizopus arrhizus*. *Phytopathology*, **55**, 159–65.

Webster, R. K., Ogawa, J. M. & Bose, E. (1970). Tolerance of *Botrytis cinerea* to 2,6-dichloro-4-nitroaniline. *Phytopathology*, **60**, 1489–92.

Webster, R. K., Ogawa, J. M. & Moore, C. J. (1968). The occurrence and behaviour of variants of *Rhizopus stolonifer* tolerant to 2,6-dichloro-4-nitroaniline. *Phytopathology*, **58**, 997–1003.

Werner, P., Lyr, H. & Casperson, G. (1978). Die Wirkung von Chloroneb, seinen Abbauprodukten, sorvie von chlorierten Phenolen auf das Wachstum und die Ultrastruktur verschiedener Pilzarten. *Archiv für Phytopathologie und Pflanzenschutz*, **14**, 301–12.

Wuu, K. D. & Grant, W. F. (1967). Chromosomal aberrations induced by pesticides in meiotic cells of barley. *Cytologia*, **32**, 31–41.

11

Antifungal activity of acylalanine fungicides and related chloroacetanilide herbicides

L. C. DAVIDSE

Laboratory of Phytopathology, Agricultural University, Wageningen, The Netherlands

Introduction

The antifungal activity of acylalanines was first recognised in 1972 when the curative and systemic properties of various lead compounds, derived from anilide herbicides, were observed against *Phytophthora infestans* (Hubele *et al.*, 1983). One of these compounds, CGA 29212 (Fig. 11.1) showed systemic, curative and protective activity at low dosage against plant pathogens in the Peronosporales, but unfortunately it was still phytotoxic. This phytotoxicity was overcome by substituting the chloroacetyl group by other acyl moieties, and this finally led to the commercial fungicides, metalaxyl and furalaxyl (Schwinn, Staub & Urech, 1977; Urech, Schwinn & Staub, 1977).

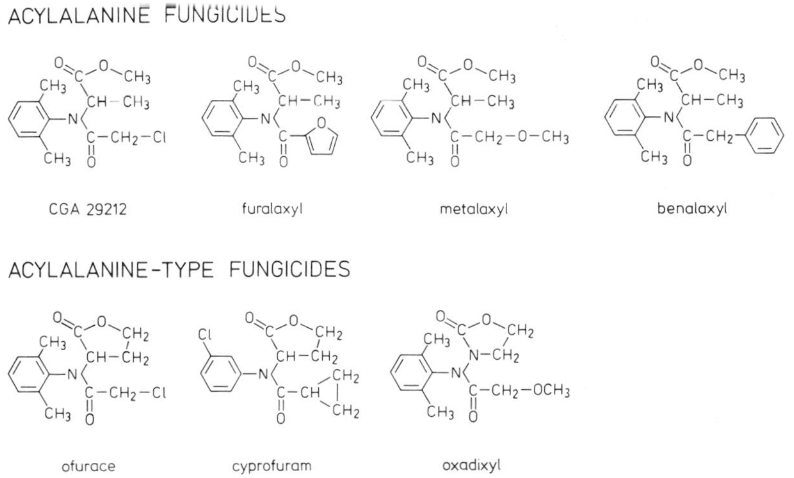

ACYLALANINE FUNGICIDES

CGA 29212   furalaxyl   metalaxyl   benalaxyl

ACYLALANINE-TYPE FUNGICIDES

ofurace   cyprofuram   oxadixyl

Fig. 11.1. Structures of acylalanine and acylalanine-type fungicides.

Antifungal activity of these compounds is dependent on chirality, the R-enantiomers being active, whereas the S-enantiomers are almost devoid of any antifungal activity (Hubele *et al.*, 1983). This contrasts with the chirality dependency of the herbicidal activity of the lead compounds, e.g. the S-enantiomers of CGA 29212 and the related commercial herbicide, metolachlor, show higher herbicidal activity than the R-enantiomers (Moser, Riks & Santer, 1982). This suggests that the primary mode of action of these compounds in plants may differ from its primary mode of action in fungi.

In the years since the discovery of furalaxyl and metalaxyl, several related compounds have been commercialised by various agrochemical companies; the structures of these fungicides are given in Fig. 11.1. Benalaxyl (Bergamaschi *et al.*, 1981) still carries the alanine methylester but, in ofurace (Kaspers & Reuff, 1979) and cyprofuram (Baumert & Busschaus, 1982), this moiety is in the cyclic butyrolactone form. In oxadixyl (Gisi *et al.*, 1983) the latter form is substituted by an oxazolidinone moiety. The last three compounds are not acylalanines; because of their structural and biological properties, they can best be described as acylalanine-type fungicides.

The acylalanine and acylalanine-type funcicides are almost selective for fungi belonging to the Peronosporales; fungi of the order Saprolegniales are not sensitive (Kerkenaar & Kaars Sijpesteijn, 1981) and, apart from the Oomycetes, only members of the Hyphochytriomycetes are inhibited by these compounds (Bruin, 1980). In a short time, the acylalanine fungicides have become important in the control of several plant-pathogenic fungi. Destructive pathogens (such as downy mildews of various crops, potato late blight, blue mould of tobacco and soil-borne *Pythium* and *Phytophthora* spp.) are amongst the main target fungi. The systemic behaviour of acylalanines allows them to be used as curative fungicides but, in practice, they are mainly used as protective fungicides.

A serious threat to the continued effectiveness of acylalanines and acylalanine-type fungicides is the rapid development of fungicide resistance in various pathogens (see Davidse, 1982). In several countries the use of acylalanines on some crops has been discontinued because of the severity of this resistance problem, e.g. during the second year of use of metalaxyl in The Netherlands, metalaxyl-resistant strains of *Phytophthora infestans* dominated the population (Davidse *et al.*, 1981b). In 1982, two years after the use of metalaxyl had been terminated, resistant populations still built up rapidly after experimental use of acylalanines

Table 11.1. *Effect of metalaxyl on precursor incorporation by Phytophthora megasperma f.sp. medicaginis in liquid culture*

Precursor	Incorporation in metalaxyl-treated cultures expressed as a per cent of incorporation in control cultures[a]	Alkali stability of product[b]
[methyl-^{3}H]Thymidine	84	98
[5,6-^{3}H]Uridine	26	2
L-[U-^{14}C]Phenylalanine	81	Not determined

[a]10 ml liquid cultures containing 20–30 mg dry weight of mycelium in a synthetic medium (Erwin & Katznelson, 1961) were pulse labelled for 15 min with 0.5 µCi of the nucleoside or 0.25 µCi of the amino acid with or without (control cultures) preincubation with metalaxyl (0.1 µg ml^{-1}) for 45 min. Acid-precipitable radioactivity was determined using standard procedures (Davidse, Hofman & Velthuis, 1983b).
[b]Percentage of radioactivity that remained associated with acid-precipitable material after incubation at 37 °C in 0.3 N KOH for 20 h.

(Davidse, Danial & Van Western, 1983a). To delay development of resistance in populations of foliar pathogens, a limited protective use of metalaxyl in combination with a second fungicide is now recommended (Staub & Sozzi, 1981). Such a strategy also reduces the risk of crop loss if resistance to metalaxyl does develop.

This chapter describes the biochemistry of the mode of action of acylalanines and the related chloroacetanilide herbicides, and discusses the resistance of fungi to these compounds. In addition, cross-resistance patterns of metalaxyl-resistant strains to these compounds are described.

Mechanism of action of metalaxyl

Studies with *Pythium splendens* (Kerkenaar, 1981), *Phytophthora palmivora* (Fisher & Hayes, 1981) and *Phytophthora megasperma* f.sp. *medicaginis* (Davidse, Gerritsma & Hofman, 1981a) have shown that metalaxyl preferentially inhibits uridine incorporation into RNA, suggesting that it interferes with RNA synthesis. Results obtained with the latter fungus are presented in Table 11.1. Complete inhibition of uridine incorporation, however, does not occur; even at metalaxyl concentrations as high as 200 µg ml^{-1}, uridine incorporation was never less than 20% of the control value.

Inhibition of [^{3}H]uridine incorporation can be caused by metalaxyl interfering with: (i) uptake of [^{3}H]uridine, (ii) metabolism of

Table 11.2. *Effects of metalaxyl and α-amanitin on RNA polymerase activity of a crude mycelial extract[a] of Phytophthora megasperma f.sp. medicaginis*

Reaction mixture[b]	Activity (%)
Control	100
−DNA	13
+Metalaxyl ($10 \, \mu g \, ml^{-1}$)	102
+α-Amanitin ($1 \, \mu g \, ml^{-1}$)	39
+α-Amanitin ($1 \, \mu g \, ml^{-1}$) + metalaxyl ($10 \, \mu g \, ml^{-1}$)	38

[a] 25 g of mycelium was homogenised in an X-Press Cell Disintegrator in 25 ml of a solution containing 50 mM Tris-HCl (pH 7.9), 10% (v/v) glycerol, 1 mM EDTA, 5 mM $MgCl_2$, 10 mM 2-mercaptoethanol, 1 mM phenylmethanesulphonyl-fluoride and 30 mM ammonium sulphate. The homogenate was centrifuged for 15 min at 12 000 g. The supernatant (18.4 mg protein ml^{-1}) was used in the assays and could be stored at −80 °C without loss of RNA polymerase activity. Activity of the preparation was 1162 units ml^{-1}.

[b] The reaction mixture contained in a total volume of 250 μl: 50 mM Tris-HCl (pH 7.9), 1 mM dithiothreitol, 1 mM ATP, GTP and CTP, 0.05 mM UTP, 1 μCi [5,6-^{3}H]UTP, 4 mM $MnCl_2$, 50 mM ammonium sulphate, 100 μg denatured calf thymus DNA and 25–50 μl of sample. Reaction mixtures were incubated at 25 °C and after 30 min acid-precipitable radioactivity was determined using standard procedures (Davidse *et al.*, 1983*b*). One unit of RNA polymerase activity is defined as the net incorporation of one picomole [^{3}H]UTP into RNA per 30 min at 25 °C.

[^{3}H]uridine into UTP, or (iii) transcription. Therefore the effects of metalaxyl on these processes in *Phytophthora megasperma* f.sp. *medicaginis* were studied (Davidse *et al.*, 1983*b*). Metalaxyl did not inhibit [^{3}H]uridine uptake; even at 200 μg ml^{-1}, [^{3}H]uridine uptake was not significantly affected. In both control and in metalaxyl-treated mycelium, [^{3}H]uridine was metabolised in a similar manner, UDP-glucose being the major metabolite. Since the latter compound is formed via UTP and glucose–phosphate, metalaxyl apparently does not inhibit conversion of [^{3}H]uridine into [^{3}H]UTP. Hence, it was concluded that inhibition of [^{3}H]uridine incorporation into RNA was caused by interference at the transcriptional level.

Crude mycelial extracts of *Phytophthora megasperma* f.sp. *medicaginis* showed a DNA-dependent RNA polymerase activity, part of which appeared to be α-amanitin-sensitive, but which was completely insensitive to metalaxyl (Table 11.2). α-Amanitin (a specific inhibitor of RNA polymerase II from many organisms), at concentrations as low as 0.05 μg ml^{-1}, caused maximum inhibition of the α-amanitin-sensitive fraction of the RNA polymerase activity. RNA polymerases could

Table 11.3. *Effects of metalaxyl and α-amanitin on partially purified RNA polymerase preparations of Phytophthora megasperma f.sp. medicaginis*

Reaction mixture[a]	Activity (%)	
	Preparation eluting from DEAE–Sephadex at about 100 mM ammonium sulphate[b]	Preparation eluting from DEAE–Sephadex at about 250 mM ammonium sulphate[c]
Control	100	100
+Metalaxyl (1 μg ml^{-1})	103	98
+α-Amanitin (0.1 μg ml^{-1})	106	7

[a]For composition, see Table 11.2, footnote *b*.
[b]Specific activity 653 units[d] per mg protein.
[c]Specific activity 7043 units[d] per mg protein.
[d]See Table 11.2, footnote *b*, for definition of units.

be partially purified from crude mycelial extracts using a procedure involving precipitation of the RNA polymerase activity with polyethylenimine, selective elution of RNA polymerases from the polyethylenimine precipitate with ammonium sulphate, ammonium sulphate fractionation, and DEAE–Sephadex chromatography (Davidse *et al.*, 1983*b*). Polymerase activity eluted from DEAE–Sephadex in two peaks at about 100 mM and 250 mM ammonium sulphate. These properties are characteristic for the RNA polymerases I and II, respectively. Preparations of both types of activity appeared to be insensitive to metalaxyl (Table 11.3), activity eluting at 250 mM ammonium sulphate was highly sensitive to α-amanitin which indicates that most of this activity was caused by RNA polymerase II. In addition, some of the latter activity might be due to RNA polymerase III because this enzyme elutes from DEAE-Sephadex at the same ionic strength as RNA polymerase II. The absence of any inhibitory effect of metalaxyl on RNA polymerase activity *in vitro* contrasts with the observed inhibition *in vivo*. This difference might result from the artificial nature of the in-vitro assay which involves denatured calf thymus DNA.

   Isolated nuclei provide an experimental system *in vitro* that resembles the situation *in vivo* as near as possible; isolated nuclei contain RNA polymerases which are tightly bound to chromatin, the natural template. Nuclei were isolated from mycelium of *Phytophthora megasperma* f.sp. *medicaginis* using a modification (Davidse *et al.*, 1983*b*) of a procedure described by Bhargava & Halvorson (1971). Using this

Table 11.4. *Effects of metalaxyl and α-amanitin on endogenous RNA polymerase activity of isolated nuclei[a] from a metalaxyl-sensitive strain[b] and a metalaxyl-resistant strain[b] of Phytophthora megasperma f.sp. medicaginis*

	Endogenous RNA polymerase activity (%)	
Reaction mixture[c]	Sensitive strain	Resistant strain
Control	100	100
+Metalaxyl (0.1 μg ml^{-1})	80	107
+Metalaxyl (10 μg ml^{-1})	64	104
+α-Amanitin (5 μg ml^{-1})	72	78
+α-Amanitin (5 μg ml^{-1}) + metalaxyl (10 μg ml^{-1})	38	76

[a]Nuclei were isolated by homogenising 50 g of mycelium and 50 ml of a solution containing 1 M sorbitol, 20% (v/v) glycerol and 5% (w/v) polyvinylpyrrolidone in liquid N_2 with a Williams homogeniser at 10 000 rpm for 2.5 min. After thawing and addition of 200 ml of the solution, the mixture was again homogenised for 30 s at 4 °C. The homogenate was centrifuged at 5500 g for 5 min. Nuclei were pelleted from the resulting supernatant at 13 200 g for 10 min and nuclear pellets were resuspended in 4–8 ml of the homogenisation medium. The preparations contained an average of 280 μg DNA ml^{-1}, with an average activity of 2.2 units per μg DNA. For definition of units see Table 11.2 footnote b.
[b]For metalaxyl sensivity of strains, see Table 11.6.
[c]The reaction mixture contained in a total of 100 μl: 50 mM Tris-HCl (pH 7.9), 1 mM dithiothreitol, 1 mM ATP, GTP and CTP, 0.01 mM UTP, 2 μCi [5,6-^{3}H]UTP, 4 mM MnCl$_2$, 50 mM ammonium sulphate and 10 μl of nuclear suspensions.

procedure, *c.* 7% from the total mycelial DNA could be recovered in crude nuclear preparations. Nuclei isolated from a metalaxyl-sensitive strain contained endogenous RNA polymerase activity that could be partially inhibited by α-amanitin as well as by metalaxyl (Table 11.4). Metalaxyl inhibition of the metalaxyl-sensitive part of the endogenous polymerase activity was complete at 10 μg ml^{-1}; the α-amanitin-sensitive part was completely inhibited by α-amanitin at 5 μg ml^{-1}. When both compounds were added together to the reaction mixture, the inhibitory effects were additive. This indicates that metalaxyl and α-amanitin inhibited two different types of polymerase activity. Since α-amanitin is known to inhibit polymerase II activity, metalaxyl cannot be an inhibitor of this polymerase.

Nuclei isolated from a metalaxyl-resistant strain of *Phytophthora megasperma* f.sp. *medicaginis* did not contain metalaxyl-sensitive polymerase activity. Endogenous nuclear RNA polymerase activity of the resistant strain was as sensitive to α-amanitin as that of the sensitive strain. Resistance to metalaxyl in this strain is apparently based on the insensitivity of the target site.

It is known that the RNA polymerases can be found in two distinct pools in isolated nuclei: one is present as a transcription complex (engaged form) while the other is free and capable of transcribing added exogenous templates but is not functionally active towards the endogenous chromatin template. A system to study the free polymerases involves the use of poly[d(A—T)] as synthetic template and actinomycin D as inhibitor of endogenous nuclear RNA synthesis. Poly[d(A—T)] is effective in directing RNA synthesis in the presence of actinomycin D because it does not contain deoxyguanosine moieties and hence does not react with this antibiotic. Results using this system are given in Table 11.5. The endogenous RNA polymerase activity of the preparation used was partial metalaxyl sensitive. Addition of poly[d(A—T)] resulted in a 2.5-fold increase of activity. In this case activity is due to the combined activity of the engaged and free forms whereas, upon addition of poly[d(A—T)] and actinomycin D, only the free form is measured. In the latter case, it is possible to determine the effect of metalaxyl on the activity of the free polymerases. The data clearly indicate that metalaxyl does not have any effect on the activity of the free polymerases. This latter observation explains the inability of metalaxyl to inhibit the activity of RNA polymerases present in crude mycelial extracts; in crude extracts the polymerases are released from the natural template and use exogenously added template.

From these data it can be concluded that the mechanism of action of metalaxyl is based on its ability to inhibit RNA synthesis by specific interference with the activity of a nuclear, α-amanitin-insensitive, RNA-polymerase–template complex. Which of the chromatin-bound RNA polymerases is inhibited, remains to be investigated by product analysis of the RNA species produced in the in-vitro system in the presence and absence of the inhibitors. Analysis of the synthesis of RNA *in vivo* in control and metalaxyl-treated mycelium has indicated that synthesis of RNA, that lacks poly(A) sequences (poly(A)$^-$RNA), is more sensitive to metalaxyl than synthesis of poly(A)$^+$RNA (Davidse *et al.*, 1983*b*). This result is consistent with the lack of an effect of metalaxyl on endogenous RNA polymerase activity mediated by polymerase II, the enzyme responsible for the synthesis of poly(A)$^+$RNA.

Table 11.5. Presence of engaged and free forms of RNA polymerases in isolated nuclei[a] from mycelium of Phytophthora megasperma f.sp. medicaginis and effects of metalaxyl on endogenous and free polymerase activity

Reaction mixture[b]	Activity (%)	Type of RNA polymerase activity
Control	100	Endogenous
+Metalaxyl (10 μg ml^{-1})	56	
+Actinomycin D (50 μg ml^{-1})	10	
+Poly[d(A—T)] (0.2 A_{260} units per ml)	249	
+Actinomycin D + poly[d(A—T)] (50 + 0.2 A_{260} units per ml)	138	Endogenous + free
+Actinomycin D + poly[d(A—T)] + metalaxyl (50 + 0.2 A_{260} units + 10 μg ml^{-1})	137	Free

[a] The preparation (see Table 11.4, footnote a) contained 306 μg DNA per ml with an activity of 2.1 units per μg DNA. For definition of units see Table 11.2, footnote b.
[b] For composition see Table 11.4, footnote c.

Cross-resistance

Acylalanine and acylalanine-type fungicides

Patterns of cross-resistance of metalaxyl-resistant strains to acylalanine and acylalanine-type fungicides have been established for a number of fungi. Field isolates of *Phytopthora infestans* with resistance to metalaxyl showed cross-resistance to benalaxyl, ofurace, cyprofuram (L. C. Davidse, unpublished results) and oxadixyl (Gisi *et al.*, 1983). Cross-resistance to benalaxyl and cyprofuram has also been observed for metalaxyl-resistant strains of *Pseudoperonospora cubensis* (Katan, 1982). Metalaxyl-resistant strains of *Phytophthora capsici*, which were obtained after mutagenic treatment, displayed cross-resistance to furalaxyl, benalaxyl, ofurace and two experimental acylalanine-type fungicides (Bruin, 1980). Nitrosoguanidine-induced metalaxyl resistance in *Phytophthora megasperma* f.sp. *medicaginis* also conferred resistance to CGA 29212, benalaxyl, ofurace and cyprofuram (L. C. Davidse & M. L. E. Van Wijk, unpublished results). Metalaxyl-resistant strains of the latter fungus were as sensitive as the wild-type strain to other fungicides active against Oomycete fungi (e.g. etridiazol, prothiocarb, propamocarb, cymoxanil, pyroxychlor and pyroxyfur). It is evident from these data that, once resistance develops to a member of the group of acylalanine and acylalanine-type fungicides, it confers resistance to all members of the group, although the degree of resistance obtained may vary.

Chloroacetanilide herbicides

Since antifungal activity of acylalanines was optimised starting with lead compounds derived from anilide herbicides, it was of interest to determine if metalaxyl-resistant strains showed cross-resistance to these compounds. Sensitivities of a wild-type and a metalaxyl-resistant strain of *Phytophthora megasperma* f.sp. *medicaginis* to the chloroacetanilide herbicides metolachlor, propachlor, dimetachlor and alachlor were determined and compared with resistance to CGA 29212 and metalaxyl (Davidse, Gerritsma & Velthuis, 1984). Metalaxyl resistance of the particular strain used was based on the insensitivity of the target site to metalaxyl. Structure of the chloroacetanilide herbicides are shown in Fig. 11.2.

In tests of colony radial-growth rate, made on asparagine–sucrose medium, both acylalanines were much more inhibitory to growth of the wild-type strain than were the chloroacetanilide herbicides (Table 11.6). Among the latter, metolachlor showed the highest antifungal activity.

The metalaxyl-resistant strain showed cross-resistance to CGA 29212, metolachlor and propachlor, but the degrees of resistance observed were much lower with the herbicides than with the acylalanine fungicides; no cross-resistance was observed to dimetachlor and alachlor. This suggests that at least part of the antifungal activity of metolachlor and propachlor is based on a metalaxyl-type of action. However, since the herbicides inhibit growth of the metalaxyl-resistant strain at a concentration at which metalaxyl is not inhibitory, they must have a second inhibitory effect which differs from that of metalaxyl. Since the herbicides have the chloroacetyl group in common (a group also present in CGA 29212, which is slightly more inhibitory to the resistant strain than metalaxyl), this group might be involved in the second mechanism of action. The chloroacetyl group seems to be required for herbicidal activity and, therefore, the second mechanism of antifungal action of the chloroacetanilide herbicides might be related to the primary inhibitory mechanism found in plants.

Mechanism of antifungal action of chloroacetanilide herbicides

Like metalaxyl, the chloroacetanilide herbicides inhibited [^{3}H]uridine incorporation into RNA by mycelium of the wild-type strain of *Phytophthora megasperma* f.sp. *medicaginis*, but much higher con-

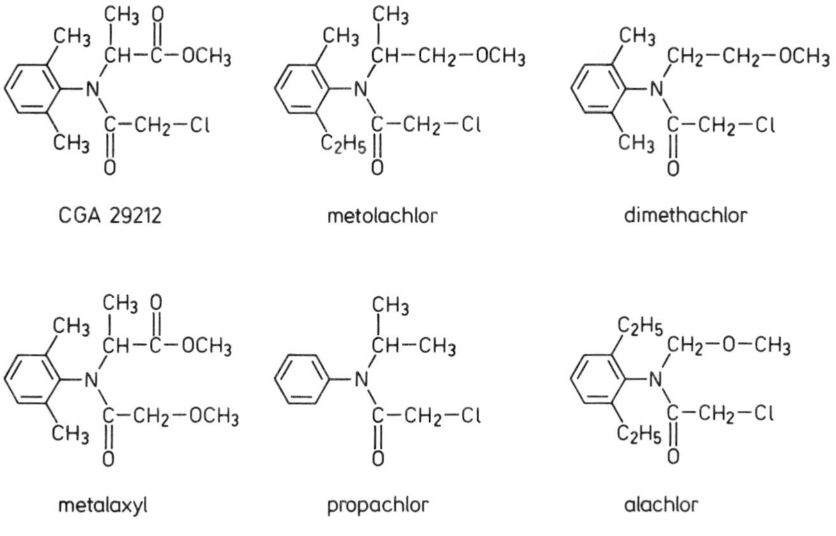

Fig. 11.2. Structures of acylalanine fungicides (CGA 29212 and metalaxyl) and chloroacetanilide herbicides (metolachlor, propachlor, dimethachlor and alachlor).

Table 11.6. *Sensitivities of a metalaxyl-sensitive and a metalaxyl-resistant strain of Phytophthora megasperma f.sp. medicaginis to acylalanine fungicides and chloroacetanilide herbicides*

	EC_{50} $(\mu M)^a$		
Compound	Sensitive strain (S)	Resistant strain (R)	Resistance factor $(EC_{50} R)/(EC_{50} S)$
Metalaxyl	0.03	1900	63 000
CGA 29212	0.09	900	10 000
Metolachlor	13	450	35
Propachlor	75	200	2.7
Dimethachlor	>400	>400	—
Alachlor	>400	>400	—

$^a EC_{50}$ values were determined by measuring colony diameter on asparagine–sucrose medium (Erwin & Katznelson, 1961), amended with the inhibitors at various concentrations and plotting the net colony diameter of triplicates against the logarithm of the concentration.

centrations were needed to obtain comparable levels of inhibition (Table 11.7). For instance, metolachlor, which among the herbicides proved to be the most inhibitory, showed a minimal inhibitory concentration (MIC) value between 0.1 and 1 $\mu g\,ml^{-1}$, whereas the MIC values for CGA 29212 and metalaxyl were below 0.01 $\mu g\,ml^{-1}$ (Davidse *et al.*, 1984). MIC values for propachlor, dimethachlor and alachlor were between 10 and 50 $\mu g\,ml^{-1}$. At high concentrations, the herbicides and CGA 29212 were almost as inhibitory to [^{3}H]uridine incorporation into RNA of the mycelium of the resistant strain, as to that of the sensitive strain (Table 11.7). Only with metalaxyl was a differential inhibition observed; this compound had little effect on the incorporation of uridine into RNA of the mycelium of the resistant strain.

In contrast with metalaxyl, the herbicides and CGA 29212 inhibited [^{3}H]uridine uptake by mycelium of *Phytophthora megasperma* f.sp. *medicaginis*; uptake by both sensitive and resistant strains was similarly affected (Table 11.7). Propachlor, which was the most inhibitory, had a MIC value between 10 and 50 $\mu g\,ml^{-1}$. Since inhibition of [^{3}H]uridine uptake obviously influences [^{3}H] incorporation into RNA, the effects on uridine incorporation observed at high concentrations of the chloroacetanilides (including CGA 29212) are not necessarily caused by specific interference with RNA synthesis. Only when inhibition of [^{3}H]uridine into RNA occurs in the absence of any effect on uptake, can it be concluded that a specific inhibition of RNA synthesis is involved. This could be demonstrated using low concentrations of CGA 29212,

Table 11.7. *Effects[a] of CGA 29212, metalaxyl, metolachlor, propachlor, dimethachlor and alachlor on [³H]uridine incorporation into RNA and its uptake by mycelium of a metalaxyl-sensitive strain and a metalaxyl-resistant strain of Phytophthora megasperma f.sp. medicaginis*

	Incorporation of [³H]uridine into RNA[b]		Uptake of [³H]uridine by mycelium[b]	
	Sensitive strain	Resistant strain	Sensitive strain	Resistant strain
CGA 29212	14	25	86	77
Metalaxyl	21	86	93	97
Metolachlor	13	10	63	50
Propachlor	3	4	16	17
Dimethachlor	28	37	74	76
Alachlor	15	12	42	33

[a]At concentrations of $200\,\mu g\,ml^{-1}$.
[b]Values are expressed as a per cent of control values. Cultures were incubated as described in Table 11.1, footnote *a*. Uptake was determined by washing the mycelia three times with 10 ml of distilled water within 30 s. Incorporation was determined using standard procedures (Davidse *et al.*, 1983*b*).

Table 11.8. *Effects of CGA 29212, metalaxyl, metolachlor, propachlor, dimethachlor and alachlor on endogenous RNA polymerase activity of nuclei[a] isolated from Phytophthora megasperma f.sp. medicaginis*

	Endogenous polymerase activity (%)	
Reaction mixture[b]	$0.1 \,\mu g\,ml^{-1}$	$10 \,\mu g\,ml^{-1}$
Control	100	100
+CGA 29212	72	60
+Metalaxyl	74	57
+Metolachlor	73	62
+Propachlor	97	94
+Dimethachlor	102	95
+Alachlor	95	103

[a] The preparation contained $248 \,\mu g$ DNA per ml with an activity of 2.3 units per μg DNA.
[b] For composition, see Table 11.4, footnote *c*.

metalaxyl or metolachlor with the metalaxyl-sensitive strain, thus indicating that the primary mechanism of action of metolachlor, like that of CGA 29212 and metalaxyl, is specific inhibition of RNA synthesis. The inhibition of uridine incorporation, observed using the metalaxyl-resistant strain and high concentrations of chloroacetanilides (including CGA 29212), evidently has another basis.

The activity of endogenous nuclear RNA polymerase was sensitive to metolachlor (Table 11.8), thus confirming the metalaxyl-type mode of action of this compound. At $0.1 \,\mu g\,ml^{-1}$, inhibition was similar to that caused by the same concentration of CGA 29212 or metalaxyl. Complete inhibition by metolachlor of the metalaxyl-sensitive fraction of this activity occurred at $1 \,\mu g\,ml^{-1}$. Propachlor, also assumed to have a metalaxyl-type of action (Table 11.6), dimethachlor and alachlor did not have any affect at $10 \,\mu g\,ml^{-1}$. The following observations indicate that the

$$\begin{array}{c} C \\ | \\ N\!-\!C\!-\!C \end{array}$$

moiety (which is present in acylalanines, metolachlor and propachlor but not in dimethachlor or alachlor) is a pre-requisite for the inhibitory mode of action of metalaxyl:

(i) among the herbicides, metolachlor and propachlor show the highest antifungal activity;

(ii) metalaxyl resistance, which is based on a lack of sensitivity of the target site to metalaxyl, is also associated with a decreased sensitivity to metolachlor and propachlor;

(iii) a metalaxyl-type of action could be demonstrated for metolachlor.

This hypothesis is compatible with the observation that substitution of the alanine methylester group of metalaxyl by a glycine methylester results in a five-fold decrease in fungicidal activity (Hubele *et al.*, 1983). The pattern of cross-resistance observed with *Phytophthora megasperma* f.sp. *medicaginis* is almost identical with that of a metalaxyl-resistant strain of *Phytophthora cactorum* to herbicidal chloroacetanilides (Leroux & Gredt, 1981). In this case, except for propachlor, cross resistance was also observed to structures having the

moiety. This moiety is probably involved in the binding of the molecule to the target site, a hypothesis which is supported by the observations that when an asymmetric C-atom is present in this group, the R-enantiomer has a high antifungal activity, whereas the S-enantiomer has a low activity or is almost inactive (Moser *et al.*, 1982; Hubele *et al.*, 1983).

Inhibition of [^{3}H]uridine uptake is evidently a feature of the second mechanism of action of the chloroacetanilides (including CGA 29212); the existence of this second mechanism was indicated by studies on cross-resistance. Suggestions about the nature of the second inhibitory mechanism can only be speculative. Interference with membrane function may be involved and this would result in a general disturbance of metabolism. Since the second mechanism of action of the chloroacetanilide herbicides might be related to the primary mechanism of action in plants the latter does not necessarily involve inhibition of RNA synthesis by interference with the activity of an RNA-polymerase–template complex. In fact, various mechanisms have been proposed for the herbicidal activity of chloroacetanilides (see Wilkinson, 1981). Different mechanisms of action in plants and fungi are also suggested by the different structural features required for herbicidal and antifungal activity. Chirality dependency of herbicidal activity is different from that of antifungal activity. The presence of the chloroacetyl group seems important for herbicidal activity but not for antifungal activity; on the other hand, high antifungal activity requires the presence of an alanine methylester moiety, or an equivalent structure, which does not seem to be essential for herbicidal activity.

It would be interesting to know if antifungal activity based on the second mechanism of action could be increased by analogue synthesis; antifungal activity is certainly more pronounced in propachlor than in the other compounds. Antifungal activity, however, would have to be increased 100–1000-fold in the analogue before it would be of any commercial value as a fungicide. At the same time, its herbicidal activity would have to be significantly reduced and this may not be possible if the primary inhibitory mechanism in plants is related to its secondary inhibitory mechanism in fungi.

Conclusion

The optimisation of antifungal activity of anilide herbicides has led to the development of highly effective acylalanines and acyl-alanine-type fungicides. At the same time, the phytotoxicity of these compounds has been reduced to a minimum. This development has been possible because the primary antifungal mechanism of these compounds differs from that of their inhibitory action on plants. The lead compounds have structural features which govern both types of action, this made possible the optimisation of antifungal activity with concurrent elimination of herbicidal activity.

The unique and highly specific interference of metalaxyl with template-bound RNA polymerase activity suggests that it can be used as a tool to study transcription in eukaryotes. In addition to its value for agriculture, metalaxyl, therefore, will also be of importance in molecular biology.

References

Baumert, D. & Busschaus, H. (1982). Cyprofuram, a new fungicide for the control of Phycomycetes. *Mededelingen Faculteit Landbouwwetenschappen, Rijksuniversiteit Gent*, **47**, 979–83.

Bergamaschi, P., Borsari, T., Caravaglia, C. & Mirenna, L. (1981). Methyl *N*-phenyl-*N*-2,6-xylyl-DL-alaninate (M 9834), a new systemic fungicide controlling downy mildew and other diseases caused by Peronosporales. *Proceedings of the 1981 British Crop Protection Conference Pest and Diseases*, pp. 11–17. Croydon: British Crop Protection Council.

Bhargava, M. M. & Halvorson, H. D. (1971). Isolation of nuclei from yeast. *Journal of Cell Biology*, **49**, 423–9.

Bruin, G. C. A. (1980). Resistance in Peronosporales to acylalanine-type fungicides. PhD thesis, University of Guelph, Canada.

Davidse, L. C. (1982). Acylalanines: resistance in downy mildews, *Pythium* and *Phytophthora* spp. In *Fungicide Resistance in Crop Protection*, ed. J. Dekker & S. G. Georgopoulos, pp. 118–27. Wageningen: Pudoc.

Davidse, L. C., Danial, D. L. & Van Westen, C. J. (1983*a*). Resistance to metalaxyl in *Phytophthora infestans* in The Netherlands. *Netherlands Journal of Plant Pathology*, **89**, 1–20.

Davidse, L. C., Gerritsma, O. C. M. & Hofman, A. E. (1981*a*). Mode d'action du metalaxyl. *Phytiatrie-Phytopharmacie*, **30**, 235–44.

Davidse, L. C., Gerritsma, O. C. M. & Velthuis, G. C. M. (1984). A differential basis of antifungal activity of acylalanine fungicides and structurally related chloroacetanilide herbicides in *Phytophthora megasperma* f.sp. *medicaginis*. *Pesticide Biochemistry and Physiology*, **21**, *in the press*.

Davidse, L. C., Hofman, A. E. & Velthuis, G. C. M. (1983*b*). Specific interference of metalaxyl with endogenous RNA polymerase activity in isolated nuclei from *Phytophthora megasperma* f.sp. *medicaginis*. *Experimental Mycology*, **7**, 344–61.

Davidse, L. C., Looyen, D., Turkensteen, L. J. & Van der Wal., D. (1981*b*). Occurrence of metalaxyl-resistant strains of *Phytophthora infestans* in Dutch potato fields. *Netherlands Journal of Plant Pathology*, **87**, 65–8.

Erwin, D. C. & Katznelson, H. (1961). Studies on the nutrition of *Phytophthora cryptogea*. *Canadian Journal of Microbiology*, **7**, 15–25.

Fisher, D. J. & Hayes, A. L. (1981). Mode of action of the systemic fungicides furalaxyl, metalaxyl and ofurace. *Pesticide Science*, **13**, 330–9.

Gisi, U., Harr, J., Sandmeier, S. & Wiedmer, H. (1983). SAN 371F, a new systemic oxazolidinone fungicide against diseases caused by Peronosporales. *Mededelingen Faculteit Landbouwwetenschappen, Rijksuniversiteit Gent*, **48**, 541–50.

Hubele, A., Kunz, W., Eckhardt, W. & Sturm, E. (1983). The fungical activity of acylalanines. In *Pesticide Chemistry, Human Welfare and the Environment*, vol. 1, *Synthesis and Structure–activity Relationships*, 5th International Congress on Pesticide Chemistry, August 1982, ed. P. Doyle & T. Fujita, pp. 233–42. Oxford: Pergamon Press.

Kaspers, H. & Reuff, J. (1979). Bekämpfung von falschen Mehltaupilzen durch ein neues organisches Fungizid (RE 20615). *Mitteilungen aus der Biologischen Bundesanstalt für Land- und Forstwirtschaft Berlin-Dahlem*, **191**, 240.

Katan, T. (1982). Cross resistance of metalaxyl-resistant *Pseudoperonospora cubensis* to other acylalanine fungicides. *Canadian Journal of Plant Pathology*, **4**, 387–8.

Kerkenaar, A. (1981). On the antifungal mode of action of metalaxyl, an inhibitor of nucleic acid synthesis in *Pythium splendens*. *Pesticide Biochemistry and Physiology*, **16**, 1–13.

Kerkenaar, A. & Kaars Sijpesteijn, A. (1981). Antifungal activity of metalaxyl and furalaxyl. *Pesticide Biochemistry and Physiology*, **15**, 71–8.

Leroux, P. & Gredt, M. (1981). Phénomènes de résistance aux fongicides antimildious: quelques résultats de laboratoire. *Phytiatrie-Phytopharmacie*, **30**, 273–82.

Moser, H., Riks, G. & Santer, H. (1982). Der Einfluss von Atropisomerie und chiralem Zentrum auf die biologische Aktivität des Metolachlor. *Zeitschrift für Naturforschung*, **87b**, 451–62.

Schwinn, F. J., Staub, T. & Urech, P. A. (1977). A new type of fungicide against diseases caused by Oomycetes. *Mededelingen Faculteit Landbouwwetenschappen, Rijksuniversiteit Gent*, **42**, 1181–8.

Staub, T. & Sozzi, D. (1981). Résistance au metalaxyl en pratique et les conséquences pour son utilisation. *Phytiatrie-Phytopharmacie*, **30**, 283–91.

Urech, P. A., Schwinn, F. J. & Staub, T. (1977). CGA 48988, a novel fungicide for the control of late blight, downy mildews and related soil-borne diseases. In *Proceedings of the 1977 British Crop Protection Conference Pest and Diseases*, pp. 623–31. Croydon: British Crop Protection Council.

Wilkinson, R. E. (1981). Metolachlor [2-chloro-*N*-2(ethyl-6-methylphenyl)-*N*-(2-methoxy-1-methylethyl)acetamide] inhibition of gibberellin precursor biosynthesis. *Pesticide Biochemistry Physiology*, **16**, 199–205.

12

Biochemical and cellular aspects of the antifungal action of ergosterol biosynthesis inhibitors[1]

H. D. SISLER* and NANCY N. RAGSDALE†

*Department of Botany, University of Maryland, College Park, MD, 20742 USA
and †Cooperative State Research Service, USDA Washington, D.C. 20250, USA

Introduction

Ergosterol biosynthesis inhibiting (EBI) fungicides constitute the largest and most important group of systemic compounds yet developed for the control of fungal diseases of plants and animals. These compounds control a broad spectrum of fungal pathogens, often exhibiting very high antifungal potency. As expected, EBI fungicides do not control plant diseases caused by fungi such as *Pythium* and *Phytophthora* species which do not synthesise sterols.

While conventionally referred to as EBI fungicides, these compounds also block sterol synthesis in Oomycetes such as *Saprolegnia ferax* which synthesise sterols other than ergosterol. In these fungi, EBI fungicides inhibit sterol biosynthetic reactions analogous to those blocked in the ergosterol pathway (Berg, 1983; Berg, Patterson & Lusby, 1983). Most EBI fungicides block sterol biosynthesis at the stage of C-14 demethylation, but compounds which block at other sites between squalene and ergosterol are also effective fungicides. While fungal resistance has not yet proved to be a serious problem with the sterol C-14 demethylation inhibitors, it would seem desirable, in any case, to develop more EBI compounds acting at other sites as a precuation against a potential resistance problem.

This chapter considers various EBI fungicides in respect to chemical structure and biochemical target site as well as the physiological and morphological consequences of their actions on fungal cells.

[1]Scientific publication No. A3556; Contribution No. 6631 of the University of Maryland Agricultural Experiment Station.

Compounds and sites of action

Sterol C-14 demethylation inhibitors

There is considerable diversity in the structures of compounds which inhibit sterol C-14 demethylation (Fig. 12.1). However, all of these compounds, with the exception of triforine, have in common a nitrogen heterocycle (pyrimidine, imidazole, triazole, pyridine or piperazine) which is substituted with a single, large lipophilic group. One of the nitrogen atoms in the heterocycle of these fungicides is believed to coordinate with the protohaem iron atom of the cytochrome P-450 enzyme(s) involved in sterol C-14 demethylation, while the lipophilic substituent interacts with nearby regions of the enzyme to increase binding affinity (Henry, 1982; Gadher *et al.*, 1983). There is considerable variation in the lipophilic substituents even among a particular class of sterol C-14 demethylation inhibitors such as the triazoles. This indicates that a relatively broad spectrum of substituents can bind effectively to the target enzyme. Nevertheless, variation in the structure of these substituents controls specificity of *N*-heterocyclic cytochrome P-450 inhibitors (Henry, 1982; Gadher *et al.*, 1983). For example, compounds which are excellent inhibitors of fungal sterol C-14 demethylase may be poor inhibitors of the same demethylase system from mammalian cells (Gadher *et al.*, 1983), and also relatively ineffective as inhibitors of the fungal cytochrome P-450 enzymes involved in certain steroid hydroxylations (Henry, 1982) or of liver enzymes involved in xenobiotic metabolism. Proper design of lipophilic substituents should result in effective inhibitors of sterol C-14 demethylase enzymes of most sterol-synthesising fungi, including strains that develop resistance through mutations which alter affinity of the target enzyme.

Pyrimidines. The pyrimidine methanol, triarimol, was the first EBI fungicide for which ergosterol biosynthesis was shown to be a target pathway (Ragsdale & Sisler, 1972, 1973). Subsequent studies made with intact cells (Ragsdale, 1975; Sherald & Sisler, 1975; Frasinel, Patterson & Dutky, 1978; Berg, 1983) or with cell-free systems (Kato & Kawase, 1976; Mitropoulos *et al.*, 1976) indicated conclusively that triarimol blocks sterol C-14 demethylation. This same mechanism has been indicated for the triarimol analogues, fenarimol and nuarimol (Buchenauer, 1977*b*; Leroux & Gredt, 1978*b*). Only the latter two analogues have been developed for use as agricultural fungicides.

While there is evidence that triarimol also blocks $\Delta^{22(23)}$ desaturation in the sterol side chain (Ragsdale, 1975; Sherald & Sisler, 1975; Frasinel

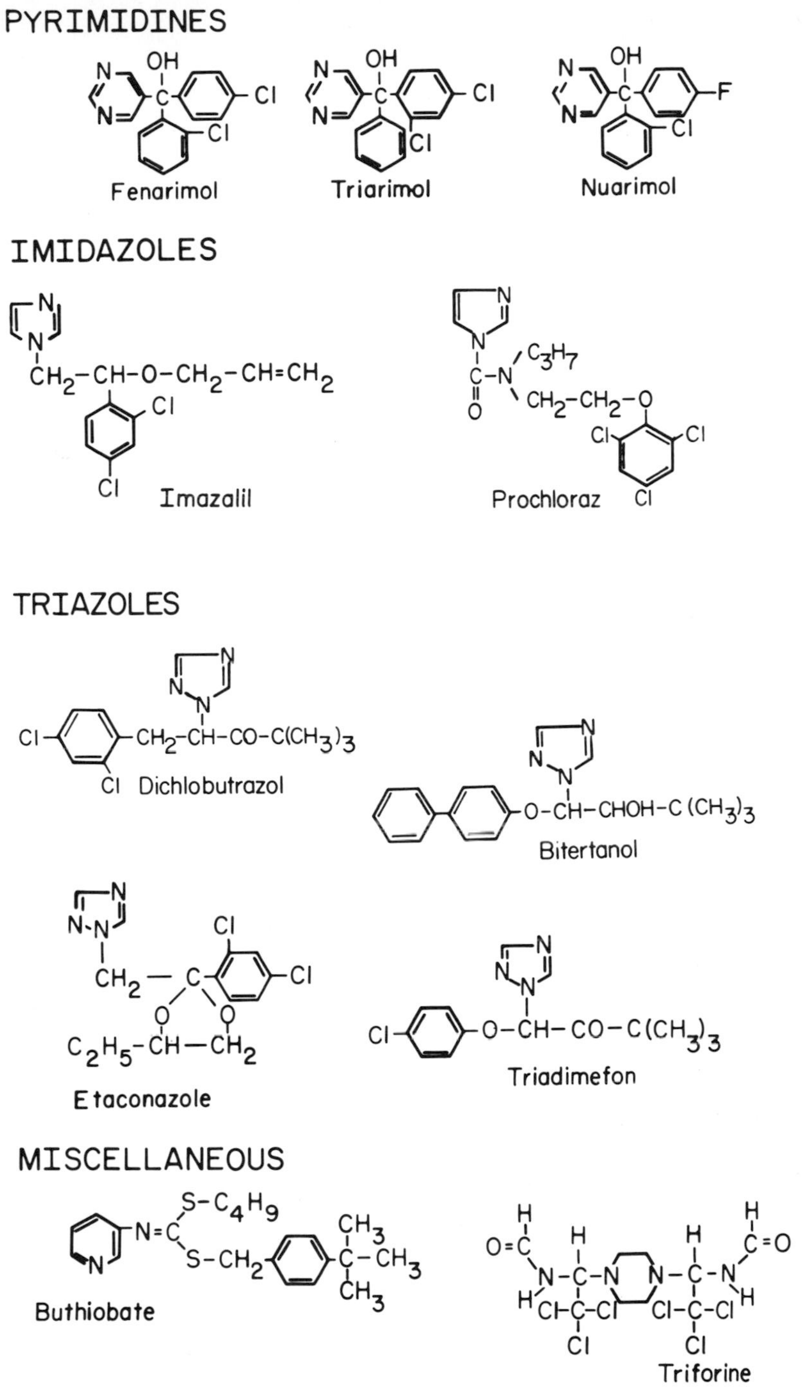

PYRIMIDINES

Fenarimol   Triarimol   Nuarimol

IMIDAZOLES

Imazalil   Prochloraz

TRIAZOLES

Dichlobutrazol

Bitertanol

Etaconazole

Triadimefon

MISCELLANEOUS

Buthiobate

Triforine

Fig. 12.1. Structures of fungicides which block ergosterol biosynthesis at the stage of C-14 demethylation (Step b to e, Figure 12.3).

et al., 1978), this effect has not been studied in any detail since it is obscured by inhibition of C-14 demethylation which occurs at an earlier stage in the pathway. It should be noted, in any case, that the Δ^{22} desaturase of yeast cells has the characteristics of a cytochrome P-450 mixed function oxidase (Hata *et al.*, 1981).

Pyrimidine methanol fungicides and closely related analogues (designed specifically as plant-growth regulators) block gibberellin (GA) biosynthesis in higher plants, thus leading to growth retardation (Shive & Sisler, 1976; Coolbaugh & Hamilton, 1976; Coolbaugh, Swanson & West, 1982*b*). Such anti-gibberellin-like activity, which sometimes proves to be an undesirable side effect in practice, has been observed for most sterol C-14 demethylation inhibitors. Whether all of the growth-retardant effects seen in higher plants are due to inhibition of GA biosynthesis, however, remains in question. It is likely that growth inhibition which is not reversible by application of exogenous GA is due to effects outside the GA biosynthetic pathway.

Imidazoles. Antifungal imidazoles are not only utilised for the control of plant fungal diseases, but are of major significance as drugs for the treatment of fungal infection in humans. Ergosterol biosynthesis in general and the sterol C-14 demethylation reaction in particular, display high sensitivity to the plant protectants in this group, which includes imazalil (Buchenauer, 1977*c*; Leroux & Gredt, 1978*a*; Siegel & Ragsdale, 1978) and prochloraz (Pappas & Fisher, 1979; Gadher *et al.*, 1983), and also to derivatives used medicinally which include miconazole (Van den Bossche *et al.*, 1978; Henry & Sisler, 1979; Marriott, 1980), ketoconazole (Marriott, 1980; Van den Bossche *et al.*, 1980; Sud & Feingold, 1981) and clotrimazole (Buchenauer, 1978*a*; Marriott, 1980). While the primary antifungal action of the aforementioned imidazole compounds is believed to be based on the inhibition of sterol C-14 demethylation, miconazole shows a somewhat less potent secondary mode of action (Sud & Feingold, 1981; Walsh & Sisler, 1982) which involves direct membrane damage (Sud & Feingold, 1981). This latter mechanism is not characteristic of EBI fungicides such as ketoconazole (Sud & Feingold, 1981), or fenarimol (Walsh & Sisler, 1982).

The fungicide, 1-dodecylimidazole is not used to control fungal diseases of either plants or animals, but is of interest as an EBI fungicide. This derivative is a potent inhibitor of sterol C-14 demethylation in *Ustilago maydis*, but is only slightly less effective as an inhibitor of 2,3-oxidosqualene cyclisation and possibly of sterol sidechain trans-methylation in the same organism (Henry & Sisler, 1979). Having only a simple C-12 aliphatic substituent, this compound lacks the biochemical specificity characteristic of the aforementioned antifungal imidazole derivatives.

Triazoles. The triazoles constitute the largest single group of EBI fungicides used for plant disease control. In addition to those shown in Fig. 12.1, triadimenol (a reduction product of triadimefon), propicona-zole, a derivative which differs from etaconazole only in having a propyl instead of an ethyl substituent on the dioxolan ring, and fluotrimazole are used in plant protection. Studies with triadimefon (Buchenauer, 1977*a*), triadimenol (Buchenauer, 1978*b*), etaconazole (Henry & Sisler, 1981; Ebert *et al.*, 1983), bitertanol (Kraus, 1979) and fluotrimazole (Buchenauer, 1978*a*) all have implicated sterol C-14 demethylation as the target site of antifungal action. Moreover, several experimental triazoles studied by Gadher and coworkers (1983) are potent inhibitors of sterol C-14 demethylation in a cell-free system from *Saccharomyces cerevisiae*.

Piperazines. Triforine (Fig. 12.1) is the only piperazine EBI fungicide in use as a practical pesticide. Patterns of sterol accumulation produced by triforine-treated cells of *Aspergillus fumigatus* (Sherald & Sisler, 1975) and *Neurospora crassa* (Sherald, Ragsdale & Sisler, 1973) indicate a specific blockage of sterol C-14 demethylation. Moreover, fungal mutants resistant to triarimol typically show resistance also to triforine (Sherald *et al.*, 1973). Triforine is generally less fungitoxic than other EBI fungicides in laboratory tests. It is ineffective at 50 μM as an inhibitor of sterol biosynthesis in a yeast cell-free system (Gadher *et al.*, 1983). The structure of triforine is unusual in comparison with those of other C-14 demethylation inhibitors in respect to the relationship of the *N*-heterocycle to the lipophilic substituents. Henry (1982) has suggested that triforine may act differently from other EBI fungicides as an inhibitor of cytochrome P-450.

Pyridines. Evidence derived from studies with whole cells and cell-free systems clearly place the pyridine derivative, buthiobate, in the category with sterol C-14 demethylation inhibitors (Kato *et al.*, 1974, 1975; Kato & Kawase, 1976). Structural characteristics of buthiobate as they influence fungitoxic activity are discussed by Tanaka *et al.* (1978).

The pyridine methanol derivative, EL241, resembles triarimol structurally and is fungitoxic to *Cladosporium cucumerinum* (Sherald *et al.*, 1973). Although the effect of this compound on sterol biosynthesis has not been ascertained, an action similar to triarimol is suggested by the fact that two triarimol-resistant mutants of *C. cucumerinum* are also resistant to EL241.

Inhibitors of sterol $\Delta^{14(15)}$ reduction and $\Delta^8 \to \Delta^7$ isomerisation

15-Azasterol antibiotic A25822B. This antibiotic (Fig. 12.2) produced by the fungus *Geotrichum flavo-brunneum*, is a potent antifungal agent. In both *Saccharomyces cerevisiae* (Hays *et al.*, 1977) and *Ustilago maydis* (Woloshuk, Sisler & Dutky, 1979), the compound selectively inhibits $\Delta^{14(15)}$-sterol reduction (which is believed to be the basis of toxicity). The azasterol is an extremely effective inhibitor of the Δ^{14}-sterol reductase with a K_i of about 2 nM (Bottema & Parks, 1978). Reduction of the $\Delta^{14(15)}$ bond is also blocked by the antibiotic in bramble cell cultures (Schmitt, Scheid & Benveniste, 1980).

Morpholines. Tridemorph (Fig. 12.2), fenpropimorph and dodemorph are fungicides of the morpholine group which are presumed to act by a common mechanism. Tridemorph inhibits reactions subsequent to the C-14 demethylation step in the sterol biosynthetic pathway. In *Ustilago maydis* it inhibits $\Delta^{14(15)}$-sterol double-bond reduction (Kerkenaar, Uchiyama & Versluis, 1981), whereas in *Botrytis cinerea* (Kato, Shoami & Kawase, 1980) and in *Saprolegnia ferax* (Berg *et al.*, 1983) and four other Oomycetes (Berg, 1983) which normally synthesise Δ^5 sterols, it blocks $\Delta^8 \to \Delta^7$ isomerisation. Bramble cells utilise the cycloartenol rather than the lanosterol pathway from squalene, and in these cells tridemorph blocks the 9β,19-cyclopropyl $\to \Delta^8$ isomerisation (cyclo-eucalenol–obtusifoliol isomerisation; Schmitt, Benveniste & Leroux,

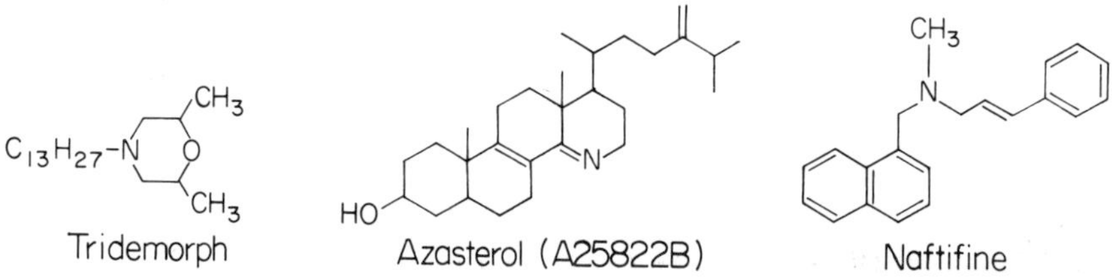

Fig. 12.2. Structures of EBI fungicides which block ergosterol biosynthesis at stages other than C-14 demethylation.

1981). The aforementioned blocks by tridemorph have been deducted from studies with whole cells, but should be verified by studies with isolated enzymes.

While several reports indicate that tridemorph inhibits fungal respiration, these results were probably obtained with the formulated product Calixin. Fungal respiration is not inhibited by purified tridemorph even at quite high concentrations (Kerkenaar & Kaars Sijpesteijn, 1979).

Trifluperidol. This compound is not used as a practical fungicide; however, it may be of interest that this hypocholesteraemic agent is an inhibitor of sterol $\Delta^8 \rightarrow \Delta^7$ isomerisation in cells of *Saccharomyces cerevisiae* (Sobus, Holmlund & Whittaker, 1977).

Squalene cyclase inhibitors

Although 1-dodecylimidazole is a strong inhibitor of sterol C-14 demethylation in *Ustilago maydis* at $0.25\,\mu g\,ml^{-1}$, at a slightly higher concentration $(1\,\mu g\,ml^{-1})$ it inhibits 2,3-oxidosqualene cyclisation (Henry & Sisler, 1979).

The hypocholesteraemic drug 3β-(β-dimethylamino ethoxy)-androst-5-en-17-one is also an inhibitor of 2,3-oxidosqualene cyclase in fungi (Field & Holmlund, 1977). In the presence of this inhibitor, *Saccharomyces cerevisiae* cells not only accumulate 2,3-oxidosqualene but also 2,3;22,23-dioxidosqualene. Thus squalene expoxidase can add oxygen across the 2,3-double bond of squalene and also across the 22,23-double bond of 2,3-oxidosqualene. Upon removal of the inhibitor, dioxidosqualene is cyclised to yield 24,25-oxidolanosterol (Field & Holmlund, 1977).

The compound 2,3-iminosqualene differs from 2,3-oxidosqualene in having an —N(H)— group instead of an oxygen atom linked to carbons 2 and 3 of squalene. It is a competitive inhibitor of the enzyme (2,3-oxidosqualene cyclase) which catalyses the conversion of 2,3-oxidosqualene to lanosterol and is an effective inhibitor of growth of *Saccharomyces cerevisiae* at 50–100 μM (Pinto & Nes, 1983).

Squalene epioxdiase inhibitor

Naftifine (Fig. 12.2) inhibits ergosterol synthesis in *Candida parapsilosis*, *Trichophyton mentagrophytes* and *Saccharomyces carlsbergensis* (Paltauf *et al.*, 1982). Cells of *C. parapsilosis* treated with

$30\,\mu g\,ml^{-1}$ naftifine accumulate large quantities of squalene which is associated with marked reduction in levels of ergosterol and strong inhibition of growth. Naftifine inhibits conversion of squalene to squalene epoxide in cell-free homogenates from *S. carlsbergensis* (Paltauf *et al.*, 1982). The evidence suggests a fungitoxic mechanism of naftifine based on inhibition of squalene epoxidase.

Effects on growth, morphology and biochemical activities
Growth

Typically EBI fungicides have little or no effect on fungal growth until after one or more cell doublings have occurred. Concentrations which are ultimately lethal or strongly inhibitory to hyphal growth often do not inhibit spore germination (Sherald *et al.*, 1973; Kato *et al.*, 1974, 1980; Buchenauer, 1979a; Sisler, Walsh & Ziogas, 1983). In *Ustilago maydis* (Ragsdale & Sisler, 1973) and *U. avenae* (Buchenauer, 1977a, 1978a, b; Kraus, 1979) only a moderate inhibition of dry-weight increase is evident even after 6–8 h of exposure to sterol C-14 demethylation inhibitors, whereas strong inhibition of sporidial multiplication is evident after 2–4 h. The delayed effects on growth are believed to be related to the time required for depletion of ergosterol levels of the cells. It is, in fact, difficult to completely inhibit growth of *U. maydis* sporidia with sterol C-14 demethylation inhibitors such as fenarimol or etaconazole (Walsh & Sisler, 1982). While strong inhibition of growth can be achieved with low concentrations of these inhibitors, a slow growth rate persists even at 50–100 fold higher concentrations. This phenomenon is undoubtedly explained by the fact that *U. maydis* sporidia survive and reproduce on the C-14 methyl sterols which accumulate when the pathway to ergosterol is blocked (Walsh & Sisler, 1982).

The ultimate effects of sterol C-14 demethylation inhibitors on growth are somewhat different in *Monilinia fructicola* and *Curvularia lunata*. Treatment of conidia with fungicides such as fenarimol and etaconazole, at concentrations which permit conidial germination or initial hyphal development, ultimately results in extrusion of cytoplasm from the cells and failure to form visible colonies (Sisler *et al.*, 1983).

In some fungal species, EBI fungicides do inhibit spore germination. Low concentrations of imazalil $(1.7–17 \times 10^{-8}\,\mathrm{M})$ delay or prevent germination of *Penicillium italicum* spores (Siegel, Kerkenaar & Kaars Sijpesteijn, 1977) while $1\,\mu g\,ml^{-1}$ etaconazole completely inhibits germination of *P. expansum* spores (Tepper & Yoder, 1982).

Morphology and cytology

Striking morphological abnormalities are produced in a number of fungal species by EBI fungicides. Treatment of *Ustilago maydis* sporidia in liquid culture with sterol C-14 demethylation inhibitors leads to the production of large sporidia that are multicelled and frequently branched (Ragsdale & Sisler, 1973; Walsh & Sisler, 1982). Sporidia fail to separate normally, suggesting a possible interference with septum formation. Similar morphological abnormalities are also produced in *U. maydis* by azasterol A25822B (Woloshuk *et al.*, 1979) and tridemorph (Kerkenaar, Barug & Kaars Sijpesteijn, 1979). Morphological effects like those produced by EBI fungicides in *U. maydis* are also produced by these inhibitors in *U. avenae* (Buchenauer, 1977*a*, 1978*a*, Kraus, 1979).

Sterol C-14 demethylation inhibitors cause distortion of germ tubes, often with excessive branching as observed in *Monilinia fructigena* (Kato *et al.*, 1974), *Monilinia fructicola* (Sisler *et al.*, 1983) and *Penicillium italicum* (Siegel *et al.*, 1977). Buthiobate and tridemorph both produce short, excessively branched germ tubes in *Botrytis cinerea* (Kato *et al.*, 1980). Morphological and cytological abnormalities of ergosterol-deficient cells resulting from fungicide treatment may be due mainly to modified activity of membrane-bound enzymes involved in wall synthesis.

Ultrastructural studies of *Ustilago avenae* sporidia grown for 8 h with nuarimol showed significant thickening of the cell wall and striking thickening of the septa (Hippe & Grossmann, 1982). Pores were present in septa of treated sporidia but were lacking in those of untreated sporidia. The plasmalemma of treated sporidia were characterised by invaginations, and extracytoplasmic vesicles were found to accumulate between the plasmalemma and cell wall. Cell-wall abnormalities are thought to arise from derangement of membrane structure and functions. In addition to effects on the cell wall and plasmalemma, nuarimol treatment resulted in increased sporidial size, decreased mitochondrial size, increased mitochondrial number, decreased nuclear size and increase in number of vacuoles and lipid droplets (Hippe & Grossmann, 1982). Ultrastructural changes similar to those described for nuarimol were produced by triadimefon in sporidia of *U. avenae* (Hippe, Buchenauer & Grossmann, 1980). Another study showed incomplete development of septa, local areas of wall thickening and accumulation of small vesicles in germ tubes of germinating spores of *Botrytis allii* treated with triadimefon (Richmond & Pring, 1979).

These effects on fungal walls most likely result from irregular deposition of wall material ràther than from specific inhibition of wall-polymer synthesis. This is indicated by the observation that buthiobate had little effect on incorporation of ^{14}C-glucose or ^{14}C-glucosamine into walls of *Monilinia fructigena* during a 2 h period following a 1 h incubation with the toxicant (Kato *et al.*, 1974). There was, in fact, an enhanced incorporation of ^{14}C-acetate into the wall fraction after incubation periods with buthiobate of more than 4 h.

Bellincampi and coworkers (1980) report increased mitotic nondisjunction resulting from treatment of germinated conidia of *Aspergillus nidulans* with fenarimol or miconazole. This effect is believed to be due to a membrane alteration which leads to a disturbance in the interaction of the spindle apparatus with the nuclear membrane.

Respiration

Studies of several sterol C-14 demethylation inhibitors in various fungi have shown that major respiratory activity is quite insensitive to toxic concentrations of these fungicides (Ragsdale & Sisler, 1973; Kato *et al.*, 1974, Sherald & Sisler, 1975; Buchenauer, 1977a; Siegel & Ragsdale, 1978; Kraus, 1979; DeWaard & Van Nistelrooy, 1980; Sisler, Ragsdale & Waterfield, 1984). Likewise, toxic concentrations of the EBI fungicide, tridemorph, did not inhibit respiration of five fungal species studied by Kerkenaar & Kaars Sijpesteijn (1979) or of *Botrytis cinerea* (Kato *et al.*, 1980). Both azide-antimycin A sensitive and insensitive respiration of *Ustilago maydis* sporidia are unaffected by $10 \,\mu g \, ml^{-1}$ fenarimol, a concentration appreciably higher than that ($2 \,\mu g \, ml^{-1}$) which strongly inhibits ergosterol biosynthesis in this fungus (Sisler *et al.*, 1984).

While it has been proposed that inhibition of mitochondrial electron transport in *Candida albicans* is the primary toxic mechanism of the sterol C-14 demethylation inhibitor, ketoconazole (Shigematsu, Uno & Arai, 1982; Uno, Shigematsu & Arai, 1982), the preponderance of evidence does not support a direct action on this pathway as the primary toxic mechanism for sterol C-14 demethylation inhibitors.

Protein, RNA and DNA syntheses

Appreciable data indicate that protein, RNA and DNA syntheses are not the primary targets of EBI fungicides. Sterol C-14 demethylation inhibitors have little or no effect on these processes initially, even though ergosterol synthesis is strongly inhibited

(Ragsdale & Sisler, 1973; Kato *et al.*, 1974; Sherald & Sisler, 1975; Buchenauer, 1977*a*; Siegel & Ragsdale, 1978; Kraus, 1979; DeWaard & Van Nistelrooy, 1979; Kato *et al.*, 1980). Likewise, tridemorph has little or no effect on protein or RNA synthesis in either *Botrytis cinerea* or *Ustilago maydis* at concentrations which strongly inhibit ergosterol biosynthesis (Kato *et al.*, 1980; Kerkenaar *et al.*, 1979). However, there is about 50% inhibition of DNA synthesis in the latter organism within 2 h of treatment with tridemorph. A concentration of $0.1\,\mu\mathrm{g\,ml}^{-1}$ of azasterol A25822B inhibits all net ergosterol synthesis in *U. maydis* within 3 h, but has no effect on RNA synthesis and only minor effects on DNA and protein synthesis during this period (Woloshuk *et al.*, 1979).

Lipids other than sterols

Various studies have shown that EBI fungicides block specific reactions in sterol biosynthesis without appreciably inhibiting the synthesis of other major lipid fractions (Ragsdale & Sisler, 1973; Kato *et al.*, 1974; Sherald & Sisler, 1975; Buchenauer, 1977*a*; 1978*a*; Kraus, 1979). A delayed accumulation of free fatty acids, which is apparently due to an ergosterol deficiency, characteristically results from the treatment of *Ustilago maydis* sporidia with sterol C-14 demethylation inhibitors (Ragsdale, 1975; Henry & Sisler, 1981). These free fatty acids are believed to be derived both from a breakdown of cell membranes and triglycerides and a disproportionality between the synthesis and utilisation of fatty acids (Sisler & Ragsdale, 1977).

Investigation of the qualitative effect of $2\,\mu\mathrm{g\,ml}^{-1}$ triarimol on fatty acids in *Ustilago maydis* showed that the percentage of 18:0 and 18:1 acids decreased while the percentage of 18:2 acids increased in free fatty acids, di- and triglycerides (Ragsdale, 1975). Triadimenol had the same effect on 18:1 and 18:2 fatty acids in *U. avenae* (Buchenauer, 1978*b*). Although the relative proportions of different fatty acids in three species of *Chlorella* were shifted by triarimol treatment, the total quantity was increased in all three species, especially the polyunsaturated acids (Frasinel *et al.*, 1978). These data indicate that triarimol does not inhibit the fatty acid desaturation system in these organisms.

Miconazole at concentrations greater than $10^{-6}\,\mathrm{M}$ interferes with the synthesis of a C-55 isoprenoid alcohol and of 2-methyl-3-prenyl-1,4-naphthoquinones (Vitamin Ks) in *Staphylococcus aureus* (Van den Bossche, Lauwers & Cornelissen, 1982). Whether EBI fungicides interfere with the synthesis of these classes of lipids in fungi at concentrations which block ergosterol biosynthesis has not been determined.

Sterol C-14 demethylation inhibitors block gibberellin biosynthesis in both higher plants and in the fungus *Gibberella fujikuroi*. Detailed investigations of this inhibition have been carried out primarily with pyrimidine methanols. These compounds block GA biosynthesis at the cytochrome P-450 dependent oxygenation of kaurene C-19 (Coolbaugh & Hamilton, 1976; Coolbaugh, Hirano & West, 1978; Coolbaugh *et al.*, 1982*b*; Coolbaugh, Heil & West, 1982*a*). The sensitive conversion in GA biosynthesis therefore is analogous to cytochrome P-450 dependent sterol C-14 demethylation in ergosterol biosynthesis.

Pyrimidine methanol derivatives regarded as growth regulators are considerably more effective as inhibitors of kaurene oxidation by microsomal systems of higher plants than are the fungicidal analogues (Coolbaugh *et al.*, 1982*a*). However, the reverse order of activity is found when the analogues are tested as inhibitors of kaurene oxidation by microsomes of the fungus *Gibberella fujikuroi* (Coolbaugh *et al.*, 1982*a*). The difference in sensitivity of the system from the fungus and the higher plant apparently reflect differences in the protein and the lipid environment surrounding the cytochrome P-450 haem group.

In the fungus, *Gibberella fujikuroi*, sensitivity of growth and GA biosynthesis to pyrimidine methanols is similar (Shive & Sisler, 1976; Coolbaugh *et al.*, 1982*a*). Since GA is not required for growth of the fungus, fungitoxicity evidently results from action at a site outside the GA-biosynthetic pathway. If fungitoxicity results from inhibition of sterol C-14 demethylation, then the cytochrome P-450 in the sterol biosynthetic pathway must be the same enzyme or at least one of similar sensitivity as that involved in kaurene oxidation.

Sterols

Sterol C-14 demethylation inhibitors. Treatment of *Ustilago maydis* sporidia (Ragsdale, 1975; Henry & Sisler, 1979, 1981; Ebert *et al.*, 1983) or *Aspergillus fumigatus* cells (Sherald & Sisler, 1975) with a sterol C-14 demethylation inhibitor results in the accumulation of eburicol, obtusifoliol and 14 α-methylfecosterol (sterols b, c and d, respectively, Fig. 12.3). Significant quantities of an unusual 6-hydroxylated sterol (14-α-methyl-ergosta-8,24(28)-dien-3β,6α-diol) also accumulate in *U. maydis* after incubation periods of 13 h or longer with etaconazole (Ebert *et al.*, 1983). With the accumulation of abnormal levels of precursor sterols in fungicide-treated cells, the total sterol level

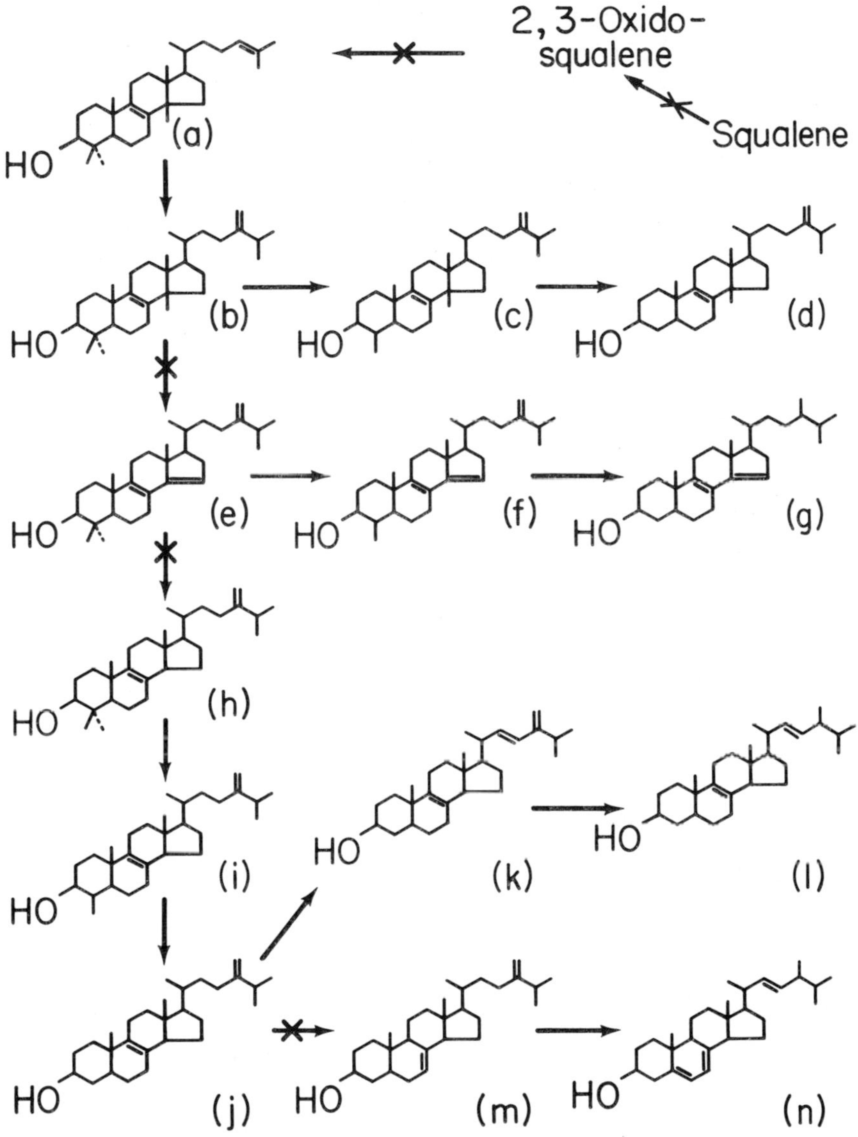

Fig. 12.3. Biosynthetic pathway of ergosterol indicating sites blocked by compounds discussed in text. Sterols (a) lanosterol, (b) eburicol, (c) obtusifoliol, (d) 14α-methylfecosterol, (e) 4,4-dimethylergosta-8,14,24(28)-triene-3β-ol, (f) 4α-methylergosta-8,14,24(28)trien-3β-ol, (g) ignosterol, (h) 4,4-dimethyl-8,24(28)-dien-3β-ol, (i) 4α-methylergosta-8,24(28)-dien-3β-ol, (j) fecosterol, (k) ergosta-8,22,24(28)trien-3β-ol, (l) ergosta-8,22-dien-3β-ol, (m) episterol, (n) ergosterol.

may equal or exceed that normally found in untreated cells. In *A. fumigatus* for example, the content of sterols in cells exposed to triarimol or triforine for 4 h is about twice that found in control cells on a dry-weight basis (Sherald & Sisler, 1975). The excess sterol is due primarily to large accumulations of eburicol and obtusfoliol (b and c, Fig. 12.3). A similar high sterol content consisting of C-14 methyl sterols occurs in a *U. maydis* mutant (Walsh & Sisler, 1982) and a mutant of *Candida albicans* (Pierce *et al.*, 1978), both of which are genetically deficient in sterol C-14 demethylation. This excess sterol production suggests that a lack of ergosterol or a similar desmethylsterol results in deregulation of sterol biosynthesis.

Accumulation of methyl sterols in the presence of sterol C-14 demethylation inhibitors has also been observed in *Monilinia fructigena* (Kato *et al.*, 1975), in *Botrytis cinerea* (Kato *et al.*, 1980), in *Ustilago avenae* (Buchenauer, 1977a, 1978a, b; Kraus, 1979) and in *Penicillium expansum* (Leroux & Gredt, 1978a).

The medicinal drugs miconazole and ketoconazole also cause accumulation of methyl sterols similar to those described in the foregoing discussion in the human pathogenic yeast, *Candida albicans* (Van den Bossche *et al.*, 1978, 1980). Likewise, four Oomycete species that normally contain primarily desmethylsterols accumulate almost exclusively the methyl sterol lanosterol (sterol a, Fig. 12.3), when grown in the presence of $10 \mu g \, ml^{-1}$ triarimol (Berg, 1983).

Sterol C-14 demethylation involves three NADPH dependent oxygenase reactions in the conversion of the 14α-methyl group from $-CH_3$ through $-CH_2OH$ to $-CHO$ and removal as HCOOH (Gibbons, Pullinger & Mitropoulos, 1979). The first oxygenation step in this sequence is catalysed by a cytochrome P-450 enzyme (Gibbons *et al.*, 1979; Gadher *et al.*, 1983). Since no accumulation of oxygenated intermediates in the demethylation process occurs in treated fungi, this step appears to be the primary target of EBI fungicides blocking sterol C-14 demethylation.

There is spectral evidence that sterol C-14 demethylation inhibitors interact with cytochrome P-450 enzymes. Type II difference spectra are produced when triazole EBI fungicides (Gadher *et al.*, 1983) or prochloraz (Rivière, 1983) interact with cytochrome P-450 of rat liver microsomes. Such spectra are indicative of direct binding of the fungicides to the haem moiety of the cytochrome. A similar Type II difference spectrum was observed for binding of ancymidol, a fenarimol analogue, to microsomes of the higher plant *Marah macrocarpus*

(Coolbaugh *et al.*, 1978). Prochloraz exhibits a high affinity (spectral dissociation constant of $1 \mu M$) for the cytochrome P-450 induced by feeding the fungicide to rats (Rivière, 1983). As mentioned earlier, binding of EBI fungicides to cytochrome P-450 apparently involves not only an interaction of the fungicide with the haem iron atom of cytochrome P-450 but also an interaction with nearby regions of the enzyme.

While attention has been focused mainly on the effect of sterol C-14 demethylation inhibitors in preventing removal of the C-14 methyl group, there is evidence that some of these inhibitors may also block insertion of the C-22 double bond in the sterol side chain. This point was mentioned earlier and is discussed by Sisler and coworkers (1984). There is no evidence that other steps except possibly the C-24(28) double bond reduction is affected by triarimol in *Ustilago maydis* (Ragsdale, 1975). Sterol C-4 demethylation reactions are apparently quite insensitive to these inhibitors and proceed in *U. maydis* even though C-14 demethylation is blocked (Ragsdale, 1975). Kato & Kawase (1976) have demonstrated that sterol C-4 demethylation in cell-free systems of *Saccharomyces cerevisiae* is not blocked by triarimol and buthiobate at concentrations which block sterol C-14 demethylation.

In other studies it was shown, for example, that $120 \mu M$ triarimol inhibits C-14 demethylation activity more than 90% in rat liver subcellular systems while HmG-Co A reductase and the introduction of the sterol C5(6) double bond are uninhibited or only slightly affected (Mitropoulos *et al.*, 1976).

Tridemorph. Treatment of *Ustilago maydis* sporidia with tridemorph leads to the accumulation of ignosterol (sterol g, Fig. 12.3), indicating a block of Δ^{14}-sterol reductase (Kerkenaar *et al.*, 1981). However, in tridemorph-treated cells of *Botrytis cinerea*, sterols j, k & l (Fig. 12.3) accumulate indicating a block of $\Delta^8 \rightarrow \Delta^7$ isomerisation (Kato *et al.*, 1980). Cultures of *Saprolegnia ferax*, *Achlya americana*, *Dictyuchus monosporous* and *Apodachlyella completa* grown with $10 \mu g \, ml^{-1}$ tridemorph contain almost exclusively Δ^8 desmethyl sterols; whereas, untreated cultures contain almost exclusively Δ^5 desmethyl sterols. Since $\Delta^8 \rightarrow \Delta^7$ isomerisation is an intermediate step in the synthesis of Δ^5 sterols, the results of these studies support an action of tridemorph on this step (Berg, 1983). The striking contrast between the sterol content of these four species when grown on triarimol or tridemorph is interesting. Tridemorph-grown cells contain primarily Δ^8-desmethyl sterols; whereas, triarimol-grown cells contained almost exclusively Δ^8-methyl sterols (Berg, 1983).

Azasterol A25822B. In both *Saccharomyces cerevisiae* (Hays *et al.*, 1977) and *Ustilago maydis* (Woloshuk *et al.*, 1979) this antibiotic blocks $\Delta^{14(15)}$ reduction, leading to accumulation of the sterol intermediate ignosterol (sterol g, Fig. 12.3), which closely resembles the antibiotic in structure (Fig. 12.2). As noted earlier, the antibiotic is an extremely potent inhibitor of Δ^{14} sterol reductase (Bottema & Parks, 1978).

Resistance

While fungal strains resistant to EBI fungicides are readily produced in the laboratory, such strains have thus far not proven to be a serious problem in the control of diseases in the field. This may be due to the fact that EBI-resistant mutants usually have reduced fitness in regard to spore germination, germ-tube elongation, mycelial growth and sporulation (Fuchs & DeWaard, 1982). DeWaard & Fuchs (1982) speculate that perhaps decreased-fitness factors are pleiotropic and coupled with the genes for resistance. As evidence they point out that the energy-dependent efflux mechanism, which is a factor in the resistance of *Aspergillus nidulans*, is fuelled at the expense of other cellular processes. DeWaard & Van Nistelrooy (1983) showed that most of their fenarimol-resistant isolates of various fungal species displayed increased sensitivity to dodine, and a limited number showed increased sensitivity to guazatine, which is chemically related to dodine. Fenarimol-resistant *Penicillium italicum* isolates, which showed cross resistance to imidazole and triazole fungicides, showed normal or increased sensitivity to fenpropimorph (DeWaard & Van Nistelrooy, 1982*a*). In the event of field resistance, this sort of information could be quite important from a practical standpoint.

The genetics of imazalil resistance was studied in *Aspergillus nidulans* by Van Tuyl (1977). Eight loci for resistance, distributed among six different linkage groups, were identified in an analysis of 21 imazalil-resistant mutants. Two cycloheximide-resistant mutants were also tolerant to imazalil. This study, therefore, showed that at least ten genes are involved in imazalil resistance. While these data might be interpreted to mean that imazalil acts at multiple, independent sites, a more likely explanation is that a multitude of cellular changes may affect its action at a single site. For example, a mutation which affects phospholipid composition might change affinity of membrane-bound cytochrome P-450 for imazalil; another mutation might alter the P-450 protein structure, and still another the energy-dependent fungicide-efflux system described by DeWaard & Van Nistelrooy (1980).

The mechanism of fungicide resistance may vary in different fungi. One of particular interest occurring in *Aspergillus nidulans* is based on an energy-dependent efflux of EBI fungicides (DeWaard & Van Nistelrooy, 1980, 1981). Uptake of fenarimol by the wild-type strain was shown to be a passive influx which then induced an energy-dependent efflux, resulting in an energy-dependent state of equilibrium. In resistant strains the efflux system is constitutive and therefore, efflux activity results in a much lower initial uptake of the fungicide. Studies by Siegel & Solel (1981) in wild-type and resistant strains of *A. nidulans* support the idea that an influx–efflux mechanism affects toxicity of the EBI fungicide, imazalil.

Various factors, such as respiratory inhibitors which decrease ATP levels, inhibit efflux activity. The influx-efflux mechanism apparently depends on intracellular ATP which, in turn, is utilised by the plasma membrane ATPase, which drives the efflux process (DeWaard & Van Nistelrooy, 1980). This is supported by the fact that N,N'-dicyclohexyl-carbodiimide, an inhibitor of plasma membrane ATPase in *Neurospora crassa*, fully inhibits fenarimol efflux by *Aspergillus nidulans*.

In some instances, resistance may be related to uptake and metabolism of the fungicide by the host plant as well as by the fungus. Therefore, alteration of a variety of factors could affect the performance of the fungicide. In a number of non-susceptible fungal species, it was shown that triforine is either not taken up or is metabolised shortly after uptake (Gasztonyi & Josepovits, 1975). Later it was shown that plant homogenates enhance triforine activity by increasing water solubility and decreasing chemical degradation (Gasztonyi & Josepovits, 1976). In the case of triadimefon, a somewhat complex situation emerges. Gasztonyi & Josepovits (1979) demonstrated that fungi sensitive to triadimefon transformed the compound to triadimenol at a high rate; whereas, species that were resistant showed little or no transformation. In addition, this study showed that plants also carry out this reaction, thus enhancing the fungitoxicity of triadimefon. Buchenauer (1979b) has shown that fungal sensitivity is not only related to the reduction of triadimefon to triadimenol, but is also affected by the production of the more active diastereomeric form. Further research on triazolylmethane fungicides by Deas & Clifford (1982) suggests that in considering the effectiveness of triadimefon applied to plants for pest control, one must take into account the rate and extent of conversion to triadimenol by the host and pathogen, inherent activities of the chemical species present, the time for an effective dose to reach the intracellular site of action, and further metabolism of the chemical by the host or pathogen to compounds more or less toxic than their precursors. A change in any one of these factors could alter the pest-control situation.

Synergism and antagonism

Although the primary action of EBI fungicides is considered to be the inhibition of ergosterol synthesis, there is no case in which the addition of ergosterol to treated cultures fully restores growth. In most situations there is a complete lack of antagonistic activity. In *Botrytis cinerea*, imazalil and triarimol toxicity is antagonised but normal growth is not restored (Leroux, Gredt & Fritz, 1976). Ergosterol mitigates toxicity of triforine to *Cladosporium cucumerinum* but has no effect on triarimol-treated cultures of this organism (Sherald *et al.*, 1973). Ergosterol also does not alleviate toxicity of triarimol to *Ustilago maydis* (Ragsdale & Sisler, 1972) or of tridemorph to *U. maydis* and *B. allii* (Kerkenaar *et al.*, 1979).

Many lipophilic compounds have been shown to alleviate toxicity. However, there is variation between organisms. Squalene, farnesol, progesterone, testosterone, β-carotene and vitamin A antagonise triarimol toxicity in *Cladosporium cucumerinum* but not in *Ustilago maydis* (Sherald *et al.*, 1973). With the exception of squalene, many of these compounds are ineffective in relieving tridemorph toxicity in *U. maydis* (Kerkenaar *et al.*, 1979). Tweens 20 and 40 have been generally effective in reducing the toxic effects of triforine, triarimol, triadimefon and imazalil in a number of organisms (Sisler *et al.*, 1984). Oleic acid and several other fatty acids have also given partial reversal of toxicity in a number of studies (Sherald *et al.*, 1973; Leroux *et al.*, 1976; Kerkenaar *et al.*, 1979).

DeWaard & Van Nistelrooy (1982*b*) carried out extensive tests to determine antagonistic or synergistic effects of chemicals on the toxicity of fenarimol to sensitive and resistant strains of *Aspergillus nidulans*. They observed antagonism by $CaCl_2$, $MgCl_2$, several non-ionic detergents, phospholipids, triglycerides, carboxin and dialkyldithiocarbamate fungicides. It was suggested that antagonism may, in some cases, result from complex formation or partitioning of the fungicide into undissolved residues of the antagonist. The variation in effectiveness of certain compounds on different fungi may be due to differences in uptake and accumulation of the antagonist. Other possible mechanisms of antagonism include competition with the toxicant and effects on membranes which reduce permeability to the toxicant or decrease toxicant affinity for the membrane-bound target enzyme. Synergists included HCl, NaOH, cationic and anionic surfactants, certain respiratory inhibitors and the fungicides chlorothalonil and folpet. DeWaard & Van Nistelrooy (1982b) postulate that synergism may be due to an increase in solubility of fenarimol in the medium, synergist accumulation in membranes stimulating uptake of the fungicide, or inhibition of an energy-dependent efflux mechanism which serves to reduce intracellular fungicide concentration.

Discussion

Although various investigations have demonstrated high sensitivity of ergosterol biosynthesis to EBI fungicides, a question still remains as to whether this inhibition constitutes the sole basis of fungitoxicity. In the case of the sterol C-14 demethylation inhibitors, the close similarity of the sterol C-14-demethylation-deficient mutant of *Ustilago maydis* to the inhibitor-treated wild type and the insensitivity of the mutant to these fungicides (Walsh & Sisler, 1982) indicate that inhibition of sterol C-14 demethylation is responsible for the observed effects on growth. Moreover, such a fungitoxic mechanism is suggested by the correlation between effectiveness of this type of EBI fungicide as inhibitor of desmethyl sterol synthesis in a yeast cell-free system and toxicity to growth of *U. maydis* (Gadher *et al.*, 1983). In view of this evidence and the high sensitivity of sterol C-14 demethylation to these fungicides, it seems likely that the cytochrome P-450 functioning in this system is the primary target of fungitoxic action. However, if sites of high sensitivity should exist in other pathways, they also are probably cytochrome P-450 dependent reactions.

In regard to azasterol A25822B, both the high sensitivity of Δ^{14} sterol reductase and the structural similarity of the substrate to the inhibitor are indications that the enzyme is the primary target of fungitoxicity.

The consequences of an ergosterol deficiency and accumulation of abnormal amounts of alternate sterols produced by EBI fungicides is difficult to assess. In view of the fact that ergosterol is a membrane component, a modification of membrane-permeability properties or activity of membrane-bound enzymes would be anticipated. Since inhibition results more in a qualitative than a quantitative change in sterols, the consequences may not be as drastic as the lethal effects expected from a total lack of sterols. Nevertheless, there are questions concerning the suitability of alternate sterols as substitutes for ergosterol and about their possible regulatory effects on other pathways.

Since *Ustilago maydis* mutant *erg-40* (Walsh & Sisler, 1982), certain mutants of *Candida albicans* (Pierce *et al.*, 1978) and *Saccharomyces cerevisiae* (Trocha, Jasne & Sprinson, 1977), which synthesise only C-14 methyl sterols, survive and reproduce, such sterols must be functional in these budding fungi even though they support only a slow, abnormal growth. This is apparently not the case in filamentous fungi such as *Monilinia fructicola* and *Curvularia lunata*, which extrude cytoplasm and fail to form visable colonies when treated with fenarimol (Sisler *et al.*, 1983). If the ergosterol-biosynthetic pathway is the primary target of fungitoxicity of the sterol C-14 demethylation inhibitors, and of azasterol A25822B and tridemorph, then it is evident that failure to make any of several transformations leads to the accumulation of intermediate sterols which are inadequate for survival or effective pathogenic activity under natural conditions. These transformations include removal of the C-14 methyl group, reduction of the C-14(15) double bond or $\Delta^8 \rightarrow \Delta^7$ isomerisation and insertion of the C-5(6) double bond.

Failure of exogenous ergosterol to alleviate toxicity of EBI fungicides is puzzling and suggests that inhibition of ergosterol biosynthesis alone is not totally responsible for fungitoxic action. On the other hand, failure to alleviate toxicity could be due to an interference of the accumulating sterol intermediates with utilisation of ergosterol or to a regulatory action of these sterols which is not reversible by ergosterol.

There is suggestive evidence that modification of the activity of membrane transport systems or membrane-bound enzymes occurs with the shift in sterol composition that follows treatment of cells with EBI fungicides. Evidence of a declining rate of uridine transport was noted in triarimol-treated sporidia of *Ustilago maydis* (Ragsdale & Sisler, 1973). Incomplete septa, cell-wall thickening, cytoplasmic extrusion and the striking morphological changes which result from EBI treatment may be due to irregularities in the activation of plasma membrane enzymes involved in wall synthesis as a consequence of changes in sterols. These and other effects based on membrane changes could account for a lethal or at least highly detrimental action on fungal cells.

Addendum

After completion of this manuscript, an article of significant interest by Taylor, Rodriguez & Parks (1983) came to our attention. In their study, antifungal activity of a 15-azasterol (Fig. 12.2), miconazole and clotrimazole was determined in a wild type, a sterol auxotropic mutant (blocked in 2,3-oxidosqualene cyclisation) and a C-14 demethylase-Δ^5-desaturase double deficient mutant of *Saccharomyces cerevisiae*. The azasterol, which blocks sterol C14(15) double-bond reduction, inhibited growth of the wild type *S. cerevisiae* but not of the double mutant or of the auxotropic strain grown on medium supplemented with ergosterol. On the other hand, the sterol C-14 demethylation inhibitors miconazole and clotrimazole were as toxic to growth of these mutants as to the wild type. These data suggest that toxicity of the azasterol is due specifically to the blocking of the sterol biosynthetic pathway, but that toxicity of miconazole and clotrimazole is not.

In these tests, appreciable toxicity of miconazole was evident only at 100 μM or higher. These concentrations produce direct membrane damage which is lethal to cells of *Saccharomyces cerevisiae* (Sud & Feingold, 1981). It seems doubtful whether this low-potency mechanism can, in general, account for the practical success of sterol C-14 demethylation inhibitors as antifungal agents.

The data of Taylor and coworkers (1983) indicate that a deficiency in sterol C-14 demethylation does not have much effect on growth of *Saccharomyces cerevisiae*. However, in *Ustilago maydis*, such a deficiency considerably reduces the growth rate and markedly alters sporidial morphology while decreasing sensitivity of the fungus to sterol C-14 demethylation inhibitors (Walsh & Sisler, 1982).

References

Bellincampi, D., Gualandi, G., La Monica, E., Poley, C. & Morpurgo, G. P. (1980). Membrane-damaging agents cause mitotic non-disjunction in *Aspergillus nidulans*. *Mutation Research*, **79**, 169–72.

Berg, L. R. (1983). The effects of triarimol and tridemorph on sterol biosynthesis of five species of Oomycetes. PhD thesis, University of Maryland.

Berg, L. R., Patterson, G. W. & Lusby, W. R. (1983). Effect of triarimol and tridemorph on sterol biosynthesis in *Saprolegnia ferax*. *Lipids*, **18**, 448–52.

Bottema, C. K. & Parks, L. W. (1978). Δ^{14}-sterol reductase in *Saccharomyces cerevisiae*. *Biochimica et Biophysica Acta*, **531**, 301–7.

Buchenauer, H. (1977a). Mode of action of triadimefon in *Ustilago avenae*. *Pesticide Biochemistry and Physiology*, **7**, 309–20.

Buchenauer, H. (1977*b*). Biochemical effects of the systemic fungicides fenarimol and nuarimol in *Ustilago avenae*. *Zeitschrift für Pflanzenkrankheiten und Pflanzenschutz*, **84**, 286–99.

Buchenauer, H. (1977*c*). Mechanism of action of the fungicide imazalil in *Ustilago avenae*. *Zeitschrift für Pflanzenkrankheiten und Pflanzenschutz*, **84**, 440–50.

Buchenauer, H. (1978*a*). Analogy in the mode of action of fluotrimazole and clotrimazole in *Ustilago avenae*. *Pesticide Biochemistry and Physiology*, **8**, 15–25.

Buchenauer, H. (1978*b*). Inhibition of ergosterol biosynthesis by triadimenol in *Ustilago avenae*. *Pesticide Science*, **9**, 507–12.

Buchenauer, H. (1979*a*). Comparative studies on the antifungal activity of triadimefon, triadimenol, fenarimol, nuarimol, imazalil and fluotrimazole. *Zeitschrift für Pflanzenkrankheiten und Pflanzenschutz*, **86**, 341–54.

Buchenauer, H. (1979*b*). Conversion of triadimefon into two diastereomers, triadimenol-I and triadimenol-II, by fungi and plants. *Abstracts of Papers, IX International Congress of Plant Protection*, No. 939. St Paul, Minn.: American Phytopathological Society.

Coolbaugh, R. C. & Hamilton, R. (1976). Inhibition of *ent*-kaurene oxidation and growth by α-cyclopropyl-α-(*p*-methoxyphenyl)-5-pyrimidine methyl alcohol. *Plant Physiology*, **57**, 245–8.

Coolbaugh, R. C., Heil, D. R. & West, C. A. (1982*a*). Comparative effects of substituted pyrimidines on growth and gibberellin biosynthesis in *Gibberella fujikuroi*. *Plant Physiology*, **69**, 712–16.

Coolbaugh, R. C., Hirano, S. S. & West, C. A. (1978). Studies on the specificity and site of action of ancymidol, a plant growth regulator. *Plant Physiology*, **62**, 571–6.

Coolbaugh, R. C., Swanson, D. I. & West, C. A. (1982*b*). Comparative effects of ancymidol and its analogs on growth of peas and *ent*-kaurene oxidation in cell-free extracts of immature *Marah macrocarpus* endosperm. *Plant Physiology*, **69**, 707–11.

Deas, A. H. B. & Clifford, D. R. (1982). Metabolism of the 1,2,4-triazolyl-methane fungicides, triadimefon, triadimenol, and diclobutrazol, by *Aspergillus niger* (Van Tiegh). *Pesticide Biochemistry and Physiology*, **17**, 120–33.

DeWaard, M. A. & Fuchs, A. (1982). Resistance to ergosterol biosynthesis inhibitors. II. Genetic and physiological aspects. *Fungicide resistance in crop protection*, ed. J. Dekker & S. G. Georgopoulos, pp. 87–100. Wageningen,: Pudoc.

DeWaard, M. A. & Van Nistelrooy, J. G. M. (1979). Mechanism of resistance to fenarimol in *Aspergillus nidulans*. *Pesticide Biochemistry and Physiology*, **10**, 219–29.

DeWaard, M. A. & Van Nistelrooy, J. G. M. (1980). An energy-dependent efflux mechanism for fenarimol in a wild-type strain and fenarimol-resistant mutants of *Aspergillus nidulans*. *Pesticide Biochemistry and Physiology*, **13**, 255–66.

DeWaard, M. A. & Van Nistelrooy, J. G. M. (1981). Induction of fenarimol-efflux activity in *Aspergillus nidulans* by fungicides inhibiting sterol biosynthesis. *Journal of General Microbiology*, **126**, 483–9.

DeWaard, M. A. & Van Nistelrooy, J. G. M. (1982*a*). Toxicity of fenpropimorph to fenarimol-resistant isolates of *Penicillium italicum*. *Netherlands Journal of Plant Pathology*, **88**, 231–6.

DeWaard, M. A. & Van Nistelrooy, J. G. M. (1982*b*). Antagonistic and synergistic activities of various chemicals on the toxicity of fenarimol to *Aspergillus nidulans*. *Pesticide Science*, **13**, 279–86.

DeWaard, M. A. & Van Nistelrooy, J. G. M. (1983). Negatively correlated cross-resistance to dodine in fenarimol-resistant isolates of various fungi. *Netherlands Journal of Plant Pathology*, **89**, 67–73.

Ebert, E., Gaudin, J., Muecke, W., Ramsteiner, K., Christian, V. & Fuhrer, H. (1983). Inhibition of ergosterol biosynthesis by etaconazole in *Ustilago maydis*. *Zeitschrift für Naturforschung*, **38**, 28–34.

Field, R. B. & Holmlund, C. E. (1977). Isolation of 2,3;22,23-dioxidosqualene and 24,25-oxidolanosterol from yeast. *Archives of Biochemistry and Biophysics*, **180**, 465–71.

Frasinel, C., Patterson, G. W. & Dutky, S. R. (1978). Effect of triarimol on sterol and fatty acid composition of three species of *Chlorella*. *Phytochemistry*, **17**, 1567–70.

Fuchs, A. & DeWaard, M. A. (1982). Resistance to ergosterol biosynthesis inhibitors. I. Chemistry and phenomenological aspects. *Fungicide resistance in crop protection*, ed. J. Dekker & S. G. Georgopoulos, pp. 71–86. Wageningen: Pudoc.

Gadher, P., Mercer, E. I., Baldwin, B. C. & Wiggins, T. E. (1983). A comparison of the potency of some fungicides as inhibitors of sterol 14-demethylation. *Pesticide Biochemistry and Physiology*, **19**, 1–10.

Gasztonyi, M. & Josepovits, G. (1975). Biochemical and chemical factors of the selective antifungal effect of triforine. I. The causes of selectivity of the contact fungicidal action. *Acta Phytopathologica Academiae Scientiarum Hungaricae*, **10**, 437–46.

Gasztonyi, M. & Josepovits, G. (1976). Biochemical and chemical factors of the selective antifungal effect of triforine. III. The role of the host plant in the selectivity of systemic action. *Acta Phytopathologica Academiae Scientiarum Hungaricae*, **11**, 141–6.

Gasztonyi, M. & Josepovits, G. (1979). The activation of triadimefon and its role in the selectivity of fungicide action. *Pesticide Science*, **10**, 57–65.

Gibbons, G. F., Pullinger, C. R. & Mitropoulos, K. A. (1979). Studies on the mechanism of lanosterol 14α-demethylation; a requirement for two distinct types of mixed-function oxidase systems. *Biochemical Journal*, **183**, 309–15.

Hata, S., Nishino, T., Komori, M. & Katsuki, H. (1981). Involvement of cytochrome P-450 in Δ^{22}-desaturation in ergosterol biosynthesis of yeast. *Biochemical and Biophysical Research Communications*, **103**, 272–7.

Hays, P. R., Parks, L. W., Pierce, H. D. & Oehlschlager, A. C. (1977). Accumulation of ergosta-8,14-dien-3β-ol by *Saccharomyces cerevisiae* cultured with an azasterol antimycotic agent. *Lipids*, **12**, 666–8.

Henry, M. J. (1982). Effect of sterol biosynthesis inhibiting (SBI) fungicides on cytochrome P-450 oxygenations in fungi. PhD thesis, University of Maryland.

Henry, M. J. & Sisler, H. D. (1979). Effects of miconazole and dodecylimidazole on sterol biosynthesis in *Ustilago maydis*. *Antimicrobial Agents and Chemotherapy*, **15**, 603–7.

Henry, M. J. & Sisler, H. D. (1981). Inhibition of ergosterol biosynthesis in *Ustilago maydis* by the fungicide 1-[2-(2,4-dichlorophenyl)-4-ethyl-1,3-dioxolan-2-ylmethyl]-1H-1,2,4-triazole. *Pesticide Science*, **12**, 98–102.

Hippe, S., Buchenauer, H. & Grossmann, F. (1980). Einfluss von Triadimefon auf die Feinstruktur der sporidien von *Ustilago avenae*. *Zeitschrift für Pflanzenkrankheit und Pflanzenschutz*, **87**, 423–6.

Hippe, S. & Grossmann, F. (1982). The ultrastructure of sporidia of *Ustilago avenae* after treatment with the fungicides nuarimol and imazalil nitrate. *Pesticide Science*, **13**, 447–51.

Kato, T. & Kawase, Y. (1976). Selective inhibition of the demethylation at C-14 in ergosterol biosynthesis by the fungicide, Denmert (S-1358). *Agricultural and Biological Chemistry*, **40**, 2379–88.

Kato, T., Shoami, M. & Kawase, Y. (1980). Comparison of tridemorph with buthiobate in antifungal mode of action. *Journal of Pesticide Science*, **5**, 69–79.

Kato, T., Tanaka, S., Ueda, M. & Kawase, Y. (1974). Effects of the fungicide, S-1358, on general metabolism and lipid biosynthesis in *Monilinia fructigena*. *Agricultural and Biological Chemistry*, **38**, 2377–84.

Kato, T., Tanaka, S. Ueda, M. & Kawase, Y. (1975). Inhibition of sterol biosynthesis in *Monilinia fructigena* by the fungicide S-1358. *Agricultural and Biological Chemistry*, **39**, 169–74.

Kerkenaar, A., Barug, D. & Kaars Sijpesteijn, A. (1979). On the antifungal mode of action of tridemorph. *Pesticide Biochemistry and Physiology*, **12**, 195–204.

Kerkenaar, A. & Kaars Sijpesteijn, A. (1979). On a difference in the antifungal activity of tridemorph and its formulated product Calixin. *Pesticide Biochemistry and Physiology*, **12**, 124–9.

Kerkenaar, A., Uchiyama, M. & Versluis, G. G. (1981). Specific effects of tridemorph on sterol biosynthesis in *Ustilago maydis*. *Pesticide Biochemistry and Physiology*, **16**, 97–104.

Kraus, P. (1979). Studies on the mechanism of action of Baycor. *Pflanzenschutz-Nachnichten* (English edition), **32**, 17–30.

Leroux, P. & Gredt, M. (1978*a*). Effect de l'imazalil sur la biosynthèse de l'ergosterol chez *Penicillium expansum* Link. *Comptes Rendus Académie des Sciences Paris, Series D*, **286**, 427–9.

Leroux, P. & Gredt, M. (1978*b*). Effets de quelques fongicides systemiques sur la biosynthèse de l'ergosterol chez *Botrytis cinerea* Pers., *Penicillium expansum* Link. et *Ustilago maydis* (DC.) Cda. *Annales de Phytopathologie*, **10**, 45–60.

Leroux, P., Gredt, M. & Fritz, R. (1976). Similitudes et différences entre les modes d'action de l'imazalile, du triadimefon, du triarimol et de la triforine. *Phytiatire–Phytopharmacie*, **25**, 317–34.

Marriott, M. S. (1980). Inhibition of sterol biosynthesis in *Candida albicans* by imidazole-containing antifungals. *Journal of General Microbiology*, **117**, 253–5.

Mitropoulos, K. A., Gibbons, G. F., Connell, C. M. & Woods, R. A. (1976). Effect of triarimol on cholesterol biosynthesis in rat-liver subcellular fractions. *Biochemical and Biophysical Research Communications*, **71**, 892–900.

Paltauf, F., Daum, G., Zuder, G., Högenauer, G., Schulz, G. & Seidl, G. (1982). Squalene and ergosterol biosynthesis in fungi treated with naftifine, a new antimycotic agent. *Biochimica et Biophysica Acta*, **712**, 268–73.

Pappas, A. C. & Fisher, D. J. (1979). A comparison of the mechanisms of action of vinclozolin, procymidone, iprodione and prochloraz against *Botrytis cinerea*. *Pesticide Science*, **10**, 239–46.

Pierce, A. M., Pierce, H. D., Unrau, A. M. & Oehlschlager, A. C. (1978). Lipid composition and polyene antibiotic resistance of *Candida albicans* mutants. *Canadian Journal of Biochemistry*, **56**, 135–42.

Pinto, W. J. & Nes, W. R. (1983). Sterochemical specificity for sterols in *Saccharomyces cerevisiae*. *The Journal of Biological Chemistry*, **258**, 4472–6.

Ragsdale, N. N. (1975). Specific effects of triarimol on sterol biosynthesis in *Ustilago maydis*. *Biochimica et Biophysica Acta*, **380**, 81–96.

Ragsdale, N. N. & Sisler, H. D. (1972). Inhibition of ergosterol biosynthesis in *Ustilago maydis* by the fungicide triarimol. *Biochemical and Biophysical Research Communications*, **46**, 2048–53.

Ragsdale, N. N. & Sisler, H. D. (1973). Mode of action of triarimol in *Ustilago maydis*. *Pesticide Biochemistry and Physiology*, **3**, 20–9.

Richmond, D. V. & Pring, R. J. (1979). Some morphological and cytological effects of fungicides. *Systemic Fungicides*, ed. H. Lyr & C. Poulter, pp. 293–298. Berlin: Akademie Verlag.

Rivière, J. L. (1983). Prochloraz, a potent inducer of microsomal cytochrome P-450. *Pesticide Biochemistry and Physiology*, **19**, 44–52.

Schmitt, P., Benveniste, P. & Leroux, P. (1981). Accumulation of 9β,19-cyclopropyl sterols in suspension cultures of bramble cells cultured with tridemorph. *Phytochemistry*, **20**, 2153–9.

Schmitt, P., Scheid, F. & Benveniste, P. (1980). Accumulation of $\Delta^{8,14}$-sterols in suspension cultures of bramble cells cultured with an azasterol antimycotic agent (A25822B). *Phytochemistry*, **19**, 524–30.

Sherald, J. L., Ragsdale, N. N. & Sisler, H. D. (1973). Similarities between the systemic fungicides triforine and triarimol. *Pesticide Science*, **4**, 719–27.

Sherald, J. L. & Sisler, H. D. (1975). Antifungal mode of action of triforine. *Pesticide Biochemistry and Physiology*, **5**, 477–88.

Shigematsu, M. L., Uno, J. & Arai, T. (1982). Effect of ketoconazole on isolated mitochondria from *Candida albicans. Antimicrobial Agents and Chemotherapy*, **21**, 919–24.

Shive, J. B. & Sisler, H. D. (1976). Effects of ancymidol (a growth retardant) and triarimol (a fungicide) on the growth, sterols and gibberellins of *Phaseolus vulgaris* (L). *Plant Physiology*, **57**, 640–4.

Siegel, M. R., Kerkenaar, A. & Kaars Sijpesteijn, A. (1977). Antifungal activity of the systemic fungicide imazalil. *Netherlands Journal of Plant Pathology*, **83**, Supplement 1, 121–33.

Siegel, M. R. & Ragsdale, N. N. (1978). Antifungal mode of action of imazalil. *Pesticide Biochemistry and Physiology*, **9**, 48–56.

Siegel, M. R. & Solel, Z. (1981). Effects of imazalil on a wild-type and fungicide resistant strain of *Aspergillus nidulans. Pesticide Biochemistry and Physiology*, **15**, 222–33.

Sisler, H. D. & Ragsdale, N. N. (1977). Fungitoxicity and growth regulation involving aspects of lipid biosynthesis. *Netherlands Journal of Plant Pathology*, **83**, Supplement 1, 81–91.

Sisler, H. D., Ragsdale, N. N. & Waterfield, W. F. (1984). Biochemical aspects of fungitoxic and growth regulatory action of fenarimol and other pyrimidine methanols. *Pesticide Science*, **15**, 167–76.

Sisler, H. D., Walsh, R. C. & Ziogas, B. N. (1983). Ergosterol biosynthesis: A target of fungitoxic action. In *Pesticide chemistry: human welfare and the environment*, vol. 3, *Mode of Action, Metabolism and Toxicology*, ed. S. Matsunaka, D. H. Hutson & S. D. Murphy, pp. 129–34. New York: Pergamon Press.

Sobus, M. T., Holmlund, C. E. & Whittaker, N. F. (1977). Effects of the hypocholesteremic agent trifluperidol on the sterol, steryl ester and fatty acid metabolsim of *Saccharomyces cerevisiae. Journal of Bacteriology*, **130**, 1310–16.

Sud, I. J. & Feingold, D. S. (1981). Heterogenity of action mechanisms among antimycotic imidazoles. *Antimicrobial Agents and Chemotherapy*, **20**, 71–4.

Tanaka, S., Kato, T., Yamamoto, S., Yoshioka, H., Taira, Z. & Watson, W. H. (1978). Crystal and liquid structures of S-*n*-butyl S'-*p-tert*-butylbenzyl *N*-3-pyridyldithiocarbonimidate (S-1358, Denmert). *Agricultural and Biological Chemistry*, **42**, 287–91.

Taylor, F. R., Rodriguez, R. J. & Parks, L. W. (1983). Relationship between antifungal activity and inhibition of sterol biosynthesis in miconazole, clotrimazole and 15-azasterol. *Antimicrobial Agents and Chemotherapy*, **23**, 515–21.

Tepper, B. L. & Yoder, K. S. (1982). Postharvest chemical control of *Penicillium* blue mold of apple. *Plant Disease*, **66**, 829–31.

Trocha, P. J., Jasne, S. J. & Sprinson, D. B. (1977). Yeast mutants blocked in removing the methyl group of lanosterol at C-14. Separation of sterols by high pressure liquid chromatography. *Biochemistry*, **16**, 4721–6.

Uno, J., Shigematsu, M. L. & Arai, T. (1982). Primary site of action of ketoconazole on *Candida albicans*. *Antimicrobial Agents and Chemotherapy*, **21**, 912–18.

Van den Bossche, H., Lauwers, W. F. & Cornelissen, F. (1982). The antimycotic miconazole: an inhibitor of the biosynthesis of polyisoprenoids in *Staphylococcus aureus* B.180. *Archives Internationales de Physiologie et de Biochimie*, **90**, B78.

Van den Bossche, H., Willemsens, G., Cools, W., Cornelissen, F., Lauwers, W. F. & Van Cutsem, J. M. (1980). *In vitro* and *in vivo* effects of the antimycotic drug ketoconazole on sterol synthesis. *Antimicrobial Agents and Chemotherapy*, **17**, 922–8.

Van den Bossche, H., Willemsens, G., Cools, W., Lauwers, W. F. J. & LeJeune, L. (1978). Biochemical effects of miconazole on fungi. II. Inhibition of ergosterol biosynthesis in *Candida albicans*. *ChemicoBiological Interactions*, **21**, 59–78.

Van Tuyl, J. M. (1977). Genetics of fungal resistance to systemic fungicides. *Mededelingen Landbouwhogeschool Wageningen*, **77–2**, 1–136.

Walsh, R. C. & Sisler, H. D. (1982). A mutant of *Ustilago maydis* deficient in sterol C-14 demethylation: characteristics and sensitivity to inhibitors of ergosterol biosynthesis. *Pesticide Biochemistry and Physiology*, **18**, 122–31.

Woloshuk, C. P., Sisler, H. D. & Dutky, S. R. (1979). Mode of action of the azasteroid antibiotic 15-aza-24-methylene-D-homocholesta-8,14-dien-3β-ol in *Ustilago maydis*. *Antimicrobial Agents and Chemotherapy*, **16**, 98–103.

13

Antifungal agents with an indirect mode of action

M. WADE

Shell Research Limited, Sittingbourne Research Centre, Sittingbourne, Kent ME9 8AG

Introduction

There are several strategies for the control of plant disease. Among these, chemical control is undoubtedly of great importance and will continue to be so for many years to come. The most obvious approach to chemical control is to devise compounds that attack the pathogen itself. These may be considered as 'conventional' antifungal agents since they directly inhibit the growth of pathogens upon or within plants. However, it should be remembered that the principal concern is with the control of disease rather than the killing of fungi and the use of 'alternative' or indirectly acting chemicals for disease control represents an equally valid strategy. Antifungal agents which act indirectly may be defined as chemicals that control disease by affecting either the virulence of the pathogen or the resistance of the host. These compounds do not directly inhibit pathogen growth.

For disease to occur, there are several requirements of both the host and the potential pathogen. Taking the pathogen first, it must have means of entering the plant. Once inside, it must have a way of avoiding or counteracting host-defence mechanisms. It must then multiply in the plant and cause the biochemical changes that are recognised as disease. Finally, for the disease to become an epidemic and of economic significance, the fungus must produce propagules for spread to other plants. As for the host, it must provide a suitable environment for pathogen growth, including the provision of essential nutrients; effective host defence mechanisms must be absent. In theory at least, chemical disruption of any one of these processes could lead to disease control. This chapter will consider each situation in turn, giving examples where possible. It should be noted that, in most cases, the literature begs many more questions than it answers and it is not usually known how a chemical acts in cases where indirect action is suspected. In addition, fungal diseases of plants are considerably more important than diseases caused by other pathogens and it so happens that all of the alternative chemical agents discussed here concern diseases caused by fungi.

Compounds acting against pathogenic processes of fungi

The first requirement of a pathogen is that it must enter the plant. Clearly, preventing this process would prevent any subsequent disease development. Many plant pathogens enter plants through the intact cuticle. Although it was originally thought that this merely involved mechanical forces, to rupture the cuticle, it is now known that the process involves hydrolytic enzymes (some of which are of the esterase type); these enzymes may act in conjunction with mechanical forces. A good example of disease control achieved by inhibiting penetration is provided by Maiti & Kolattukudy (1979). Diisopropyl fluorophosphate, $[(CH_3)_2CHO]_2P(O)F$, is a very potent inhibitor of esterases such as acetylcholinesterase. Maiti & Kolattukudy found that it is an inhibitor of cutinesterase and, when applied to epicotyls of pea plants, prevented infection of the plants by *Fusarium solani*; this pathogen normally enters plants through the cuticle. They suggested that disease control was due to inhibition of the cutinases produced by the fungus, but better evidence for this hypothesis comes from experiments in which they raised specific antisera to the purified cutinases of the pathogen. These antisera also prevented infection by the pathogen, but neither diisopropyl fluorophosphate, nor the antisera inhibited fungal growth directly. Although it is unlikely that it would be practicable to use antisera for disease control in the field, the experiments indicate that, if suitable synthetic inhibitors of cutinases can be found, these may be of considerable practical value.

Once inside the plant, a pathogen must counteract, or avoid in some way, the defence mechanisms of the host. The way in which fungi do this is a grey area of plant pathology at the moment but is the subject of much research, being directly relevant to the question of the biochemical basis of host and cultivar specificity. It is tempting to think that chemicals could be designed to interfere with these host-defence mechanisms, but no compound has yet been shown to work in this way. Perhaps it will become feasible to design such molecules as our understanding of defence mechanisms is increased.

Having counteracted host-defence mechanisms in some way,

pathogens then multiply in their hosts to a greater or lesser extent, depending on the type of disease involved. 'Conventional' therapeutic chemicals act on this phase of the disease process and they will not be considered here.

Disease is recognised by the metabolic disorders that are induced in the plant following infection. In many diseases, symptoms are attributable to highly potent toxins produced by the pathogen and, in some cases, these have been characterised chemically. Since the object is to control disease rather than to kill the pathogen, it is logical that control could be brought about by inhibiting the action or production of these toxins. Whether or not this approach could be successful depends at least in part on whether the toxin is the principal component responsible for virulence of the pathogen. There are not many good examples of this approach. Horsfall & Zentmyer (1942) found that several basic chemicals such as 8-hydroxyquinoline and diaminoazobenzene (1) counteracted the effects of the toxins produced by the Dutch Elm disease

(1) H_2N—〈 〉—N=N—〈 〉—NH_2 Diaminoazobenzene

fungus, and Howard (1942) similarly found that diaminoazobenzene acts as an antidote of the toxin produced by *Phytophthora cactorum*, the cause of a damaging disease of maple trees. Howard was able to obtain a marked improvement in the growth of trees naturally infected with this fungus by administering this compound.

Possibly a more specific antitoxic effect concerns the eyespot disease of sugarcane caused by *Helminthosporium sacchari*. The symptoms of eyespot can be entirely attributed to a low molecular weight toxin, helminthosporoside (2). Strobel (1973) has suggested that this toxin is a

(2)   Helminthosporoside

2-hydroxycyclopropyl-α-galactoside, although there is some doubt about the validity of this claim. The toxin binds to a receptor protein in the plasma membrane of susceptible clones of sugarcane. A similar protein is present in the plasma membrane of resistant clones, but due to a small difference in its tertiary structure, this protein is unable to bind the toxin and those clones that are resistant to the fungus are also resistant to the toxin. The receptor protein may be associated with the transport of α-galactoside and Strobel has found that if leaves of cane plants are pretreated with certain α-galactosides (e.g. raffinose, mellibiose or methyl-α-galactoside), these leaves become resistant to the effects of the toxin. Strobel suggested that there is a competitive interaction between the saccharides and the toxin. A much more permanent resistance to the toxin was obtained by pretreating leaves with the detergent Triton X-100. The relevance of this latter observation to practical disease control does not appear to have been explored, but it seems quite feasible that where a disease can be attributed almost entirely to a toxin, control could be achieved by counteracting that toxin.

The commercial fungicide tricyclazole (5-methyl-[1,2,4]-triazole [3,4-b]-benzothiazole) (3) appears to interfere with the mechanisms of

(3)   Tricyclazole

pathogenicity of the fungus. Tricyclazole is a systemic fungicide used to control the blast disease of rice caused by the fungus *Pyricularia oryzae*. It is highly specific for this fungus, and is not able to control several other diseases of rice. Although the compound is highly active against the fungus on rice plants, it has only very weak effects *in vitro*. The lack of correlation between disease control and fungitoxicity *in vitro* suggests three possible modes of action: (i) the compound is changed by the plant; (ii) it affects the resistance of the host to infection; (iii) it affects fungal pathogenicity.

Sisler and his co-workers discovered that the formation of melanin pigments by *Pyricularia oryzae* is inhibited *in vitro* by concentrations of tricyclazole well below those required for inhibition of fungal growth *in vitro*. In fungi, melanin is a complex polymer of naphthalene derivatives. They have found that tricyclazole inhibits the reduction of 1,3,8-trihydroxynaphthalene (4) to vermelone (5), a step in melanin biosynthesis (Tokousbalides & Sisler, 1979). Sisler is firmly of the opinion that the fungicidal properties of tricyclazole are due to its effects

(4)   1,3,8-Trihydroxynaphthalene

(5)

Vermelone

on the pathogenicity of the fungus. This is based on the unusually high sensitivity of melanin biosynthesis in *P. oryzae* to tricyclazole and on the fact that there are mutants of *P. oryzae* – the so-called buff mutants – that lack melanin and are also non-pathogenic. Tokousbalides & Sisler (1978, 1979) suggested several ways that a defect in melanin bio-synthesis could affect pathogenicity; these included inhibition of the production of related phytotoxic metabolites. Recently, however, Sisler has stressed the importance of melanin in rigidifying the cell wall and appressoria of *P. oryzae*. Without melanin, the appressorial infection peg does not have sufficient tensile strength to penetrate the cuticle and infection does not take place.

Besides low molecular weight toxins, many pathogens utilise hydro-lytic enzymes to aid their progress through host tissue. Pectolytic enzymes, for example, are important in many diseases and cause tissue maceration and cell death. The possibility of inhibiting such enzymes to control disease has been considered. Grossmann (1968) looked at a number of compounds for their ability to inhibit pectic and cellulolytic enzymes *in vitro* and found one such compound, rufianic acid (**6**)

(6)

Rufianic acid

capable of doing this. It was particularly active against pectin methyl-esterase activity but also inhibited pectolytic activity. When applied to the roots of tomato plants, rufianic acid caused a significant (though not complete) suppression of the *Fusarium* and *Verticillium* wilt diseases (Martin & Grossmann, 1972). This was associated with reduced activi-ties of pectic enzymes in the xylem sap and reduced amounts of the fungus in the treated plants. Although it is tempting to conclude that rufianic acid inhibited pectic enzymes in the plants and that this resulted in reduced virulence of the fungi, Grossmann pointed out that this was in no way proven and other explanations for the activity of rufianic acid might exist. A particular problem was that the content of rufianic acid in xylem sap was insufficient to inhibit pectolytic enzymes. Also rufianic acid has some direct fungitoxic properties although, with wilt pathogens, these are weak.

Compounds preventing the dispersal of pathogens

With the diseases of aerial parts of plants caused by fungal pathogens, plant to plant transfer of the pathogen is brought about by the production of spores, in particular those produced vegetatively. Plant pathology for practical purposes is concerned with diseases of populations of plants. Thus, there is the possibility of controlling disease by preventing the production of fungal spores. Unfortunately, no compounds that are truly effective in this respect have been found to date, although many 'conventional' compounds inhibit sporulation probably by inhibition of fungal growth. It should also be questioned, especially in relation to airborne diseases, whether disease control based on this specific mode of action would be practicable.

Compounds causing changes in host physiology that reduce disease

So far in this chapter, consideration has been given to those features of the pathogen necessary for disease and the ways in which it might be possible to interfere with them. An equally valid approach is to consider the properties of the host required for disease to occur. These are of course complementary to those of the pathogen.

In the disease situation, host and pathogen co-exist intimately in a delicate balance particularly with the obligate or biotrophic pathogens. It is not surprising therefore, that compounds that cause small changes in host physiology may also affect the susceptibility of plants to disease. There are many examples of such compounds, but what is perhaps more surprising is that the majority of the plant-growth-regulator effects reported against disease concern the wilt pathogens, fungi that are not obligately parasitic.

Corden & Dimond (1959) looked at a homologous series of auxin-active compounds based on naphthalene acetic acid (**7**). These compounds gave good control of the *Fusarium* wilt disease of tomato and the degree of control was related to the activity of these derivatives as inhibitors of root growth. The explanation of disease control was based

CH₂COOH

(7)   Naphthalene acetic acid

on a hypothesis for auxin action that was in vogue at that time. They suggested that the effects on disease and root growth had a common basis through increased calcium bonding to the pectic substances of the cell walls. Reduced methylation of the pectic substances allowed increased formation of calcium pectate bridges. This decreased cell wall plasticity and, therefore, inhibited root growth. Calcium pectate is resistant to the action of pectolytic enzymes of fungi and this could explain why the treated plants were also more resistant to disease (Edgington, Corden & Dimond, 1961). Disease control was in fact dependent on an adequate calcium supply in the tomato plants (Corden & Edgington, 1960).

More recent studies on the biochemical basis for auxin action have shown that the effect of certain auxins on the degree of methylation of pectins of the cell wall is incidental to their effects on plant growth (Cleland, 1963). It is still possible, however, that the effects on disease are related to pectin methylation since certain auxin-active compounds can promote pectin methylesterase activity while others may promote pectin methylation. Ethionine ($C_2H_5SCH_2CH_2CH(NH_2)CO_2H$) is also known to inhibit the methylation of pectins and induces resistance of avocado to *Phytophthora cinnamomi* (Zentmyer, Moje & Mircetich, 1962). Whether the mechanism of induced resistance is the same or not has not been investigated. With wilt diseases, there is clear evidence that ethionine adversely affects pathogen virulence rather than host susceptibility. In a study of the action of ethionine against common scab of potato caused by *Streptomyces scabies*, McIntosh, Burrell & Bateman (1977) found the compound sufficiently active against the pathogen to account for disease control. There are other examples of growth-regulating compounds affecting wilt diseases. They include auxin-transport inhibitors such as the morphactins (Buchenauer & Grossmann, 1969; **8**) and a number of growth retardants such as Pydanon (Buchenauer & Erwin, 1976; **9**).

(8)

COOH

Morphactins

(9)

HO   CH₂COOH

Pydanon

Ethylene can induce changes in peroxidase and polyphenol oxidase activities (Gentile & Matta, 1975), pectic enzymes (demonstrated *in vitro*, Gertman & Fuchs, 1974), induction of phytoalexins (see for example, Henfling, Lister & Kuc, 1978) and in enzymes which degrade fungal cell walls (Pegg, 1976), leading to increased resistance to wilts. Similarly, the application of the ethylene generator Ethephon ($ClCH_2CH_2$—PO_3H) to tomato plants induced resistance to subsequent infection with *Verticillium albo-atrum* (Retig, 1974).

As already mentioned, however, there are rather fewer examples of growth-regulating compounds affecting disease other than wilts. One example is the control of powdery mildews by the cytokin-reactive compound kinetin (Dekker, 1963; Cole & Fernandes, 1970; **10**).

(**10**)   Kinetin

Compounds that act on host-resistance mechanisms

There are two ways in which one can consider using host-resistance mechanisms for disease control. First, there is the activation of the host-resistance mechanisms themselves. Secondly, there is the use of the host's own defensive compounds directly.

Disease-resistance mechanisms of plants are imperfectly understood but it is clear that, as with disease-resistance in animals, several mechanisms are involved. Among the better studied of plant disease-resistance mechanisms is the production of phytoalexins, low molecular weight antibiotics produced by plants in response to infection. There are many plant–pathogen interactions in which the production of phytoalexins can adequately explain the expression of natural resistance. Other mechanisms include hypersensitive cell death, lignification, callose deposition, papillae formation, etc., around sites of pathogen penetration.

Many chemicals are known to elicit the production of phytoalexins in the absence of a pathogen. For example, certain glucans obtained from fungal cell walls are highly potent as phytoalexin inducers and non-specific in their activity (Cline, Wade & Albersheim, 1978). This might suggest that they would be useful agents for plant-disease control.

Unfortunately, however, a problem is that the production of phyto-alexins is very frequently accompanied by the death of host cells, the so-called hypersensitive response. The elegance of the natural resistance response lies in the fact that this hypersensitive response happens only at the precise time and place where it is needed to counteract a fungus. The sacrifice by the plant is minimal. Clearly, it would not be a good idea to trigger such a reaction indiscriminately throughout the plant.

Ideally, we need a compound that alters the plants in a subtle way such that, on challenge by a pathogen, it would behave in the same way as a genetically resistant plant, and host-resistance mechanisms are activated only at the precise times and places where needed.

One such class of molecules, the dichlorocyclopropanes, that appear to fulfil these requirements was discovered during the random screening of compounds for activity against the rice blast disease. The most active compound, and the one on which most work has been conducted, is WL28325 (**11**). A comprehensive account of these studies has been

presented previously (Langcake, Cartwright & Ride, 1981). Briefly, WL28325 is a systemic prophylactic compound highly specific for the rice blast disease. In this respect, it resembles tricyclazole. It was found that WL28325 is not sufficiently fungitoxic to account for disease control and it did not appear to give rise to fungitoxic metabolites within the plant. The most noticeable feature of the infection process in treated plants was the highly localised death of host cells in the vicinity of the invading fungus. This could be seen as a necrotic browning reaction and is very similar to the hypersensitive response of genetically resistant cultivars of rice to *Pyricularia oryzae*. At the same time, it was possible to detect the production of antifungal compounds; these were later identified as the momilactones A and B (Cartwright, Langcake & Ride, 1980; Cartwright *et al.*, 1981) which are now considered to be phyto-alexins of rice. The production of these compounds is enhanced in infected plants treated with WL28325. These effects could be brought about either by effects of the compound on the host-resistance mechan-isms, 'priming' them ready for fungal attack, or on the pathogenicity of the fungus; this was a problem that originally faced Sisler and co-workers in relation to the action of tricyclazole. In contrast to the case of tricyclazole, it is thought that the compound acts on the plant rather than on the fungus since it is possible to mimic the effects of the fungus on treated plants with picolinic acid, a compound that is reputed to be a metabolic product of *P. oryzae* (Langcake *et al.*, 1981).

Another example of a compound that appears to overcome the host's defence mechanism is the systemic fungicide tris-*O*-ethyl phosphonate (TEPA; aliette) (**12**). This compound has very low activity *in vitro* but it

(**12**)

$$\left[\begin{array}{c} H_5C_2O \\ \\ H \end{array} \underset{O^-}{\overset{O}{>}} P \underset{}{\overset{}{\Big\backslash}} \right]_3 \quad Al^{3+}$$

Aliette (TEPA)

is very effective in plant tissue against downy mildew (*Plasmopara viticola*) on vine and *Phytophthora* species on tomato. TEPA stimulates defence reactions and the synthesis of phytoalexins in these crops (Bompeix *et al.*, 1980). Bompeix has shown that the anti-phycomycete activity of TEPA in plants can be annulled by the shikimic acid pathway inhibitors, glyphosate and aminooxyacctic acid (Bompeix, Fettouche & Saindrenan, 1981). The shikimic acid pathway is the source of phenols required in phytoalexin synthesis.

Phenylthiourea (PTU) (**13**) gives complete protection to cucumber plants from *Cladosporium cucumerinum* by causing a strong lignification reaction around sites of fungal penetration. Genetically resistant

(**13**) —NH.CS.NH_2 Phenylthiourea (PTU)

plants respond to *C. cucumerinum* in the same way. In treated tissue, the levels of PTU required to protect the plant are 25–50 times less than those required to stop fungal growth *in vitro* (Sijpesteijn, 1969). PTU inhibits polyphenol oxidase, thus making more phenolics available for peroxidase action and lignin synthesis.

Probenazole (3-allyloxy-1,2-benzisothiazole-1,1-dioxide; **14**), a commercial rice blast fungicide (Watanabe *et al.*, 1977), is converted to two

(**14**) $OCH_2 \quad CH{=}CH_2$ Probenazole

active principles; *N*-D-glucopyranosylbenzisothiazole-1,1-dioxide and benzisothiazole-1,1-dioxide (Uchiyama *et al.*, 1973). Neither are inhibitory to *Pyricularia oryzae in vitro*. Although probenazole inhibits appressorium formation by *P. oryzae*, other effects on growth are much more pronounced in the presence of host tissue. Blast lesions on treated rice are similar to hypersensitive lesions of resistant cultivars.

An alternative to activating resistance mechanism, is the direct use of phytoalexins as chemicals for disease control. This has been tried on several occasions, for example by Ward, Unwin & Stoessl (1975) with the phytoalexins from pepper known as capsidiol (**15**) and by Carter,

(15)   Capsidiol

Chamberlain & Wain (1978) with synthetic analogues of vignafuran (**16**), a phytoalexin from cowpeas. Langcake & Pryce (1977) tried the

(16)   Vignafuran

effects of the grapevine phytoalexin ε-viniferin (**17**) against the downy mildew disease of grapevine. In each case, some disease control was

(17)   ε-Viniferin

obtained. However, phytoalexins would seem to have limited value. There are several problems. Firstly, the compounds themselves are not readily accessible, in many cases they cannot be easily synthesised. Secondly, they are not transported around plants (i.e. they are not systemic). Thirdly, they are molecules that are metabolically and sometimes environmentally unstable. Finally, phytoalexins are generally of rather low fungitoxicity, especially in comparison with synthetic fungicides. Nevertheless, it is possible that such problems could be overcome by analogue synthesis.

Contribution of host-defence mechanisms to the effectiveness of fungicides

There are some fungicides that have potent antifungal effects *in vitro* yet which also induce responses in the host plant very similar to the natural-resistance response. It is difficult to exclude the possibility that these host responses contribute to the effectiveness of some fungicides. There are several examples of this, but only two are mentioned here. The acylalanines are systemic fungicides recently introduced for the control of Oomycete fungi that cause downy mildews and other diseases (Davidse, Chapter 11, this volume). Metalaxyl (**18**) is an example of one

(**18**)   Metalaxyl

of these fungicides. It has potent activity *in vitro* against Oomycete fungi such as *Phytophthora infestans*, the causal agent of the potato late blight disease. In plants, however, it seems to be, if anything, even more active against the fungus than *in vitro*, but no effects are seen until the fungus has penetrated the host and has started to form haustoria (Crute, 1978; Staub, Dahmen & Schwinn, 1978; Cohen, Reuvini & Eyal, 1979). Haustoria are specialised fungal structures that obtain nutrients from the host by penetrating the host cells. Ward *et al.* (1980) looked at the effects of metalaxyl on the *Phytophthora* disease of soybean. They found lesions on treated plants which were indistinguishable from lesions on plants inoculated with an avirulent race of the pathogen (that is, in a situation of natural resistance). They also studied the production of phytoalexins in the two situations, and found that the levels of phytoalexins that accumulated were very similar in the two cases. On its own, metalaxyl did not promote the formation of phytoalexins. There are two possibilities here. Phytoalexins may accumulate because the fungicide kills the fungus, releasing fungal products that non-specifically stimulate the production of the defence compounds. On the other hand, the fungicide may interfere with either the pathogenicity of the fungus or the resistance mechanisms of the host. Either way, phytoalexins would accumulate. Since the fungicide is active against the fungus directly, it is likely that the effects on host-resistance mechanisms are incidental or secondary. It is possible, however, that they make a significant contribution to the effectiveness of the fungicide.

The hypersensitive necrosis induced in vines treated with 2-cyano-*N*-(ethylaminocarbonyl)-2(methoxyimino) acetamide (curzate; **19**) following infection by the downy mildew fungus was associated with

$$(19) \quad \underset{\underset{NOH}{\overset{\|}{C}}}{\overset{\overset{O}{\overset{\|}{C}}}{NC-C-C-NH-C-NH-C_2H_5}} \quad \text{Curzate}$$

enhanced synthesis of vine phytoalexins (P. Langcake, unpublished results). Neither metalaxyl (Ward *et al.*, 1980) nor curzate, in the absence of infection, induced phytoalexin production in soybean and vines respectively.

Conclusions

This review started out by showing the processes in both pathogens and in their hosts that are essential for disease to occur. It is proposed that each of these processes is amenable to chemical attack and that such chemicals could be useful in the control of plant disease. Some selected examples of compounds that may control plant disease by acting on these disease processes have been described. One notable feature, however, is that few of the compounds mentioned have so far found extensive commercial use. It is legitimate to ask, therefore, what are the possible advantages of indirectly acting chemicals over 'conventional' ones? There seem to be several.

First, the processes of disease and host resistance are amenable to scientific study and are receiving considerable attention in many laboratories throughout the world. This holds the prospect that it will shortly be possible to design chemicals for disease control based on this new-found knowledge.

Second, the processes of disease and host resistance are specific to the disease situation. The implication of this is that any chemical that acts specifically against pathogen virulence or host resistance is unlikely to affect other non-target organisms. In brief, these compounds would be considerably more environmentally acceptable.

Third, since the compounds are not likely to be intrinsically toxic to the host plant, they may be introduced into the plant without detrimental effects and this is a prerequisite for systemic and therapeutic compounds that have to penetrate host tissue.

Fourth, because indirectly acting chemicals are not directly fungitoxic, they should not exert the selection pressure in the environment that may lead to the occurrence of fungicide-resistant strains. It is open to debate, however, whether activated host-resistance mechanisms will be overcome by physiological races of pathogens in the way that genetic resistance has frequently failed. It probably depends on the precise stages in the resistance mechanisms which are affected by the compounds.

It is too early to say, however, whether indirectly acting compounds will have significant impact on practices in disease control. However, it is worth pointing out that the majority of agrochemicals companies have seen fit to rely entirely on screens that involve real diseases on real plants. This shows that they are not prepared to ignore the contribution of indirect mechanisms to fungicidal effectiveness, particularly now that the usefulness of tricyclazole, probenazole and aliette, for instance, have proved the point. Further examples of indirectly acting chemicals for disease control are being found and will continue to be found, whether by rational design or by empirical screening.

References

Bompeix, G., Fettouche, F. & Saindrenan, P. (1981). Mode d'action du Phoséthyl Al. *Phytiatrie-Phytopharmacie*, **30**, 257–72.

Bompeix, G., Ravise, A., Raynall, G., Fettuche, F. & Durand, M. C. (1980). Stimulation des réactions de défense et de synthese de phytoalexines chez la vigne et la tomate sous l'influence du tris-*O*-ethyl phosphonate d'aluminium. In *Abstracts of 18ème Colloque de la Société Française de Phytopathologie*, Toulouse, April 1980.

Buchenaeur, H. & Erwin, D. C. (1976). Effect of the plant growth regulator pydanon on *Verticillium* wilt of cotton and tomato. *Phytopathology*, **66**, 1140–3.

Buchenauer, H. & Grossmann, F. (1969). Effects of morphactins on various tomato diseases. *Deutsche Botanische Gesellschaft, Berichte*, **3**, 149–59.

Carter, G. A., Chamberlain, K. & Wain, R. L. (1978). Investigations on fungicides. XX. The fungitoxicity of analogues of the phytoalexin 2-(2'-methoxy-4'-hydroxyphenyl)-6-methoxybenzofuran (vignafuran). *Annals of Applied Biology*, **88**, 57–64.

Cartwright, D., Langcake, P., Pryce, R. J., Leworthy, D. P. & Ride, J. P. (1981). Isolation and characterisation of two phytoalexins from rice as momilactones A and B. *Phytochemistry*, **20**, 535–7.

Cartwright, D., Langcake, P. & Ride, J. P. (1980). Phytoalexin production in rice and its enhancement by a dichlorocyclopropane fungicide. *Physiological Plant Pathology*, **17**, 259–67.

Cleland, R. (1963). Independence of the effects of auxin on cell wall methylation and elongation. *Plant Physiology*, **38**, 12–18.

Cline, K., Wade, M. & Albersheim, P. (1978). Fungal glucans which elicit phytoalexin accumulation in soybean also elicit the accumulation of phytoalexins in other plants. *Plant Physiology*, **62**, 918–21.

Cohen, Y., Reuvini, M. & Eyal, H. (1979). The systemic antifungal activity of Ridomil against *Phytophthora infestans* on tomato plants. *Phytopathology*, **69**, 645–9.

Cole, J. S. & Fernandes, D. L. (1970). Changes in the resistance of tobacco leaf to *Erysiphe cichoracearum* D.C. induced by topping, cytokinins and antibiotics. *Annals of Applied Biology*, **66**, 239–43.

Corden, M. E. & Dimond, A. E. (1959). The effect of growth-regulating substances on disease resistance and plant growth. *Phytopathology*, **49**, 68–72.

Corden, M. E. & Edgington, L. V. (1960). A calcium requirement for growth-regulator-induced resistance to *Fusarium* wilt of tomato. *Phytopathology*, **50**, 625–6.

Crute, I. R. (1978). Studies on new systemic fungicides active against *Bremia lactucae*. In *Abstracts of Third International Congress of Plant Pathology*, Munich, p. 384.

Dekker, J. (1963). Effect of kinetin on powdery mildew. *Nature*, **197**, 1027–8.

Edgington, L. V., Corden, M. E. & Dimond, A. E. (1961). The role of pectic substances in chemically induced resistance to *Fusarium* wilt of tomato. *Phytopathology*, **51**, 179–82.

Gentile, I. A. & Matta, A. (1975). Production of and some effects of ethylene in relation to *Fusarium* wilt of tomato. *Physiological Plant Pathology*, **5**, 27–35.

Gertman, E. & Fuchs, Y. (1974). Changes in pectin methyl esterases (PME) activity caused by ethylene applied at different temperatures. *Plant Cell Physiology*, **15**, 501–5.

Grossmann, F. (1968). Studies on the therapeutic effects of pectolytic enzyme inhibitors. *Netherlands Journal of Plant Pathology*, **74**, (suppl. 1), 91–103.

Henfling, J. W. D., Lister, N. & Kuc, J. (1978). Effect of ethylene on phytuberin and phytoberol accumulation in potato tuber slices. *Phytopathology*, **68**, 857–62.

Horsfall, J. G. & Zentmyer, G. A. (1942). Antidoting of toxins of plant diseases. *Phytopathology*, **32**, 22–3.

Howard, F. L. (1942). Antidoting of toxins of *Phytophora cactorum* as a means of plant disease control. *Science*, **94**, 345.

Langcake, P., Cartwright, D. & Ride, J. P. (1981). The dichlorocyclopropanes and other fungicides with indirect mode of action. In *Systemic Fungicides and Antifungal Compounds*, proceedings of a symposium, Reinhardsbrunn, DDR, ed. H. Lyr & C. Potter. Berlin: Akademie-Verlag.

Langcake, P & Pryce, R. J. (1977). A new class of phytoalexins from grapevine. *Experientia*, **33**, 151–2.

McIntosh, A. H., Burrell, M. M. & Bateman, G. L. (1977). Mechanism of action of foliar sprays of daminozide and ethionine against potato common scab. In *Proceedings of the 1977 British Crop Protection Conference*, vol. 1, pp. 87–93.

Maiti, I. B. & Kolattukudy, P. E. (1979). Prevention of fungal infection of plants by specific inhibition of cutinase. *Science*, **205**, 567–8.

Martin, J. & Grossmann, F. (1972). Inhibition of pectic and cellulolytic enzymes of *Rhizoctonia solani* and the influence of some inhibitors of the disease process. II. Effect of some inhibitors *in vivo*. *Phytopathologische Zeitschrift*, **75**, 87–110.

Pegg, G. F. (1976) Glucanohydrolases of higher plants: a possible defence mechanism against parasitic fungi. In *Cell Wall Biochemistry Related to Host-plant Pathogen Interactions*, ed. B. Solheim & J. Rahe, pp. 305–45. Oslo: Universitetsforlaget.

Retig, N. (1974). Changes in peroxidase and polyphenoloxidase associated with natural and induced resistance of tomato to *Fusarium* wilt. *Physiological Plant Pathology*, **4**, 145–50.

Sijpesteijn, A. K. (1969). Mode of action of phenylthiourea, a therapeutic agent for cucumber scab. *Journal of the Science of Food and Agriculture*, **20**, 403–5.

Staub, T., Dahmen, H. & Schwinn, F. (1978). Effects of Ridomil on the development of

Plasmopara viticola and *Phytophthora infestans* on their host plants. *Zeitschrift für Pflanzenkrankheiten und Pflanzenschutz*, **87**, 83–91.

Strobel, G. A. (1973). The helminthosporoside-binding protein of sugar cane. *Journal of Biological Chemistry*, **248**, 1321–8.

Tokousbalides, M. G. & Sisler, H. D. (1978). Effect of tricyclazole on growth and secondary metabolism in *Pyricularia oryzae*. *Pesticide Biochemistry and Physiology*, **8**, 26–32.

Tokousbalides, M. C. & Sisler, H. D. (1979). Site of inhibition by tricyclazole in the melanin biosynthetic pathway of *Verticillium dahliae*. *Pesticide Biochemistry and Physiology*, **11**, 64–73.

Uchiyama, M., Abe, H., Sato, R., Shimura, M. & Watanabe, T. (1973). Fate of 3-allyloxy-1,2-benzisothiazole-1,1-dioxide (Oryzemate) in rice plants. *Agricultural and Biological Chemistry*, **37**, 737–45.

Ward, E. W. B., Lazarovits, G., Stossel, P., Barrie, S. D. & Unwin, C. H. (1980). Glyceollin production associated with control of *Phytophthora* rot of soybeans by the systemic fungicide metalaxyl. *Phytopathology*, **70**, 738–40.

Ward, E. W. B., Unwin, C. H. & Stoessl, A. (1975). Experimental control of late blight of tomatoes with capsidiol, the phytoalexin from peppers. *Phytopathology*, **65**, 168–9.

Watanabe, T., Igarashi, H., Matsumoto, K., Seki, S., Mase, S. & Sekizawa, Y. (1977). The characteristics of probenazole (Oryzemate) for the control of rice blast. *Journal of Pesticide Science*, **2**, 291–6.

Zentmyer, G. A., Moje, W. & Mircetich, S. M. (1962). Ethionine as a chemotherapeutant. *Phytopathology*, **52**, 34.

14

Molecular and cellular aspects of the interaction of benzimidazole fungicides with tubulin and microtubules

T. G. BURLAND* and K. GULL†

*McArdle Laboratory, University of Wisconsin, Madison, WI 53706, USA
†Biological Laboratory, University of Kent, Canterbury, Kent CT2 7NJ, UK

Introduction

Numerous electron microscopic studies of fungi have documented the presence of microtubules within the cells of these eukaryotic organisms. In these studies the microtubules are visualised as filamentous structures of variable length, but with a constant diameter of 24 nm. They are often found in the cytoplasm where they may function in organelle positioning and movement (Howard & Aist, 1977, 1980; Oakley & Morris, 1980). Microtubules constitute the major structural components of the mitotic and meiotic spindles of fungal nuclei, these spindles often being enclosed within an intact nuclear envelope. However, microtubules are also present as components of highly ordered organelles such as the centrioles and flagella of certain cells in the life cycles of the Mastigomycetes and Myxomycetes. The study of microtubule structure, organisation and function in fungal cells has been aided by the discovery that a group of benzimidazole fungicides act by binding to the protein subunit of microtubules – tubulin. In this chapter we discuss the organisation of microtubule formation in fungal cells and the biochemical nature of fungal tubulins. We then address aspects of the mechanism of action and selective toxicity of the benzimidazole fungicides, together with an analysis of the genetics of resistance.

Microtubule-organising centres in fungal cells

The polymerisation of microtubules in cells appears to be restricted to specific, localised regions. These areas of the cell may have a definable and recognisable ultrastructural form. However, in some cases they are merely represented by areas of electron-dense, amorphous material (Tucker, 1979). The members of this collection of different structures have been referred to, in generic terms, as microtubule-organising centres (MTOCs). The various MTOCs of fungal cells have been extensively characterised ultrastructurally (Beakes, 1981; Beckett, 1981; Heath, 1981; Zickler, 1981). However, there are only a few experimental accounts of the activity of these structures (Byers, Shriver & Goetsch, 1978; Hyams & Borisy, 1978; Havercroft, Quinlan & Gull, 1981*b*; Kuriyama *et al.*, 1982; Roobol, Havercroft & Gull, 1982).

In *Saccharomyces cerevisiae*, all the microtubules of the cell appear to be nucleated by a specialised MTOC located in the nuclear envelope; this structure has been variously termed the spindle plaque or spindle pole body (SPB). Its position in the nuclear membrane appears to facilitate its nucleation of both cytoplasmic and spindle microtubules (Zickler, 1981).

The SPB of *Saccharomyces cerevisiae* has been shown to nucleate the polymerisation of microtubules *in vitro*. Hyams & Borisy (1978) prepared sphaeroplasts of yeast and then lysed these sphaeroplasts on an air/water interface. These lysed and spread preparations were then picked up on electron microscope grids and incubated on preparations of mammalian microtubule protein under conditions which did not allow the self-assembly of the microtubule protein. Examination of the grids in the electron microscope revealed that yeast SPBs were often associated with long microtubules, indicating that they were capable of initiating the assembly of microtubules *in vitro*. There is no biochemical information regarding the identity of the elements of the SPB which are necessary for microtubule assembly. However, Hyams & Borisy were able to show that the structure of the SPB and its ability to initiate the assembly of microtubules *in vitro* was essentially unchanged by treatment with DNase I, RNase A or phospholipase A. The proteolytic enzyme trypsin did destroy both the structural organisation of the SPB and its ability to influence microtubule assembly.

In the cell, the yeast SPB has two important properties. It appears to nucleate the assembly of a very specific number of microtubules and secondly, the ends of these microtubules proximal to the SPB are closed. Byers *et al.* (1978) were able to show that when SPBs from lysed sphaeroplasts were incubated with mammalian tubulin, the microtubules polymerised *in vitro* also possessed this closed end proximal to the

SPB. Byers and colleagues were also able to show that the number of microtubules arising on each SPB *in vitro* achieved saturation early in the incubation, in spite of the continued elongation of the microtubules that were present; this again suggested that each SPB possesses a limited number of discrete sites for microtubule assembly.

The main MTOC of *Physarum polycephalum* myxamoebae is an amorphous, osmiophilic structure attached to the nucleus and to the two centrioles (Fig. 14.1). The nucleus–MTOC–centriole complex can be isolated from the cell. The anterior centriole is linked to the MTOC by a bridge structure and the MTOC is then linked to the nucleus by a group of unusually stable microtubules (Havercroft *et al.*, 1981*b*; Roobol *et al.*, 1982). When the nucleus–MTOC–centriole complex was incubated with purified tubulin from *Physarum* myxamoebae or mammalian cells, the growth of 45–70 microtubules was initiated onto the MTOC. The number and length of nucleated microtubules was proportional to the tubulin concentration. Pretreatment of the MTOC complex with DNase

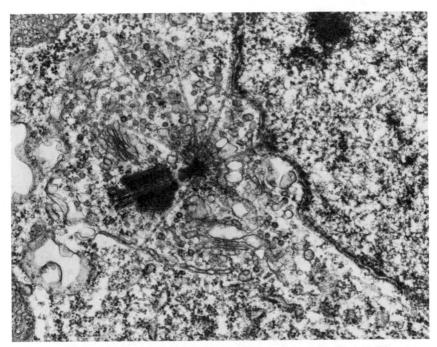

Fig. 14.1. Centriole–MTOC complex in a myxamoeba of *Physarum polycephalum*. Note the cytoplasmic microtubules which radiate from the electron-dense MTOC which is located between the nucleus and the anterior centriole. The area also includes numerous vesicles and dictyosomes.

I, RNase A or antitubulin antibody did not affect its nucleation capacity. However, after pretreatment with trypsin the MTOC and any nucleation capacity were destroyed. The structural organisation of the microtubule arrays nucleated by this MTOC in the myxamoeba have been reconstructed by Wright and his co-workers (Wright, Mir & Moisand, 1980).

These and other reports of the MTOCs of fungal cells emphasise some important generic properties of these organelles. MTOCs often nucleate specific numbers of microtubules. They often nucleate microtubules which have specific alignments and the microtubules are often organised into defined arrays. Also, in most cells there is a high fidelity in the protofilament number of the microtubules that are nucleated. In most, but not all cells microtubules have 13 protofilaments. Unfortunately, although we have a large number of ultrastructural descriptions of the MTOCs of cells, we have no biochemical account of the identity of the molecule(s) responsible for nucleation of microtubules.

Tubulin polypeptides

The major component of microtubules is tubulin which exists as a heterodimer of one α and one β subunit. Mammalian tubulin can be purified with relative ease and has been extensively characterised (Dustin, 1978; Roberts & Hyams, 1979). Mammalian cell tubulin has the important characteristic of possessing a specific binding site for the plant alkaloid colchicine. There is one colchicine-binding site per tubulin dimer and radioactive colchicine can be used as an experimental 'tag' for the tubulin molecule. However, early attempts to identify tubulin in fungi such as *Schizosaccharomyces pombe* (Burns, 1973) and *Physarum polycephalum* (Jockusch, Brown & Rusch, 1971) using colchicine-binding activity as a marker were unsuccessful. Colchicine-binding activity has been reported for *Saprolegnia ferax* and *Allomyces moniliformis* (Olson, 1973; Heath, 1975). This activity may well be due to impurities in the [^{3}H]-colchicine used at that time (Davidse & Flach, 1977). Since more critical work has been possible with purified tubulins, it seems that insensitivity to colchicine may well be a general property of fungal tubulins (discussed later).

Davidse & Flach (1977) were able to show that strains of *Aspergillus nidulans* possessed proteins which had similar electrophoretic mobilities to mammalian tubulin on one dimensional polyacrylamide gels. Unfortunately, these proteins were not purified to homogeneity and all attempts to assemble them *in vitro* into microtubules were unsuccessful.

A most important result, however, was that Davidse and his colleagues were able to show that these *A. nidulans* proteins were present in cell extracts that possessed binding activity for methyl benzimidazol-2-yl carbamate (MBC). A number of laboratories have used copolymerisation in an attempt to identify fungal tubulins. These copolymerisation experiments involve mixing radiolabelled fungal cell extracts with purified mammalian brain microtubule protein. The protein is then taken through two or more cycles of assembly and disassembly *in vitro* and the product is run on SDS polyacrylamide gels. The presence of any copurifying fungal tubulins is detected by autoradiography. There are some technical problems involved in this approach, but it has been used to give an initial identification of tubulins in *Dictyostelium discoideum* (Capuccinelli, Martinotti & Hames, 1978); *Saccharomyces cerevisiae* (Water & Kleinsmith, 1976; Clayton, Pogson & Gull, 1979; Kilmartin, 1981); *A. nidulans* (Sheir-Neiss *et al.*, 1976) and *Physarum polycephalum* (Roobol, Pogson & Gull, 1980*a*). This approach is, of course, limited only to the identification of the fungal tubulins and it does not constitute a purification protocol. However, the purification of tubulin has now been achieved with three fungi – *P. polycephalum* (Roobol, Pogson & Gull, 1980*b*), *S. uvarum* and *S. cerevisiae* (Kilmartin, 1981). These fungal tubulins have now been characterised in some detail, particularly in comparison with mammalian cell tubulin. Although other fungal tubulins have not been purified, a more extensive characterisation of the *Dictyostelium* tubulins has been performed using immunological techniques, and the molecular cloning of the tubulin genes of *S. cerevisiae* (Neff *et al.*, 1983) and *S. pombe* (Toda, Adachi & Yanagida, 1984) has provided valuable information on the amino acid sequence of fungal tubulins.

Tubulin polypeptides of Physarum and Dictyostelium

Physarum polycephalum is a slime mould which displays several different arrangements of microtubules in the different phases of its life cycle. The unicellular myxamoebae possess cytoplasmic microtubules along with mitotic spindle and centriolar microtubules. The myxamoebae can transform under appropriate conditions to a flagellate where microtubules are additionally used as the components of the flagellar axonemes. The myxamoebae can also be transformed into the large, syncytial plasmodium which does not possess cytoplasmic microtubules or centrioles. The only microtubules present in this stage of the *Physarum* life cycle are the microtubules of the mitotic and (potentially) the meiotic spindles (Havercroft & Gull, 1983). Only the myxamoebal and plasmodial forms in the life cycle are proliferative cells; however, the flagellate can return to the myxamoebal cell type.

Initially, we developed a method to enable us to purify tubulin from the myxamoebae by assembly of microtubules *in vitro* (Roobol *et al.*, 1980*b*; Roobol & Gull, 1982). The tubulins purified by these procedures have been subjected to extensive biochemical characterisation (Clayton *et al.*, 1980; Roobol *et al.*, 1980*b*; Quinlan *et al.*, 1981; Clayton & Gull, 1982). In summary, one dimensional SDS polyacrylamide gels of *Physarum* tubulin have shown it to consist of one β subunit that is similar to the mammalian β-tubulin, and an α subunit that is dissimilar to mammalian α-tubulin in that it migrates ahead of the β subunit. The difference between the *Physarum* tubulin and mammalian tubulin is also reflected in their sensitivity to antimicrotubule agents. *Physarum* tubulin is insensitive to colchicine, even at concentrations well above those which totally abolish the polymerisation of mammalian tubulin *in vitro* (Quinlan *et al.*, 1981). The *Physarum* tubulins are, however, sensitive to some of the benzimidazole drugs (discussed later). On two dimensional (2D) gels these myxamoebal tubulins separate into just two spots which have been designated $\alpha 1$ and $\beta 1$. Recently, we have been able to characterise the tubulin polypeptides expressed in the plasmodium of *Physarum*. In this phase of the life cycle we have shown that four tubulin species are separable by 2D gel electrophoresis (Fig. 14.2). These are

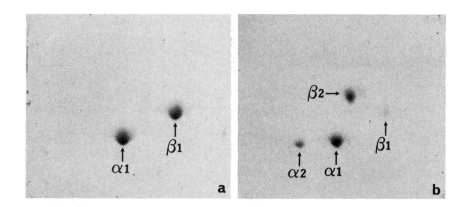

Fig. 14.2. Two dimensional gel analysis of tubulins purified from (*a*) myxamoebae and (*b*) plasmodium of *Physarum polycephalum*. The myxamoebae express two tubulin species, $\alpha 1$ and $\beta 1$ whilst the plasmodium expresses four; $\alpha 1$, $\alpha 2$, $\beta 1$ and $\beta 2$.

the $\alpha 1$, $\alpha 2$, $\beta 1$ and $\beta 2$ (Burland *et al.*, 1983; Roobol *et al.*, 1983). Translation *in vitro* of the myxamoebal and plasmodial RNAs indicates that there are distinct mRNAs and therefore, probably separate genes for the $\alpha 1$-, $\alpha 2$-, $\beta 1$- and $\beta 2$-tubulins. Consequently, in *Physarum* there is a cell-type-dependent expression of different members of a tubulin multigene family.

Dictyostelium discoideum is a cellular slime mould whose sole proliferative phase is a small, uninucleate amoeba. These amoebae possess cytoplasmic microtubules and during mitosis, microtubules in the mitotic spindle. The *Dictyostelium* amoebal tubulin has never been purified although recently White, Tolbert & Katz (1983) have produced a careful identification using well characterised antibodies. They separated *Dictyostelium* proteins using two dimensional gels and noted that there were no spots in the position that is characteristic for mammalian tubulins [the position coordinates being defined by a polypeptide's isolectric point (first dimension), and molecular weight (second dimension)]. The 2D gels were subjected to Western blotting and probed with two monoclonal antibodies specific for α-tubulin (Kilmartin, Wright & Milstein, 1982) and a polyclonal antibody that recognised both α- and β-tubulin. This approach detected the two *Dictyostelium* tubulin spots as being of similar molecular weight to mammalian tubulin but having a much more basic isoelectric point, the pI of the mammalian tubulins being 5.7–6.0 whilst the *Dictyostelium* tubulins had pIs in the range 6.2–6.7. Also, as found previously with *Physarum* the *Dictyostelium*, α-tubulin migrated faster than the β-tubulin in the second dimension. This reversed order of migration of the α- and β-tubulins of *Physarum* and *Dictyostelium* emphasises the importance of establishing identity of tubulins with rigorous criteria such as one and 2D peptide mapping and well characterised monoclonal antibodies.

Tubulin polypeptides of yeasts

The tubulins of *Saccharomyces uvarum* and *S. cerevisiae* have been purified by Kilmartin (1981) and shown to be capable of assembly *in vitro* into microtubules. Kilmartin's approach was to produce a partial purification by DEAE Sephadex chromatography and then to concentrate the column fractions containing the yeast tubulin. Warming this highly concentrated column eluate to 25°C produced microtubules, and this in-vitro assembly yielded a highly purified preparation of yeast tubulin. Two dimensional polyacrylamide gel analysis of these preparations revealed two major subunits – the α- and β-tubulins of yeast. This ability to purify the tubulin and assemble it *in vitro* into microtubules has a number of advantages in the subsequent analysis of the tubulins. Kilmartin was able to show that yeast tubulin possessed a number of characteristics that differentiated it from mammalian tubulin including its sensitivity to the benzimidazole fungicides and as *Physarum*, its insensitivity to colchicine (discussed later).

The tubulin polypeptides of *Saccharomyces cerevisiae* and *S. pombe* have recently been more extensively characterised by molecular cloning techniques. Neff *et al.* (1983) have identified the β-tubulin gene of *S. cerevisiae* encoded at the *TUB 2* locus. Cloning of this gene has allowed its nucleotide sequence to be determined, and so the predicted amino acid sequence to be compared with that of two vertebrate β-tubulins from pig and chicken (Valenzuela *et al.*, 1980; Krauhs *et al.*, 1981). The correlation between the predicted amino acid sequence of *S. cerevisiae* β-tubulin and that of the animal β-tubulins is very close, 70% of amino acid residues in the chicken sequence being identical with the yeast. *S. cerevisiae* β-tubulin is 12 residues longer than its chicken brain counterpart, the predicted molecular weights being 51 073 for the yeast and 49 935 for chicken. Gene disruption experiments showed that the *S. cerevisiae* β-tubulin gene encoded at the *TUB 2* locus is essential for the growth of a haploid strain and cannot be substituted by a cryptic copy of this gene elsewhere in the genome. This and other experiments suggest that there is only one β-tubulin gene in a haploid *S. cerevisiae* genome.

Toda *et al.* (1983) have studied the α- and β-tubulin genes of *Schizosaccharomyces pombe* by molecular cloning techniques. The initial experiments in this work included the isolation of cold-sensitive mutants. A number of these *NDA2* mutants possessed pleiotropic phenotypes at the restrictive temperature of 22°C – including arrest of nuclear division, absence of mitotic spindles and abnormal nuclear positioning. Some of these mutants were also supersensitive to thiabendazole and MBC at the permissive temperature. Toda *et al.* (1983) subsequently identified the *NDA2* gene of *S. pombe* as one of two α-tubulin genes in this organism by using transformation, nucleotide sequencing and integration of cloned sequences into the chromosome. The (NDA2)1 cloned sequence was found to be derived from the *NDA2* gene, but Toda and colleagues could not determine the chromosomal location of the second α-tubulin sequence (NDA2)2 that had been cloned. The nucleotide sequences of the two α-tubulin clones have been determined and showed no extensive homology in the 5′ and 3′ flanking regions. In the coding regions of the two clones, however, there was considerable homology – about 70% overall. The (NDA2)1 sequence possessed a 90 bp intervening sequence although no intervening sequence was present in the (NDA2)2 sequence. The two *S. pombe* α-tubulin clones had sequences which showed high homology to pig α-tubulin, although the overall size varied slightly. Pig α-tubulin has 451 amino acids, four smaller than (NDA2)1 and two larger than (NDA2)2. 107 substitutions and four additions were found in (NDA2)1 in comparison with pig α-tubulin, giving 76% homology. (NDA2)2 was calculated to have the same degree of homology (76%) when compared with pig α-tubulin – 106 substitutions and two deletions. Extensive changes were detected in the regions 35–44 and 444–52, indicating that these two regions may be hot spots of evolutionary variation for α-tubulin. The tubulins are considered to be evolutionarily conserved proteins; however, this work has shown that there are over 100 changes between the *S. pombe* and pig α-tubulin amino acid sequences whilst there are only three amino acid substitutions between the chicken and pig sequences (Ponstingl *et al.*, 1981).

It is clear that the purification of the *Physarum* and *Saccharomyces* tubulins has allowed differences between fungal and mammalian tubulins to be characterised biochemically. Recent progress using molecular cloning techniques and the determination of nucleotide sequence of the tubulin genes has started to produce a fascinating insight into the evolution of the α- and β-tubulin polypeptides. In each case, knowledge of the interaction of the benzimidazole fungicides with tubulin has been crucial to the development of the research.

Benzimidazole fungicides

Derivatives of benzimidazoles (Fig. 14.3) were introduced commercially as fungicides and anthelmintics. Of the fungicides, benomyl has probably been the most studied with regard to biological activity; the activity of benomyl *in vivo* can be largely ascribed to its hydrolysis product methyl benzimidazol-2-yl carbamate (MBC; Seiler, 1975).

Actions in vivo

Benomyl is effective as a systemic fungicide against a broad range of pathogenic and nonpathogenic fungi. Hastie (1970) observed that levels of benomyl as low as 1 μM made diploid strains of *Aspergillus nidulans* mitotically unstable, causing them to break down to haploids and aneuploids. This suggested that benomyl caused abnormal chromosome segregation at mitosis. Higher levels (*c.* 4 μM) of MBC, the active breakdown product of benomyl, inhibited mitosis, DNA synthesis and growth of *A. nidulans* (Davidse, 1973); since the inhibition of mitosis preceded inhibition of DNA synthesis and growth, it was deduced that inhibition of mitosis was the primary effect of MBC. A direct effect of MBC on the functioning of the mitotic spindle could account for both mitotic instability and inhibition of mitosis. In *Saccharomyces cerevisae* and *Ustilago maydis*, inhibition of cytokinesis by benomyl and by MBC was noted in addition to inhibition of mitosis (Hammerschlag & Sisler, 1973). Again, it was reasoned that the primary

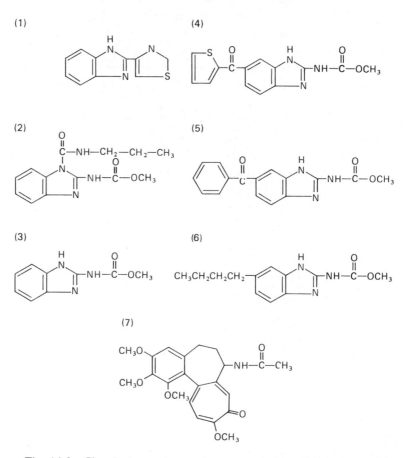

Fig. 14.3. Chemical structures of some antimicrotubule agents: (1) thiabendazole; (2) benomyl; (3) methyl benzimidazol-2-yl carbamate: (4) nocodazole; (5) mebendazole; (6) parbendazole; (7) colchicine (from Gull, 1981).

effect of the drug was inhibition of mitosis, and the inhibition of cytokinesis was a result of the mitotic block.

The production of mitotic instability at low drug levels and the inhibition of mitosis at higher drug levels are common effects of several benzimidazoles on a number of organisms. In the slime moulds *Dictyostelium discoideum* and *Polysphondylium pallidum*, low levels of benomyl, thiabendazole, cambendazole and nocodazole cause mitotic breakdown of diploid myxamoebae into haploids (Williams & Barrand, 1978; Welker & Williams, 1980). Higher levels of thiabendazole, cambendazole and nocodazole inhibit mitosis (Cappuccinelli, Fighetti & Rubino, 1979; Williams, 1980). High benzimidazole levels also lead to the production of diploids from haploid myxamoebae of *D. discoideum* and *P. pallidum* (Williams & Barrand, 1978; Welker & Williams, 1980). Presumably, the diploid myxamoebae are generated in three steps: the first step, a block in mitosis; the second step, a block in cytokinesis (as a consequence of the mitotic block); and the third step, replication of the chromosomes that failed to segregate during the block to division, thus generating a viable diploid cell. Such a sequence of events would suggest that the action of benzimidazoles is specific to mitosis, allowing other essential processes (including chromosome replication) to continue normally. However, the possibility that benzimidazoles generate diploid cells by causing two haploid cells to fuse cannot yet be ruled out.

Benzimidazoles are active on both vegetative phases of the life cycle of the slime mould *Physarum polycephalum*. Growth of the uninucleate myxamoebae is inhibited by benomyl, MBC, parbendazole and nocodazole (Wright *et al.*, 1976; Mir & Wright, 1977; Quinlan *et al.*, 1981; T. G. Burland, unpublished). Under moist conditions, *P. polycephalum* myxamoebae normally develop flagella, but flagellar development is inhibited by both MBC and nocodazole (Mir & Wright, 1977). As in the mitotic spindle, microtubules are a major structural component of flagella, and it seems likely that it is through action on microtubules that benzimidazoles inhibit flagellar development as well as mitosis. Growth of the multinucleate plasmodial phase of *P. polycephalum* is much less sensitive to benzimidazoles than is growth of myxamoebae. However, at concentrations higher than needed to inhibit myxamoebal growth, MBC, nocodazole and parbendazole each lead to a vast increase in plasmodial nuclear size; these giant nuclei are probably polyploid (Wright *et al.*, 1976; T. G. Burland, unpublished), arising by continued nuclear replication in the absence of nuclear division. Wright *et al.* (1976) noted that many of the giant nuclei produced following MBC treatment contained abnormal microtubule structures, consistent with a direct effect of MBC on microtubules. Sporulation of *P. polycephalum* plasmodia is also inhibited by nocodazole, but Chapman & Coote (1982) suggested that this may be a consequence of the inhibition by nocodazole of the mitotic divisions that immediately precede sporulation.

Yeasts also are sensitive to benzimidazoles. MBC induces mitotic chromosome loss and generates diploids from haploids at high frequency in *Saccharomyces cerevisiae* (Wood, 1982). In *S. cerevisiae* and *Schizosaccharomyces pombe*, the effects of benomyl and MBC on cell cycle progression have been studied. In both organisms these benzimidazoles allow cell cycle progression through S phase, and arrest the cell cycle at or very shortly before mitosis (Quinlan, Pogson & Gull, 1980; Fantes, 1982; Walker, 1982; Wood & Hartwell, 1982).

All of the above effects of benzimidazoles on lower eukaryotes *in vivo* could be accounted for by a direct and specific action of the drugs on microtubules. Benzimidazoles seem to have analogous effects on higher eukaryotes. For example, treatment of cultured mammalian cells with nocodazole leads to the disassembly of cytoplasmic and mitotic spindle microtubules (De Brabander *et al.* 1976, 1981). Similarly, parbendazole leads to the depolymerisation of the microtubules in cultured Vero cells (Havercroft, Quinlan & Gull, 1980*a*). Thus benzimidazoles act on microtubules from a very broad range of eukaryotes, suggesting substantial conservation through evolution of their site(s) of action.

Actions in vitro

Since benzimidazoles seem to act specifically on microtubules, it should be fruitful to study the drug–microtubule interaction *in vitro*. Ideally, these studies should be performed on a system where the assembly of tubulin subunits into microtubules can be controlled *in vitro*. Until recently, this assembly was possible only with tubulins from mammalian brain.

Hoebeke, Van Nijen & De Brabander (1976) studied the effects of nocodazole on the assembly *in vitro* of rat brain tubulin. They found that nocodazole prevented the assembly of microtubules, but did not promote disassembly. The apparent discrepancy with actions *in vivo*, where benzimidazoles lead to disassembly of microtubules, is presumably a result of a dynamic equilibrium *in vivo* where tubulin subunits are continually being added to and removed from microtubules (e.g., see Fulton & Simpson, 1979); inhibition of polymerisation would thus lead to a net depolymerisation of microtubules. Evidence was obtained (Hoebeke *et al.*, 1976) that nocodazole binds to tubulin. When radiolabelled nocodazole was mixed with rat brain tubulin, radioactivity and tubulin subsequently co-eluted from a Sephadex G50 column. Binding of radiolabelled nocodazole to tubulin was also confirmed by equilibrium dialysis (Hoebeke *et al.*, 1976). An important finding, based on the colchicine-binding assay of Sherline, Bodwin & Kipnis (1974), was that nocodazole competitively inhibited the binding of colchicine to purified rat brain tubulin (Hoebeke *et al.*, 1976). Colchicine, a known specific antimicrotubule drug active in higher eukaryotes, had been ineffective as an antimicrotubule agent in lower eukaryotes, so competitive inhibition of colchicine binding by nocodazole strongly suggested that benzimidazoles may be effective in this role in lower eukaryotes. Inhibition of assembly of mammalian brain tubulin into microtubules *in vitro* has since been noted for a number of benzimidazoles. Nocodazole, oxibendazole, parbendazole, mebendazole and fenbendazole all inhibit bovine brain microtubule assembly more effectively than does colchicine; benomyl, cambendazole, MBC and thiabendazole inhibit assembly less effectively (Friedman & Platzer, 1978).

Although no microtubule assembly *in vitro* was available for *Aspergillus nidulans*, Davidse & Flach (1977) were able to show that radiolabelled MBC formed a complex with a tubulin-like protein from *A. nidulans* extracts. The MBC-protein complex was sufficiently stable to elute in a single peak from a Sephadex G200 column, at an apparent molecular weight (110 000) characteristic of $\alpha\beta$-tubulin. The reciprocal of the binding constant for the MBC-protein reaction (2.2 μM) was remarkably close to the concentration of MBC (4.5 μM) needed to exert a 50% inhibition of mycelial growth; furthermore, the altered MBC–protein-binding constants in strains resistant and supersensitive to MBC correlated well with the altered sensitivity of growth to MBC in these strains (Davidse & Flach, 1977), suggesting that MBC acted specifically on the tubulin-like protein of *A. nidulans*. Davidse & Flach (1977) also obtained evidence that nocodazole and colchicine competitively inhibited the binding of MBC to this protein.

Roobol *et al.* (1980*a*) were able to achieve microtubule assembly *in vitro* with tubulins from *Physarum polycephalum* myxamoebae, thus opening the way for studies *in vitro* of lower eukaryote microtubule assembly. Quinlan *et al.* (1981) observed that both parbendazole and nocodazole inhibited microtubule assembly *in vitro*, and that the relative effectiveness of these drugs on microtubule assembly correlated with their effectiveness at inhibiting myxamoebal growth. As Hoebeke *et al.* (1976) had noted for nocodazole binding to rat brain tubulin, Quinlan *et al.* (1981) observed that binding of parbendazole to *P. polycephalum* tubulin was readily reversible. This property of benzimidazole action will be of great value for studies *in vivo*, where microtubule structure and function can be analysed by treatment with antimicrotubular drugs. The antimicrotubule action of benzimidazoles in lower eukaryotes has been confirmed in yeast, where benomyl has been shown to inhibit microtubule assembly in the in-vitro system recently developed by Kilmartin (1981). Both of these studies on *Physarum* and *Saccharomyces* purified tubulins clearly showed that colchicine was not a very effective inhibitor of polymerisation.

Genetics of resistance to benzimidazoles

Genetic analysis of fungal resistance to benzimidazoles has been most extensive in *Aspergillus nidulans*. Hastie & Georgopoulos (1971) isolated nine mutants resistant to benomyl, and all were cross-resistant to MBC and to thiabendazole (TBZ). Two mutants were analysed, and each carried a single mutation conferring benzimidazole resistance; the two mutations were, however, mapped to unlinked loci, *ben1* and *ben2* (Hastie & Georgopoulos, 1971). Van Tuyl (1977) was able to isolate mutants resistant to benzimidazoles in ten different species, from the genera *Aspergillus, Aureobasidium, Cercosporella, Cladosporium, Fusarium, Penicillium, Phialophora, Rhodotorula* and *Ustilago*. Genetic analysis of the *A. nidulans* mutants showed that mutation at any one of three loci could confer MBC resistance on the mycelium. However, most of the mutations mapped to one locus, *benA*, and Van Tuyl (1977) showed that *benA* was the same locus as *ben1* identified by Hastie & Georgopoulos (1971). The level of drug resistance seemed to be higher among *benA* mutants than *benB* or *benC*, although mutations at all three loci were semi-dominant. By contrast, in *Aspergillus niger* and *Ustilago maydis* (Van Tuyl, 1977) and in *Neurospora crassa* (Borck & Bramer, 1974), only a single locus was identified that could mutate to MBC resistance.

Because of the known action of benzimidazoles on microtubules, it was reasoned that at least some of the mutations conferring MBC resistance would be in tubulin genes. In *Aspergillus nidulans*, Sheir-Neiss, Lai & Morris (1978) identified the tubulins of the wild-type among proteins resolved by two dimensional (2D) gel electrophoresis. Of 26 benomyl-resistant *benA* mutants studied, 18 had electrophoretically altered β-tubulins, whereas the α-tubulins were all normal (Sheir-Neiss *et al.*, 1978). This indicated that the *benA* mutations were in a β-tubulin structural gene. Analysis of diploids heterozygous at *benA* showed co-dominant expression of normal and mutant β-tubulins, confirming that the mutations were not in a gene whose product simply modified β-tubulin. The observation that benzimidazole resistance can be acquired through structural alteration of β-tubulin greatly strengthens the view that benzimidazoles act directly on tubulin. The products encoded by the *benB* and *benC* loci in *A. nidulans* have not yet been identified.

A few *benA* mutations that confer benomyl resistance of *Aspergillus nidulans* also make the mycelium temperature-sensitive for growth, presumably by altering the β-tubulin structure such that microtubules become unstable at nonpermissive temperature (Oakley & Morris, 1981). This confirms that the cellular target for benzimidazoles, the *benA* gene product, is essential for cell growth. One of the mutations causing temperature-sensitive growth, *benA33* (Oakley & Morris, 1981), results in a block in mitosis when mycelia are grown at nonpermissive temperatures. This shows that the *benA*-encoded β-tubulin is essential for mitosis. Oakley & Morris (1981) went on to show that at nonpermissive temperatures, the mitotic spindle microtubules are unable to depolymerise in *benA33* strains.

Mutations that suppress MBC resistance have been studied in *Aspergillus nidulans*. Several mutations that suppress *benA*-encoded MBC resistance (or that suppress the temperature-sensitive growth phenotype of some *benA* mutants) also map to the *benA* locus (Morris, Lai & Oakley, 1979). At least some of these suppressor mutations in *benA* are not true revertants since they also lead to novel structural changes in β-tubulin, detectable on 2D gels. In addition, other suppressors of *benA*-encoded MBC resistance map to *tubA*, a structural gene for α-tubulin (Morris *et al.*, 1979). This shows that alterations in α-tubulin as well as in β-tubulin structure can alter sensitivity to benzimidazoles. This being the case, it is not yet clear why some of the mutants selected for resistance to benzimidazoles seem not to carry mutations in α-tubulin structural genes.

Benzimidazole resistance has also been analysed genetically in yeasts. Yamamoto (1980) isolated several mutants of *Schizosaccharomyces pombe* resistant to TBZ, and all were cross-resistant to MBC. The frequency of mutation to resistance was higher at 36 °C than at 26 °C, and different classes of mutants were isolated at the two temperatures.

Mutants isolated at 26 °C were resistant to TBZ at all growth temperatures, and all the mutations mapped to a single locus, *ben1*. Resistant mutants isolated at 36 °C, however, were not resistant to TBZ at lower temperatures, and the mutations mapped to two other loci, *ben2* and *ben3* (Yamamoto, 1980). The unusual temperature-dependent drug-resistance phenotypes may be related to the fact that wild-type *S. pombe* is naturally more resistant to TBZ at 36 °C than at lower temperatures.

Roy & Fantes (1982) also isolated *Schizosaccharomyces pombe* mutants resistant to benzimidazoles (benomyl), but they sought a particular class, one that was cold-sensitive for cell division. Of 150 benomyl-resistant mutants isolated at 35 °C, seven became blocked in mitosis when shifted down to 20 °C; these mutants mapped to a single locus, *ben4* unlinked to *ben1*, *2* and *3* (Roy & Fantes, 1982). At low temperature, the *ben4* mutants were blocked in cell cycle progression rather than in growth, with the cell cycle block point shortly before mitosis. Roy & Fantes (1982) suggested that this phenotype was consistent with a mutation in a β-tubulin structural gene. However, unlike most other mutants resistant to benomyl or to MBC (including the *Aspergillus nidulans benA* mutants), these *ben4* mutants of *S. pombe* were not cross-resistant to another benzimidazole, TBZ, and six of the seven *ben4* mutants were supersensitive to the spindle poison isopropyl-*N*-(3-chlorophenyl)carbamate (CIPC). It thus seems quite possible that the *ben4* mutants represent a novel class of benzimidazole-resistant mutants that carry mutations in a gene other than a β-tubulin gene – perhaps in a gene encoding α-tubulin or a microtubule-associated protein.

In *Physarum polycephalum*, mutations conferring resistance of myx-amoebae to MBC (and cross-resistance to parbendazole) have been mapped to four unlinked loci, *benA, B, C* and *D* (T. G. Burland, T. Schedl & W. F. Dove, unpublished data). There are at least three DNA sequences in the haploid *P. polycephalum* genome that are homologous to β-tubulin (T. Schedl, unpublished); because these sequences are located on restriction fragments that show natural variation in length, it is possible to investigate by simple Mendelian genetics whether any of the MBC-resistance mutations map to the β-tubulin DNA sequences (see Schedl & Dove, 1982 for an example of this technique). The *benA* and *ben D* loci are each linked to a different β-tubulin DNA sequence, suggesting that *benA* and *benD* each define a different β-tubulin structural gene (T. G. Burland, T. Schedl & W. F. Dove, unpublished). Since the *P. polycephalum* tubulins have been identified (Burland *et al.*,

1983), it has been possible to show that one mutation in *benD*, *benD210*, results in the production of a novel, electrophoretically altered β-tubulin polypeptide in the myxamoebae, confirming *benD* as a structural gene for β-tubulin (T. G. Burland, T. Schedl & W. F. Dove, unpublished). Myxamoebae carrying the *benD210* mutation also express a normal β-tubulin (which is presumably the product of the *benA* gene), indicating that at least two β-tubulin genes are expressed in the myxamoebae. Thus mutation in only one of two expressed β-tubulin genes is sufficient to confer MBC resistance on the cells. Remarkably, the *benD210* mutation is recessive to its wild-type allele, yet confers MBC resistance despite the concurrent expression of the other normal 'MBC-sensitive' β-tubulin (T. G. Burland, T. Schedl & W. F. Dove, unpublished). One possible explanation of this result is that microtubules (MTs) are assembled using both the expressed β-tubulins and that it is not necessary for all of the β-tubulin subunits of a MT to be mutant in order to make the MT resistant to MBC. Alternatively, MTs may be assembled using either one β-tubulin or the other, but not both; in this case, only those MTs containing mutant β-tubulin would function in the presence of MBC. Either way, it would seem that the two myxamoebal β-tubulins (one encoded by *benD*, the other presumably encoded by *benA*) are functionally redundant, at least in terms of functions necessary for cell survival in the presence of MBC.

The *benA* and *benD* genes show different patterns of expression over the *Physarum polycephalum* life cycle; *benA* mutations confer MBC resistance only on the myxamoebae, whereas *benD* mutations confer MBC resistance on both myxamoebae and plasmodia (T. G. Burland, T. Schedl & W. F. Dove, unpublished). As expected from this result, the mutant β-tubulin expressed in *benD210* myxamoebae is also expressed in plasmodia, while the presumptive *benA* encoded β-tubulin is absent in plasmodia.

The nature of the structural genes defined by MBC resistance mutations in the *benC* and *benB* loci of *Physarum polycephalum* is unknown, although it is clear that *benC*, which is expressed only in myxamoebae, is not linked to any of the β-tubulin DNA sequence loci (T. G. Burland, T. Schedl & W. F. Dove, unpublished). It is suspected that the *benB* locus controls MBC uptake; *benB* mutations are expressed in both myxamoebae and plasmodia, and plasmodia carrying *benB* mutations are morphologically abnormal (T. G. Burland, unpublished), suggesting a membrane alteration (plasmodia have no cell wall). Since there are no cytoplasmic MTs in plasmodia (Havercroft & Gull,

1983), it seems unlikely that tubulin mutations could lead to morphological abnormality.

In summary, mutations conferring resistance to benzimidazoles have been characterised in several fungi. In at least some of these organisms, multiple loci can mutate to give a drug-resistant phenotype. In both *Aspergillus nidulans* and *Physarum polycephalum*, it has been shown that benzimidazole resistance can arise by mutations in β-tubulin genes that lead to alteration of β-tubulin structure, observations that strengthen conclusions from studies *in vivo* and *in vitro* that benzimidazoles act specifically on tubulin. That the only altered gene products in benzimidazole-resistant mutants so far identified are β-tubulins might seem to suggest that benzimidazoles bind specifically to β-tubulin and not to α-tubulin. However, β-tubulin interacts intimately with other molecules in the cell, notably α-tubulin and microtubule-associated proteins (MAPs). It seems possible that alterations not only in the benzimidazole binding site(s) but also in allosteric sites on any of the tubulins or MAPs could also modify benzimidazole binding. Thus, mutation in any of the genes encoding these proteins may alter benzimidazole sensitivity. Indeed, some mutations in an α-tubulin gene of *A. nidulans* increase benomyl sensitivity (i.e. suppress resistance; Morris *et al.*, 1979). Thus it is not at all clear from the genetic evidence where the benzimidazole binding site(s) lies. The genetic evidence does show, however, that gene products other than β-tubulins are involved in benzimidazole action. Some of these gene products may simply control drug uptake, but others are likely to involve the intracellular site(s) of action. Identification of the gene products of the *benB* and *benC* loci of *A. nidulans*, the *ben1, 2, 3* and *4* loci of *Schizosaccharomyces pombe* and the *benC* and *benB* loci of *P. polycephalum* will surely enhance our knowledge of the cellular sites of action of benzimidazoles and of the interaction of those sites with other gene products.

Acknowledgements. We thank our colleagues W. F. Dove, T. Schedl, A. Roobol, E. C. A. Paul and M. Wilcox for comments on the manuscript and permission to quote unpublished data. T. G. Burland is supported by Program-Project Grant CA-23076 in Tumor Biology to W. F. Dove and by Core Grant CA-07175 to the McArdle Laboratory from the National Cancer Institute. Work in K. Gull's laboratory is supported by grants from The Wellcome Trust, The Science and Engineering Research Council, The Agricultural Research Council and The Royal Society.

References

Beakes, G. (1981). Ultrastructure of the Phycomycete nucleus. In *The Fungal Nucleus*, ed. K. Gull & S. G. Oliver, pp. 1–35. Cambridge University Press.

Beckett, A. (1981). Ultrastructure and behaviour of nuclei and associated structures within the meiotic cells of Euascomycetes. In *The Fungal Nucleus*, ed. K. Gull & S. G. Oliver, pp. 37–63. Cambridge University Press.

Borck, K. & Braymer, H. D. (1974). The genetic analysis of resistance to benomyl in *Neurospora crassa*. *Journal of General Microbiology*, **85**, 51–6.

Burland, T. G., Gull, K., Schedl, T., Boston, R. S. & Dove, W. F. (1983). Cell type-dependent expression of tubulins in *Physarum*. *Journal of Cell Biology*, **97**, 1852–9.

Burns, R. G. (1973). (^{3}H)-Colchicine binding: failure to detect any binding to soluble proteins from various lower organisms. *Experimental Cell Research*, **81**, 285–92.

Byers, B., Shriver, K. & Goetsch, L. (1978). The role of spindle pole bodies and modified microtubule ends in the initiation of microtubule assembly in *Saccharomyces cerevisiae*. *Journal of Cell Science*, **30**, 331–53.

Cappuccinelli, P., Fighetti, M. & Rubino, S. (1979). A mitotic inhibitor for chromosomal studies in slime moulds. *FEMS Microbiology Letters*, **5**, 25–7.

Capuccinelli, P., Martinotti, G. & Hames, B. D. (1978). Identification of cytoplasmic tubulin in *Dictyostelium discoideum*. *FEBS Letters*, **91**, 153–7.

Chapman, A. & Coote, J. G. (1982). Sporulation competence in *Physarum polycephalum* CL and the requirement for DNA replication and mitosis. *Journal of General Microbiology*, **128**, 1489–1501.

Clayton, L. & Gull, K. (1982). Tubulin and the microtubule organising centres of *Physarum polycephalum* myxamoebae. In *Microtubules in Microorganisms*, ed. P. Cappuccinelli & N. R. Morris, pp. 179–201. New York & Basel: Marcel Dekker.

Clayton, L., Pogson, C. I. & Gull, K. (1979). Microtubule proteins in the yeast *Saccharomyces cerevisiae*. *FEBS Letters*, **106**, 67–70.

Clayton, L., Quinlan, R. A., Roobol, A., Pogson, C. I. & Gull, K. (1980). A comparison of tubulins from mammalian brain and *Physarum polycephalum* using SDS-polyacrylamide gel electrophoresis and peptide mapping. *FEBS Letters*, **115**, 301–5.

Davidse, L. C. (1973). Antimitotic activity of methylbenzimidazol-2-yl carbamate (MBC) in *Aspergillus nidulans*. *Pesticide Biochemistry and Physiology*, **3**, 317–25.

Davidse, L. C. & Flach, W., (1977). Differential binding of benzimidazol-2-yl carbamate to fungal tubulin as a mechanism of resistance to this antimitotic agent in mutant strains of *Aspergillus nidulans*. *Journal of Cell Biology*, **72**, 174–93.

De Brabander, M., Geuens, G., Nuydens, R., Willebrords, R. & De Mey, J. (1981). Microtubule stability and assembly in living cells: the influence of metabolic inhibitors, taxol and pH. *Cold Spring Harbor Symposium of Quantitative Biology*, **46**, 227–40.

De Brabander, M. J., Van De Veire, R. M. L., Aerts, F. E. M., Borgers, M. & Janssen, P. A. J. (1976). The effects of methyl[5-(2-thienylcarbonyl)-1H-benzimidazol-2-yl]carbamate, (R17934; NSC 238159), a new synthetic antitumoral drug interfering with microtubules, on mammalian cells cultured *in vitro*. *Cancer Research*, **36**, 905–16.

Dustin, P. (1978). *Microtubules*. Berlin: Springer-Verlag.

Fantes, P. A. (1982). Dependency relations between events in mitosis in *Schizosaccharomyces pombe*. *Journal of Cell Science*, **55**, 383–402.

Friedman, P. A. & Platzer, E. G. (1978). Interaction of anthelmintic benzimidazoles and benzimidazole derivatives with bovine brain tubulin. *Biochimica et Biophysica Acta*, **544**, 605–14.

Fulton, C. & Simpson, P. A. (1979). Tubulin pools, synthesis and utilization. In *Microtubules*, ed. K. Roberts & J. S. Hyams, pp. 117–74. London: Academic Press.

Gull, K. (1981). Microtubules and microtubule proteins in fungi. In *The Fungal Nucleus*, ed. K. Gull & S. G. Oliver, pp. 113–32. Cambridge University Press.

Hammerschlag, R. S. & Sisler, H. D. (1973). Benomyl and methyl-2-benzimidazole carbamate (MBC): biochemical, cytological and chemical aspects of toxicity to *Ustilago maydis & Saccharomyces cerevisiae. Pesticide Biochemistry and Physiology*, **3**, 42–54.

Hastie, A. C. (1970). Benlate-induced instability of *Aspergillus* diploids. *Nature*, **226**, 771–2.

Hastie, A. C. & Georgopoulos, S. G. (1971). Mutational resistance to fungitoxic benzimidazole derivates in *Aspergillus nidulans. Journal of General Microbiology*, **67**, 371–3.

Havercroft, J. C. & Gull, K. (1983). Demonstration of different patterns of microtubule organization in *Physarum polycephalum* myxamoebae and plasmodia using immunofluorescence microscopy. *European Journal of Cell Biology*, **32**, 67–74.

Havercroft, J. C., Quinlan, R. A. & Gull, K. (1981a). Binding of parbendazole to tubulin and its influence on microtubules in tissue-culture cells as revealed by immunofluorescence microscopy. *Journal of Cell Science*, **49**, 195–204.

Havercroft, J. C., Quinlan, R. A. & Gull, K. (1981b). Characterisation of a microtubule organising centre from *Physarum polycephalum* myxamoebae. *Journal of Ultrastructure Research*, **74**, 313–21.

Heath, I. B. (1975). Colchicine and colcemid binding components of the fungus *Saprolegnia ferax. Protoplasma*, **85**, 177–92.

Heath, I. B. (1981). Nucleus-associated organelles in fungi. *International Review of Cytology*, **69**, 191–221.

Hoebeke, J., Van Nijen, G. & De Brabander, M. (1976). Interaction of oncodazole (R17934), a new antitumoral drug, with rat brain tubulin. *Biochemical and Biophysical Research Communications*, **69**, 319–24.

Howard, R. J. & Aist, J. R. (1977). Effects of MBC on hyphal tip organisation, growth and mitosis of *Fusarium acuminatum*, and their antagonism by D_2O. *Protoplasma*, **92**, 195–210.

Howard, R. J. & Aist, J. R. (1980). Cytoplasmic microtubules and fungal morphogenesis: ultrastructural effects of methyl benzimidazole-2-yl carbamate determined by freeze substitution of hyphal tip cells. *Journal of Cell Biology*, **87**, 55–64.

Hyams, J. S. & Borisy, G. G. (1978). Nucleation of microtubules *in vitro* by isolated spindle pole bodies of the yeast *Saccharomyces cerevisiae. Journal of Cell Biology*, **78**, 401–14.

Jockusch, B. M., Brown, D. F. & Rusch, H. P. (1971). Synthesis and some properties of an actin-like nuclear protein in the slime mould *Physarum polycephalum. Journal of Bacteriology*, **108**, 705–14.

Kilmartin, J. V. (1981). Purification of yeast tubulin by self-assembly *in vitro. Biochemistry*, **20**, 3629–33.

Kilmartin, J. V., Wright, B. & Milstein, C. (1982). Rat monoclonal antitubulin antibodies derived by using a new nonsecreting rat cell line. *Journal of Cell Biology*, **93**, 576–82.

Krauhs, E., Little, M., Kempf, F., Hofer-Warbinek, R., Ade, W. & Ponstingl, H. (1981). Complete amino-acid sequence of β tubulin from porcine brain. *Proceedings of the National Academy of Sciences, USA*, **75**, 4962–6.

Kuriyama, R. C., Sato, Y., Fukui, S. & Nishibayashi, S. (1982). *In vitro* nucleating of microtubules from microtubule-organising center prepared from cellular slime mould. *Cell Motility*, **2**, 257–72.

Mir, L. & Wright, M. (1977). Action of antimicrotubular drugs on *Physarum polycephalum*. *Microbios Letters*, **5**, 39–44.

Morris, N. R., Lai, M. H. & Oakley, C. E. (1979). Identification of a gene for α-tubulin in *Aspergillus nidulans*. *Cell*, **16**, 437–42.

Neff, N. F., Thomas, J. H., Grisafi, P. & Botstein, D. (1983). Isolation of the β-tubulin gene from yeast and demonstration of its essential function *in vivo*. *Cell*, **33**, 211–19.

Oakley, B. R. & Morris, N. R. (1980). Nuclear movement is β-tubulin dependent in *Aspergillus nidulans*. *Cell*, **19**, 255–62.

Oakley, B. R. & Morris, N. R. (1981). A β-tubulin mutation in *Aspergillus nidulans* that blocks microtubule function without blocking assembly. *Cell*, **24**, 837–45.

Olson, L. W. (1973). A low molecular weight colchicine binding protein from the aquatic phycomycete *Allomyces neo-moniliformis*. *Archiv für Mikrobiologie*, **91**, 281–97.

Ponstingl, H., Krauhs, E., Little, M. & Kempf, T. (1981). Complete amino-acid sequence of α-tubulin from porcine brain. *Proceedings of the National Academy of Sciences, USA*, **78**, 2757–61.

Quinlan, R. A., Pogson, C. I. & Gull, K. (1980). The influence of the mitotic inhibitor methyl benzimidazole-2-yl carbamate (MBC) on nuclear division and the cell cycle in *Saccharomyces cerevisiae*. *Journal of Cell Science*, **46**, 341–52.

Quinlan, R. A., Roobol, A., Pogson, C. I. & Gull, K. (1981). A correlation between *in vivo* and *in vitro* effects of the microtubule inhibitors colchicine, parbendazole and nocodazole on myxamoebae of *Physarum polycephalum*. *Journal of General Microbiology*, **122**, 1–6.

Roberts, K. & Hyams, J. S. (1979). *Microtubules*. New York & London: Academic Press.

Roobol, A. & Gull, K. (1982). Purification of microtubule proteins from *Physarum polycephalum* by *in vitro* microtubule assembly. In *Microtubules in Microorganisms*, ed. P. Cappuccinelli & N. R. Morris, pp. 227–33. New York & Basel: Marcel Dekker.

Roobol, A., Havercroft, J. C. & Gull, K. (1982). Microtubule nucleation by the isolated microtubule organising centre of *Physarum polycephalum* myxamoebae. *Journal of Cell Science*, **55**, 365–81.

Roobol, A., Pogson, C. I. & Gull, K. (1980a). Identification and characterisation of microtubule proteins from myxamoebae of *Physarum polycephalum*. *Biochemical Journal*, **189**, 305–12.

Roobol, A., Pogson, C. I. & Gull, K. (1980b). *In vitro* assembly of microtubule proteins from myxamoebae of *Physarum polycephalum*. *Experimental Cell Research*, **130**, 203–15.

Roobol, A., Wilcox, M., Paul, E. C. A. & Gull, K. (1983). Identification of tubulin isoforms in the plasmodium of *Physarum polycephalum* by *in vitro* microtubule assembly. *European Journal of Cell Biology*, **33**, 24–8.

Roy, D. & Fantes, P. A. (1982). Benomyl resistant mutants of *Schizosaccharomyces pombe* cold-sensitive for mitosis. *Current Genetics*, **6**, 195–201.

Schedl, T. & Dove, W. F. (1982). Mendelian analysis of the organization of actin sequences in *Physarum polycephalum*. *Journal of Molecular Biology*, **160**, 41–57.

Seiler, J. P. (1975). Toxicology and genetic effects of benzimidazole compounds. *Mutation Research*, **32**, 151–68.

Sheir-Neiss, G., Lai, M. H. & Morris, N. R. (1978). Identification of a gene for β-tubulin in *Aspergillus nidulans. Cell*, **15**, 639–47.

Sheir-Neiss, G., Nardi, R. V., Gealt, M. A. & Morris, N. R. (1976). Tubulin-like proteins from *Aspergillus nidulans. Biochemical and Biophysical Research Communications*, **96**, 285–90.

Sherline, P., Bodwin, L. K. & Kipnis, D. M. (1974). A new colchicine binding assay for tubulin. *Analytical Biochemistry*, **62**, 400–7.

Toda, T., Adachi, Y. & Yanagida, M. (1984). Two cloned sequences complementing pleiotropic *nda2* mutations of the fission yeast *Schizosaccharomyces pombe* code for α-tubulin. *Journal of Molecular Biology*, (in press).

Tucker, J. N. (1979). Spatial organisation of microtubules. In *Microtubules*, ed. K. Roberts, J. S. Hyams, pp. 315–57. London & New York: Academic Press.

Valenzuela, P., Quiroga, M., Zaldirar, J., Rutter, W., Kirschner, M. & Cleveland, D. (1980). Nucleotide and corresponding amino-acid sequences encoded by α and β tubulin mRNAs. *Nature*, **289**, 650–5.

Van Tuyl, J. M. (1977). Genetics of fungal resistance to systemic fungicides. PhD thesis, University of Wageningen, The Netherlands.

Walker, G. M. (1982). Cell cycle specificity of certain antimicrotubular drugs in *Schizosaccharomyces pombe. Journal of General Microbiology*, **128**, 61–71.

Water, R. D. & Kleinsmith, L. J. (1976). Identification of α and β tubulin in yeast. *Biochemical and Biophysical Research Communications*, **70**, 704–8.

Welker, D. L. & Williams, K. L. (1980). Mitotic arrest and chromosome doubling using thiabendazole, cambendazole, nocodazole and benlate in the slime mould *Dictyostelium discoideum. Journal of General Microbiology*, **116**, 397–407.

White, E., Tolbert, E. M. & Katz, E. R. (1983). The identification of tubulin in *Dictyostelium discoideum:* characterisation of some unique properties. *Journal of Cell Biology*, **97**, 1011–19.

Williams, K. L. (1980). Examination of the chromosomes of *Polysphondylium pallidum* following metaphase arrest by benzimidazole derivates and colchicine. *Journal of General Microbiology*, **116**, 409–15.

Williams, K. L. & Barrand, P. (1978). Parasexual genetics in the slime mould *Dictyostelium discoideum:* haploidization of diploid strains using benolate. *FEMS Microbiology Letters*, **4**, 155–9.

Wood, J. S. (1982). Genetic effects of methyl benzimidazole-2-yl-carbamate on *Saccharomyces cerevisiae. Molecular and Cellular Biology*, **2**, 1064–79.

Wood, J. S. & Hartwell, L. H. (1982). A dependent pathway of gene functions leading to chromosome segregation in *Saccharomyces cerevisiae. Journal of Cell Biology*, **94**, 718–76.

Wright, M., Moisand, A., Tollon, Y. & Oustrin, M. L. (1976). Mise en evidence de l'action du methylbenzimidazole-2-yl-carbamate (MBC) et du methyl[5(-2 thienyl carbonyl)1H benzimidazole-2-yl-carbamate] (R17934) sur le noyau de *Physarum polycephalum* (Myxomycetes). *Comptes Rendue Academie Sciences, Paris*, **283**, 1361–4.

Wright, M., Mir, L. & Moisand, A. (1980). The structure of the proflagellar apparatus of the amoebae of *Physarum polycephalum*: relationship to the flagellar apparatus. *Protoplasma*, **103**, 69–81.

Yamamoto, M. (1980). Genetic analysis of resistant mutants to antimitotic benzimidazole compounds in *Schizosaccharomyces pombe. Molecular and General Genetics*, **180**, 2131–4.

Zickler, D. (1981). Ultrastructure of the yeast nucleus. In *The Fungal Nucleus*, ed. K. Gull & S. G. Oliver, pp. 63–85. Cambridge University Press.

15

The molecular basis for the antifungal activities of N-substituted azole derivatives. Focus on R 51 211

H. VANDEN BOSSCHE, G. WILLEMSENS,
P. MARICHAL, W. COOLS and W. LAUWERS*

*Department of Comparative Biochemistry and *Analytical Department, Janssen Pharmaceutica, Research Laboratories, B-2340 Beerse, Belgium*

Introduction

The triazole derivative, R 51 211 – WHO proposed name, itraconazole – is orally active in the treatment of experimental aspergillosis, cryptococcosis, histoplasmosis, trichophytosis, candidosis and sporotrichosis (Van Cutsem *et al.*, 1983). The first clinical pilot studies with R 51 211 indicate that this new antimycotic has promising features in the treatment of fungal infections (Cauwenbergh, 1983).

Antimycotic imidazole and triazole derivatives interfere at low concentrations ($\geqslant 10^{-11}$ M) with the microsomal cytochrome-P-450-dependent lanosterol 14α-demethylase system, resulting in the accumulation of 14α-methylsterols and decreased availability of ergosterol (Vanden Bossche *et al.*, 1978, 1980, 1984). Although other biochemical targets are affected at higher drug concentrations, e.g. inhibition of fatty acid desaturase (Vanden Bossche *et al.*, 1981) and direct interactions with membrane components (Vanden Bossche *et al.*, 1982; Brasseur *et al.*, 1983) have been observed at concentrations $\geqslant 10^{-8}$M and $\geqslant 10^{-5}$ M respectively, it is hypothesised that the antimycotic activity may be related to the interaction with the cytochrome-P-450-dependent 14α-demethylase system.

In this paper the effects of R 51 211 on the growth of *Candida albicans* and on cytochrome P-450 (cyt. P-450) and sterol synthesis in yeast and mammalian subcellular fractions are compared with those obtained with other antimycotic azole derivatives, especially miconazole and ketoconazole. The chemical structures of R 51 211, miconazole and ketoconazole are given in Fig. 15.1.

Materials and methods
Growth
Strain, inocula and media. Inocula of *Candida albicans* ATCC 28516 (=RV 4688) and ATCC 44859 (=RV 12377) were prepared as previously described (Vanden Bossche, Willemsens & Van Cutsem, 1975; Vanden Bossche *et al.*, 1978). The casein-hydrolysate–yeast-extract–glucose medium (CYG medium) contained 5 g casein hydrolysate, 5 g yeast extract and 5 g glucose per litre. Eagle's minimum essential medium (EMEM; this medium contained penicillin and streptomycin), containing glutamine and (where indicated) foetal calf serum, was prepared as previously described (Vanden Bossche *et al.*, 1980). The outgrowth of mycelium was assessed by microscopic observation. The number of yeast cells per unit volume of CYG medium was determined with a Coulter Counter (Vanden Bossche *et al.*, 1978).

Sterol biosynthesis
Subcellular fraction. Subcellular fractions of *Candida albicans* and of rat liver were obtained as described previously (Willemsens,

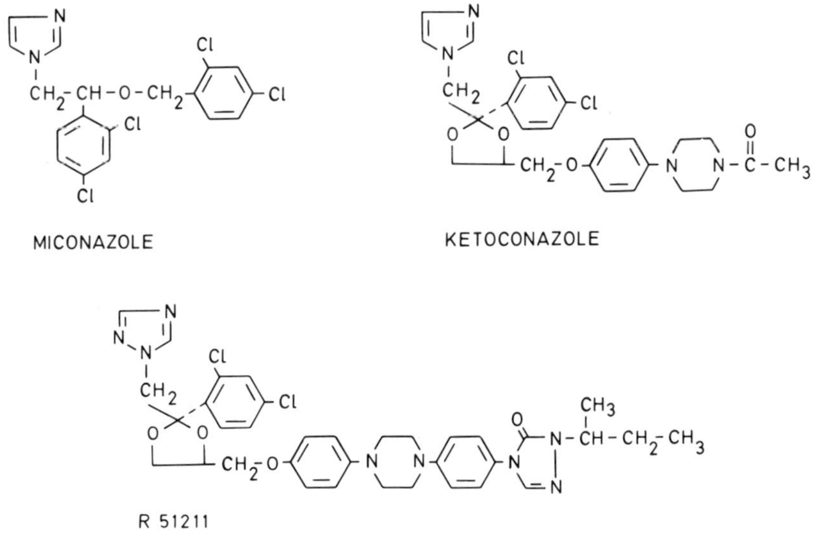

Fig. 15.1. Chemical structures of miconazole, ketoconazole and R 51 211.

Cools & Vanden Bossche, 1980). For the studies presented here, the supernatant of a $10\,000\,g$ centrifugation was used (S $10\,000$ fraction).

[^{14}C]Mevalonate incorporation. Sterol synthesis in yeast and rat liver subcellular fractions was studied by the method of Mitropoulos *et al.* (1976). In summary, the incubation mixture contained 8 ml of the S $10\,000$ fraction (protein content $= 15\,\text{mg ml}^{-1}$), $16\,\text{mM}$ fructose-1,6-diphosphate, $1.6\,\text{mM}$ NAD, $4\,\text{mM}$ MgCl$_2$ and $2.5\,\mu$Ci [^{14}C]mevalonate (DBED-salt); final volume $= 10\,\text{ml}$. Miconazole, ketoconazole and R 51 211 were dissolved in dimethylsulphoxide (DMSO; miconazole) or DMSO $+ 15\,\mu$l $6\,\text{M}$ HCl (ketoconazole and R 51 211). $50\,\mu$l drug and/or solvent were added to the incubation mixture. The incubation was carried out in 100 ml Erlenmeyer flasks in a Heto® reciprocating shaker at $37\,°$C for 3 h. The incubation mixtures were gassed with $0.85\,\text{l}$ of air min^{-1}.

Intact cells. Growth conditions, incubation circumstances, disruption of cells, extraction and saponification of lipids, thin layer chromatography (TLC), determination of radioactivity derived from [^{14}C] acetate and identification of chemical structures of the sterols by gas-chromatographic–mass-spectrometric analysis were as described previously (Vanden Bossche *et al.*, 1978, 1980).

Cytochrome P-450

Strain, inocula and media. Inocula of *Saccharomyces cerevisiae* (B.19328-1) were prepared as previously described (Vanden Bossche *et al.*, 1978). Cells were grown aerobically in a reciprocating shaker at $37\,°$C for 64 h, and 0.5 ml portions were used to inoculate 100 ml of CYG medium. Cells were grown for 24 h at $30\,°$C without shaking and 20 ml (*c.* 5×10^8 cells) was used to inoculate 5 l of PYG medium; PYG medium contained 1 g polypeptone, 1 g yeast extract and 4 g glucose per 100 ml (Ishidate *et al.*, 1969).

Isolation of yeast microsomes. *Saccharomyces cerevisiae*, grown at $30\,°$C on a magnetic stirrer for 18 h in Erlenmeyer flasks containing 5 l PYG medium, were harvested by centrifugation. 20 g of cells were suspended in 20 ml of $0.65\,\text{M}$ sorbitol and 20 g glass beads (diameter 0.4–0.5 mm) were added. The cells were broken in a Braun cell homogeniser (model MSK) using 12 bursts of 1 min (frequency: $4000\,\text{min}^{-1}$). This and all further operations were carried out at approximately $4\,°$C. The homogenate was centrifuged for 5 min at 1500 g. The cell-free supernatant was centrifuged for 20 min at 10 000 g using a Beckman Spinco L2-65B. The supernatant was centifuged as before and the final supernatant was centrifuged for 90 min at 100 000 g. The precipitate was resuspended in 0.65 M sorbitol and recentrifuged as before. The final pellet was resuspended in 0.1 M potassium phosphate buffer (pH 7.5) and designated microsomal fraction.

Isolation of microsomes from mammalian tissues. Rat liver. Livers of normal or phenobarbital-treated female Wistar rats were minced in 0.9% NaCl and homogenised in 4 vol (w/v) of 1.15% KCl using a Potter–Elvehjem homogeniser with teflon pestle (Yoshida *et al.*, 1977). The microsomal fraction was obtained as described for the yeast microsomes. Sodium phenobarbital (PB) was administered intraperitoneally at 80 mg kg^{-1} d^{-1} for 6 d. Rats were sacrificed 24 h after the last dose; food was withheld during the last 24 h.

Rabbit liver. Male New Zealand white rabbits (2.5 kg) were injected intraperitoneally with 3-methylcholanthrene (MC), dissolved in olive oil, for 8 d at a daily dose of 25 mg kg^{-1} of body weight (Yoshida *et al.*, 1977) or with 80 mg PB kg^{-1} d^{-1} for 8 d. Rabbits were sacrificed 24 h after the last dose. Livers were removed and microsomes prepared as described above.

Liver, adrenals and testes from male and female mongrel dogs. Dogs were anaesthetised with Nembutal® (45 mg kg^{-1} given intravenously). After the carotid artery was cut, liver, adrenals and testes were dissected out and thoroughly washed in ice-cold 0.9% NaCl. The microsomal fraction of the livers was obtained as described above. The adrenals and testes were minced for 5 s in 1.15% KCl using an Ultra-Turax prior to homogenisation with the Potter–Elvehjem homogeniser.

Analytical methods. The cyt. P-450 content was determined by measuring the reduced carbon monoxide (CO) difference spectrum using an extinction coefficient of 91 cm^{-1} mM^{-1} (Omura & Sato, 1964). In summary: both the sample and reference cells contained a microsomal suspension (1 mg of protein per ml of 0.1 M phosphate buffer pH 7.5, when a yeast microsomal suspension was used) that had been treated with dithionite. The sample was then saturated with CO and the difference spectrum was measured 1 or 5 min after the addition of drug and/or solvent. The absorbance increment between 448 (yeast micro-somes) or 450 (mammalian microsomes) and 490 nm was used for the calculation of the cyt. P-450 content. The effects of miconazole, ketoconazole, R 51 211 and/or DMSO (final concentration 0.3%) were determined on microsomal fractions from liver, adrenals and testes, diluted with 0.1 M potassium phosphate buffer (pH 7.5) to obtain an absorbance difference similar to that obtained with a yeast microsomal fraction containing 1 mg protein per ml, i.e. about 0.1 nmol mg^{-1}.

To trace the spectral transitions of the Soret band of cyt. P-450, associated with the addition of ketoconazole or miconazole, yeast microsomes were diluted to 1 mg protein per ml in 0.1 M potassium phosphate buffer (pH 7.5). The rat liver microsomes were diluted to 0.5 mg protein ml^{-1}. The suspension was divided into the reference and sample cells. A base-line of equal light absorbance was established, ketoconazole was added to the sample and DMSO to the reference cell, and the resulting difference spectrum recorded. The spectral transitions of the Soret band of cyt. P-450 induced by the antimycotics and the reduced CO-difference spectra were traced with a Cary® double beam spectrophotometer, model 1605, using 1 cm path-length quartz cuvettes. Miconazole, ketoconazole and R 51 211 were dissolved in DMSO. Protein was determined by the method of Lowry *et al.* (1951).

Results and discussion
Growth studies
Inhibition of the outgrowth of mycelium was observed when ketoconazole or R 51 211, at concentrations as low as 10^{-9} M, were added to Eagle's minimum essential medium (EMEM) containing glutamine and foetal calf serum and inoculated with *Candida albicans* strain ATCC 44859 (Fig. 15.2). Complete inhibition was obtained at 3 × 10^{-9} M of R 51 211 and at 4 × 10^{-9} M of ketoconazole; with micon-azole complete inhibition of outgrowth was only reached at 5 × 10^{-7} M. Miconazole also differs from R 51 211 and ketoconazole with regard to its effects on yeast budding (Fig. 15.3). 50% inhibition of growth of *C. albicans* (ATCC 44859) was achieved at 7 × 10^{-8} M, i.e. at a lower concentration than that needed to block the mycelial outgrowth. R 51 211 and ketoconazole induced 50% inhibition of yeast growth at 3.8 × 10^{-8} and 3.5 × 10^{-8} M. This is almost ten times higher than the concentrations needed to inhibit mycelial outgrowth.

The higher activity of miconazole in CYG medium as compared with that in EMEM + serum may result from interference with serum

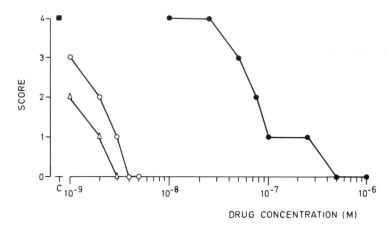

Fig. 15.2. Effects of R 51 211 (open triangles), ketoconazole (open circles), and miconazole (filled circles) on the mycelial outgrowth of *Candida albicans* (ATCC 44859) in EMEM supplemented with glutamine and foetal calf serum. Temperature 37 °C. Incubation time: 24 h. Scores – 4: mycelial outgrowth >90%; 3: 75–90% mycelium; 2: 50–75% mycelium; 1: <50% mycelium; 0: yeast form only. C: control.

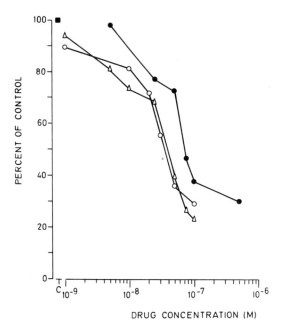

Fig. 15.3. Effects of R 51 211 (open triangles), ketoconazole (open circles), and miconazole (filled circles), on yeast budding. *Candida albicans*: strain ATCC 44859. Incubation time: 24 h, temperature: 37 °C. C: control.

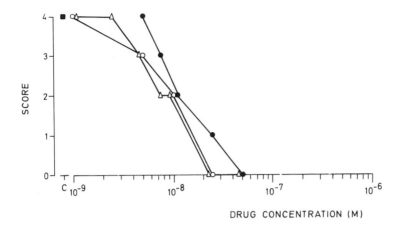

Fig. 15.4. Effects of R 51 211 (open triangles), ketoconazole (open circles) and miconazole (filled circles) on the mycelial outgrowth of *Candida albicans* (ATCC 44859) in EMEM supplemented with glutamine only. Scores are given in the legend to Fig. 15.2. C: control.

components (proteins?). This can be deduced from results obtained with *C. albicans* (ATCC 44859) grown in EMEM not supplemented with serum (Fig. 15.4). In fact, an almost equal activity with all three azole derivatives was found. It should be noted that R 51 211 and ketoconazole were less active in the absence of serum (Figs. 15.2, 15.4).

Sterol biosynthesis

Miconazole and ketoconazole inhibit ergosterol biosynthesis in a variety of yeasts and fungi resulting in an accumulation of the 14α-methylsterols: lanosterol,   24-methylene-dihydrolanosterol, 4,14-dimethylzymosterol,   obtusifoliol,   14-methylfecosterol   and 14-methyl,24-methylene-ergosterol (Vanden Bossche *et al.*, 1978, 1980). Miconazole and ketoconazole share this property with other *N*-substituted imidazoles and triazoles, e.g. clotrimazole, imazalil, econazole, etaconazole, parconazole and terconazole (Siegel & Ragsdale, 1978; Marriot, 1980; Ebert *et al.*, 1983; Vanden Bossche *et al.*, 1984). Preliminary results indicated that the triazole derivative, R 51 211, also inhibits ergosterol biosynthesis in *Candida albicans* (Vanden Bossche *et al.*, 1984).

The results summarised in Fig. 15.5 further demonstrate R 51 211's ability to inhibit ergosterol biosynthesis in *Candida albicans*. In this experiment *C. albicans* (ATCC 44859) was grown for 24 h in the mycelium-promoting EMEM. 50% inhibition was achieved at

3.2×10^{-9} M, which is significantly lower than the concentrations needed with ketoconazole (8×10^{-9} M) or miconazole (about 10^{-7} M). The lower activity of miconazole, as compared with the two other azole derivatives, correlates extremely well with the higher doses of miconazole needed to inhibit the outgrowth of mycelium in media supplemented with serum (Fig. 15.2).

The three azole derivatives also affect ergosterol biosynthesis in the yeast form of *Candida albicans*. As shown in Fig. 15.6, miconazole, ketoconazole and R 51 211 added at the onset of the experimental growth phase (after 4 h incubation) greatly affect ergosterol synthesis in the yeast form of *C. albicans* (strain ATCC 28516), even with only 1 h of contact. 50% inhibition of ergosterol biosynthesis was achieved with about 5.5×10^{-10}, 7×10^{-10} and 4×10^{-9} M of R 51 211, miconazole and ketoconazole respectively.

The effects of R 51 211 (Fig. 15.7) and ketoconazole (Fig. 15.8) on sterol biosynthesis in *Candida albicans* (ATCC 44859) grown for 24 h in

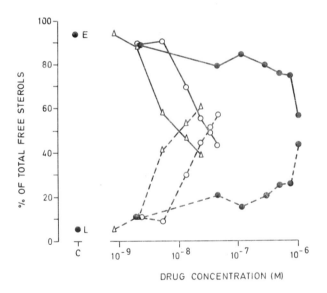

Fig. 15.5. Effects of R 51 211 (open triangles), ketoconazole (open circles) and miconazole (filled circles) on the incorporation of radioactivity derived from [^{14}C]acetate into ergosterol (solid line) and in 14α-methylsterols, i.e. the sum of 4,4′,14- and 4,14-methylsterols (broken line) in *Candida albicans* (strain ATCC 44859) grown for 24 h in EMEM supplemented with glutamine and foetal calf serum. Incubation time: 24 h, temperature: 37°C. The ergosterol fraction (E) contains the 4-desmethylsterols. L = 14α-methylsterols.

CYG (yeast form) or EMEM + serum (mycelial form) are shown. The different sterols were separated by TLC after saponification, so that both esterified and non-esterified sterols are included. These figures clearly show that both azole derivatives affect ergosterol biosynthesis at much lower concentrations in *C. albicans* grown in EMEM + serum than in cells grown in CYG medium. This correlates with their greater ability to block mycelial outgrowth (Fig. 15.2) than yeast budding (Fig. 15.3).

In all the experiments summarised so far, an accumulation of 14α-methylsterols was observed concomitant with the decrease in ergosterol. This shift was even more pronounced than shown in Figs. 15.5, 15.7 and 15.8. Preliminary results were obtained by using [14C] acetate as substrate, TLC to separate the different sterols, contact photography (Kodak X-omatic cassette, Kodak film X-Omat S) to localise the different sterols, and gas chromatography combined with mass spectrometry to identify them. These results revealed for example that R 51 211, at concentrations $\geq 2.5 \times 10^{-9}$ M, induced in *Candida*

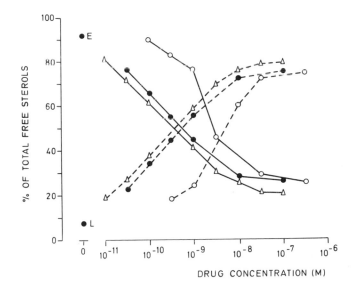

Fig. 15.6. Effects of the azole derivatives R 51 211 (open triangles), ketoconazole (open circles), and miconazole (filled circles) on ergosterol biosynthesis (solid lines) and accumulation of 14α-methylsterols (broken lines). *Candida albicans* (strain ATCC 28516) were grown for 4 h in CYG medium, [14C] acetate and drugs were added and the cells collected 1 h later. Temperature: 37 °C. E = ergosterol; L = 14α-methylsterols.

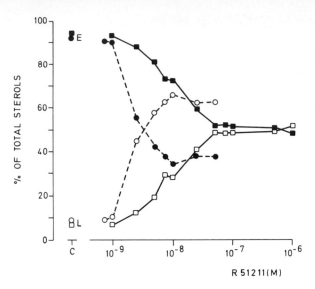

Fig. 15.7. The effects of R 51 211 on sterol synthesis in *Candida albicans* (ATCC 44859) grown for 24 h in CYG medium (solid lines) or EMEM + serum (broken lines), both supplemented with [^{14}C] acetate. The different sterols are separated by TLC after saponification. Radioactivity derived from [^{14}C] acetate in ergosterol (filled circles), and in 14α-methylsterols (open circles). E: ergosterol (filled circles and squares); L: lanosterol (open circles and squares).

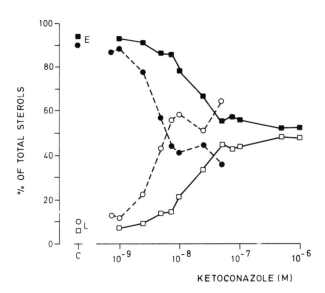

Fig. 15.8. The effects of ketoconazole on sterol synthesis in *Candida albicans* (ATCC 44859) grown for 24 h in CYG medium (solid lines) or EMEM + serum (broken lines). Further details are given in the legend to Fig. 15.7. C: control.

albicans (strain ATCC 44859) an accumulation of 14α-methylfecosterol and 14α-methyl,24-methylene-ergosterol also. No ergosterol was found in the presence of 5×10^{-9}M R 51 211 or 7.5×10^{-9}M ketoconazole. Both 14α-methylsterols have, in the TLC system used so far, an Rf-value almost similar to that of ergosterol so that the radioactivity accumulated in these 14α-methylsterols was added to that of ergosterol.

The accumulation of 14α-methylsterols, e.g. lanosterol, together with a decreased availability of ergosterol greatly disturbs membrane properties, affecting (for example) the permeability of membranes and growth of *Saccharomyces cerevisiae* (Vanden Bossche *et al.*, 1984). Both the accumulation of 14α-methylsterols and the decreased ergosterol content may thus create unbalanced conditions in the cell and, according to Roberts *et al.* (1983), from these a generalised activation of chitin synthetase zymogen may originate. Chitin is the major component of the primary septum, separating bud and mother cell in the yeast form and playing an important role in septum and primary-wall formation in the hyphal form of *Candida albicans* (Gooday, 1978; Rogers, Perkins & Ward, 1980). Chitin synthetase (E.C. 2.4.1.16) has been found distributed in the plasma membrane in a latent state (Duran, Cabib & Bowers, 1979). The activation of chitin synthetase must be regulated to obtain the proper temporal and spatial induction of septum formation and/or formation of the primary wall. Cabib (see Roberts *et al.*, 1983) proposed that, under normal conditions of growth, an activator is directed to the specific site where chitin synthesis is to occur. This activator may be a protease. It is possible that in the unbalanced conditions produced in the cell by ergosterol biosynthesis inhibition, protease in excess of that normally involved is liberated from, for example, the osmotically fragile vacuole with a concomitant uncoordinated activation of chitin synthetase.

Our preliminary results indicate that in ketoconazole- and R 51 211-treated *C. albicans*, chitin-like material is irregularly distributed, whereas in control cells it is only found at the septum site and at the point just below the apex of the hypha (Vanden Bossche *et al.*, 1983; and P. Marichal, unpublished results). Roberts *et al.* (1983) suggest that a high chitin content may result in greater lysis. It is of interest to note that chitin synthesis is more important in mycelial outgrowth than in yeast budding. Therefore the greater effect of R 51 211 on mycelial outgrowth, as compared with yeast budding, may be the result of a greater effect on ergosterol biosynthesis (Fig. 15.7) and/or the greater sensitivity of mycelial outgrowth to a disturbance of chitin synthesis.

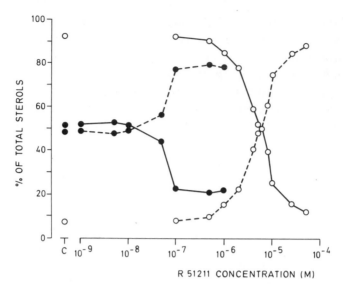

Fig. 15.9. Effect of R 51 211 on the [^{14}C] mevalonate incorporation into 4-desmethylsterols, e.g. ergosterol (solid lines) and 14α-methyl-sterols, i.e. the sum of 4,4′,14- and 4,14-methylsterols (broken lines) of *Candida albicans* (filled circles) and rat liver (open circles) subcellular fractions. Incubation time: 3 h, temperature: 37 °C. C: control.

The accumulation of lanosterol also disturbs artificial membranes having a phospholipid composition similar to that of mammalian membranes (P. Marichal, unpublished results). Therefore it was impor-tant to determine R 51 211's effect on cholesterol synthesis in mamma-lian membrane systems. The selective toxicity of R 51 211 has been established using subcellular fractions (S 10 000) of *Candida albicans* (ATCC 28516) and rat liver (Fig. 15.9). Ergosterol synthesis in the yeast fraction is greater than 85 times more sensitive to R 51 211 than cholesterol synthesis in a similar fraction obtained from rat liver. This suggests that R 51 211 should have a lower toxicity than ketoconazole and miconazole, for which ergosterol synthesis is 28 and 67 times more sensitive than cholesterol synthesis (Willemsens *et al.*, 1980). Thus, for R 51 211, as for ketoconazole and miconazole, it can be concluded that the selective toxicity observed in both animals and humans, is at least partly related to a greater sensitivity of ergosterol synthesis as compared with cholesterol synthesis.

Cytochrome P-450

In contrast with the 14α-demethylation, 4-demethylation is insensitive to carbon monoxide (CO; Gibbons, Pullinger & Mitropoulos, 1979) indicating that the microsomal CO-binding pigment, cytochrome P-450, is involved in the 14-demethylation only. Evidence is available for the presence of cyt. P-450 as a component of the enzyme system required to initiate oxidation of the 14α-methyl group of lanosterol in yeast microsomes (Aoyama & Yoshida, 1978a, b).

The azole derivative-induced accumulation of the 14α-methylsterols: lanosterol, 24-methylene dihydrolanosterol, obtusifoliol, 4,14-dimethylzymosterol, 14-methylfecosterol and 14-methyl,24-methylene ergosterol indicates that 14α-demethylation but not the 4-demethylation is affected by azole derivatives. These results led to the working hypothesis that azole derivatives may affect ergosterol biosynthesis via interactions with cyt. P-450.

Table 15.1 shows the contents of cyt. P-450 in microsomal fractions from *Saccharomyces cerevisiae*, rat, rabbit and dog liver, testis and adrenal gland. It should be noted that the absorption maximum of the reduced CO-compound of the yeast microsomal fraction was situated at 448 nm, resembling the cyt. P-450 from liver microsomes of 3-methylcholanthrene-pretreated rabbits (Hashimoto & Imai, 1976), but different from the peak found at 450 nm in normal and phenobarbital-pretreated animals.

Addition of miconazole, ketoconazole or R 51 211 to a yeast and rat liver microsomal suspension resulted in a dose-dependent decrease of the absorption difference between 448 and 490 nm (yeast) and between 450 and 490 nm (liver). As shown in Fig. 15.10 the cyt. P-450 of rat liver microsomes is much less sensitive to R 51 211, ketoconazole and miconazole; the difference in sensitivity is most pronounced with R 51 211. The results presented in Table 15.1 further indicate that the cyt. P-450 of mammalian microsomes is much less sensitive to these antimycotics than that present in *Saccharomyces cerevisiae* microsomes. For example, a 50% decrease in the peak height of the Soret band of the reduced CO compound of yeast microsomal cyt. P-450 was achieved with 2×10^{-8} M of R 51 211, whereas to obtain a similar effect on the most sensitive mammalian system tested, i.e. dog testis microsomes, 10^{-5} M of this triazole was needed. Although the spectral properties of cyt. P-450 from *S. cerevisiae* resemble those of liver cyt. P-450 from methylcholanthrene-pretreated rabbits (Sato & Omura, 1979), the latter cyt. P-448 is also much less sensitive to the three azole derivatives tested.

The difference in sensitivity between mammalian and yeast microsomal cyt. P-450 correlates extremely well with the much greater effects

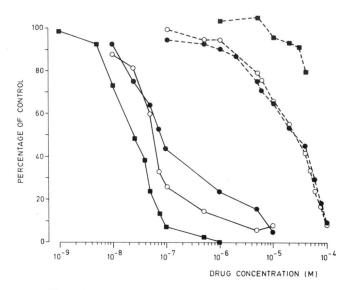

Fig. 15.10. Effect of miconazole (filled circles), ketoconazole (open circles) and R 51 211 (filled squares) on yeast (solid lines) and rat liver (broken lines) microsomal cyt. P-450. For yeast microsomes the $\Delta A_{448-490nm}$ and for liver microsomes the $\Delta A_{450-49nm}$ were determined and as percentages of the control value plotted against drug concentrations.

of miconazole (Vanden Bossche *et al.*, 1978), ketoconazole (Vanden Bossche *et al.*, 1980) and R 51 211 (Vanden Bossche *et al.*, 1984 and this chapter) on ergosterol biosynthesis in yeast, as compared with cholesterol biosynthesis in different mammalian systems (Vanden Bossche *et al.*, 1980; Willemsens *et al.*, 1980; Heeres, De Brabander & Vanden Bossche, 1981). Therefore, we speculate that the azole-derivative-induced inhibition of the 14α-demethylase system results from an effect on the microsomal cyt. P-450 of yeast.

An interference of azole derivatives with cyt. P-450 could be expected since it has been shown that imidazole derivatives interact with iron-haem proteins and display an extremely high affinity for cyt. P-450 (Wilkinson, Hetnarski & Yellin, 1972), causing a type II difference spectrum with maxima at about 430 nm (409–445 nm) and minima at about 393 nm (Hajek, Cook & Novak, 1982; Hajek & Novak, 1982). Type II difference spectra were also obtained in this study with miconazole and ketoconazole (Figs. 15.11*a* and 15.11*b*). R 51 211 induced a similar difference spectrum in the Soret region, i.e. maximum at about 430 nm and minimum at 398 nm. This suggests that the

Table 15.1. *Effects of azole derivatives on cytochrome P-450 in microsomal fractions of Saccharomyces cerevisiae and mammalian cells*[a]

Species	Cytochrome P-450[b] (nmol mg^{-1} protein)	I_{50}-values (10^{-7} M)[c]		
		miconazole	ketoconazole	R 51 211
Saccharomyces cerevisiae	0.09 ± 0.04 (12)	0.8 (5)	0.6 (5)	0.2 (3)
Rat liver: basal	0.54 ± 0.12 (10)	230.0 (3)	250.0 (3)	>300.0 (4)[d]
PB-induced	1.15 ± 0.14 (5)	39.0 (3)	31.0 (3)	>300.0 (3)
Rabbit liver: MC-induced	1.54 ± 0.02 (3)	120.0 (1)	82.0 (1)	>300.0 (1)
PB-induced	2.54 ± 0.20 (4)	110.0 (1)	120.0 (1)	>300.0 (1)
Male dog: liver	0.32 ± 0.04 (3)	560.0 (2)	560.0 (3)	>300.0 (2)
adrenal	0.34 ± 0.10 (5)	24.0 (4)	14.0 (4)	>300.0 (2)
testis	0.10 ± 0.04 (7)	10.0 (3)	13.0 (4)	100.0 (3)
Female dog: liver	0.26 ± 0.09 (4)	300.0 (1)	530.0 (3)	>300.0 (2)
adrenal	0.34 ± 0.11 (3)	41.0 (2)	12.0 (3)	>300.0 (1)

[a]The effects on mammalian cell cyt. P-450 were determined on microsomal fractions diluted with buffer to obtain an absorbence difference similar to that obtained with yeast microsomal fractions, containing 1 mg protein ml^{-1}, i.e. about 0.1 nmol mg^{-1}. PB = sodium phenobarbital; MC = 3-methylcholanthrene.

[b]The results presented are mean values ± standard deviation followed by the number of experiments in parentheses.

[c]By a weighed non-linear regression procedure, a sigmoidal dose-response model was fitted to the individual observations and the corresponding I_{50}-values determined. I_{50}-value: drug concentration needed to obtain a 50% decrease in the Soret peak height of the reduced carbon monoxide compound of cyt. P-450.

[d]At the limit of solubility (under the present experimental conditions).

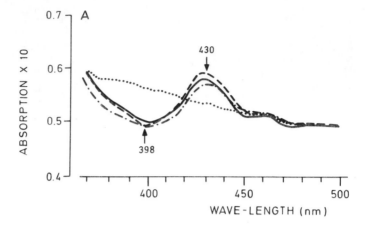

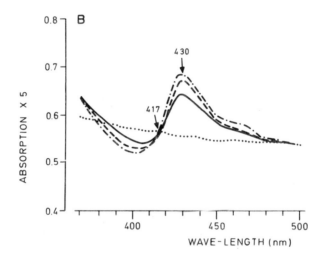

Fig. 15.11. Spectral transitions of the Soret band of yeast (*a*) and rat liver (*b*) microsomal cyt. P-450. *Saccharomyces cerevisiae* microsomes and rat liver microsomes were diluted to 1 mg and 0.5 mg protein ml^{-1} in 0.1 M potassium phosphate buffer (pH 7.5). The microsomal suspensions were divided between the reference and sample cuvettes. After establishment of the base-line (dotted line) DMSO was added to the reference cuvette and in the experiments using yeast microsomes 5×10^{-6} M (dash-dotted line), 10^{-5} M (solid line) or 2×10^{-4} M (broken line) of ketoconazole was added to the sample cuvette. The resulting difference spectrum was recorded. In the experiments using rat liver microsomes (Fig. 15.11*b*), 10^{-5} M (solid line), 5×10^{-5} M (broken line) or 10^{-4} M (dash-dotted line) of ketoconazole was used.

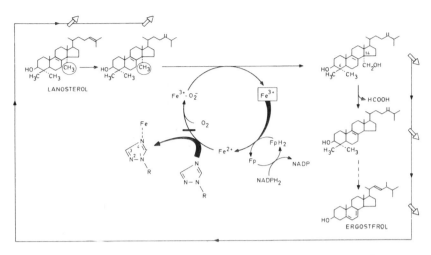

Fig. 15.12. Interaction of R 51 211 with the 14α-demethylation system. It is suggested that the triazole nitrogen (N_4) binds to the catalytic haem iron at the site of dioxygen binding, a step prior to the 'activation' of oxygen ($Fe^{3+} - O_2^-$). Fe^{3+} and Fe^{2+} represent respectively the oxidised and reduced forms of cyt. P-450; Fp: flavoprotein, R: the side chain of R 51 211 of which the chemical structure is shown in Fig. 15.1. Arrows pointing upwards indicate accumulation; arrows pointing downwards indicate decreased synthesis; the heavy horizontal line indicates inhibition.

observed interaction of these derivatives with cyt. P-450 is due to binding of the nitrogen (N_3 in the imidazole and N_4 in the triazole ring) to the catalytic haem iron atom at the site occupied by the exchangeable sixth ligand. In the resting form of ferric cyt. P-450, this ligand is most likely water (Dawson, Anderson & Sono, 1982). Upon reduction of the iron, the sixth ligand becomes the site of dioxygen binding (White & Coon, 1980), a step prior to the 'activation' of oxygen, required to initiate oxidation of the 14α-methyl group of lanosterol.

In Fig. 15.12 the interaction of R 51 211 with the haem iron is presented schematically. The resting form of ferric cyt. P-450 (Fe^{3+}) is first reduced (Fe^{2+}). In the presence of the triazole derivative, the nitrogen is bound to the haem iron occupying as such the site of dioxygen binding, the step prior to the formation of a superoxide–Fe^{3+} complex ($Fe^{3+}—O_2^-$) needed to hydroxylate the 14α-methyl group of 24-methylenedihydrolanosterol, lanosterol and other 14α-methyl-sterols. This results in a decreased synthesis of ergosterol and accumulation of 14-methylated sterols.

Conclusion

Evidence is available for a direct interaction of azole derivatives with microsomal cyt. P-450. Yeast microsomal cyt. P-450 is much more sensitive to these compounds than cyt. P-450 of mammalian cell microsomes. From the azole derivatives tested, R 51 211 is not only the most active against yeast cyt. P-450, it is much more selective too. As shown in Fig. 15.13, it is deduced that from this selective interaction a selective inhibition of 14α-demethylation in yeast and fungal cells results. From the decreased availability of ergosterol and the concomitant accumulation of 14α-methylsterols, a deterioration of membrane activities and an uncoordinated activation of chitin synthetase might result and lead to growth inhibition and cell death.

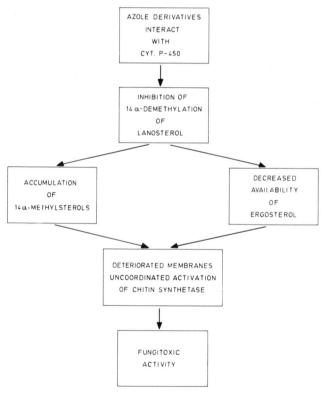

Fig. 15.13. Schematic presentation of the hypothesis on the mechanism underlying the antifungal properties of azole derivatives.

Acknowledgements. The authors are grateful to Dr Paul A. J. Janssen for his constant interest. Grateful appreciation is expressed to B. Wouters for typing the text and to J. Van Mierlo for drawing the figures. This work was supported in part by a grant from the 'Instituut voor Aanmoediging van het Wetenschappelijk Onderzoek in Nijverheid en Landbouw'.

References

Aoyama, Y. & Yoshida, Y. (1978*a*). Interaction of lanosterol to cytochrome P-450 purified from yeast microsomes: evidence for contribution of cytochrome P-450 to lanosterol metabolism. *Biochemical and Biophysical Research Communications*, **82**, 33–8.

Aoyama, Y. & Yoshida, Y. (1978*b*). The 14α-demethylation of lanosterol by a reconstituted cytochrome P-450 system from yeast microsomes. *Biochemical and Biophysical Research Communications*, **85**, 28–34.

Brasseur, R., Vanden Bosch, C., Vanden Bossche, H. & Ruysschaert, J. M. (1983). Mode of insertion of miconazole, ketoconazole and deacetylated ketoconazole in lipid layers. A conformational analysis. *Biochemical Pharmacology*, **32**, 2175–80.

Cauwenbergh, G. (1983). Clinical research with new orally active antimycotics: a first review. In *Proceedings 13th International Congress of Chemotherapy* (Vienna, 1983), ed. K. H. Spitzy & K. Karrer, SS4.8/1–12. Vienna: Verlag H. Egermann.

Dawson, J. H., Anderson, L. A. & Sono, M. (1982). The sixth ligand of ferric cytochrome P-450. In *Cytochrome P-450, Biochemistry, Biophysics and Environmental Implications*, ed. E. Hietanen, M. Laitinen & O. Hänninen, pp. 523–9. Amsterdam: Elsevier Biomedical Press.

Duran, A., Cabib, E. & Bowers, B. (1979). Chitin synthetase, distribution on the yeast plasma membrane. *Science*, **203**, 363–5.

Ebert, E., Gaudin, J., Mecke, W., Ramsteiner, K., Vogel, C. & Fuhrer, H. (1983). Inhibition of ergosterol biosynthesis by etaconazole in *Ustilago maydis*. *Zeitschrift für Naturforschung*, **38c**, 28–34.

Gibbons, G. F., Pullinger, C. R. & Mitropoulos, K. A. (1979). A requirement for two distinct types of mixed function-oxidase systems. *Biochemical Journal*, **183**, 309–15.

Gooday, G. W. (1978). The enzymology of hyphal growth. In *The Filamentous Fungi. 3. Development Biology*, ed. J. E. Smith & D. R. Berry, pp. 51–77. London: Edward Arnold.

Hajek, K., Cook, N. I. & Novak, R. F. (1982). Mechanism of inhibition of microsomal drug metabolism by imidazole. *Journal of Pharmacology and Experimental Therapeutics*, **223**, 97–104.

Hajek, K. & Novak, R. F. (1982). Spectral and metabolic properties of liver microsomes from imidazole-pretreated rabbits. *Biochemical and Biophysical Research Communications*, **108**, 664–72.

Hashimoto, C. & Imai, Y. (1976). Purification of a substrate complex of cytochrome P-450 from liver microsomes of 3-methylcholanthrene-treated rabbits. *Biochemical and Biophysical Research Communications*, **68**, 821–7.

Heeres, J., De Brabander, M. & Vanden Bossche, H. (1981). Ketoconazole: Chemistry and basis for selectivity. *Current Chemotherapy and Immunotherapy*, ed. P.

Periti & G. G. Grassi, pp. 1007–9. Washington, D.C.: The American Society for Microbiology.

Ishidate, K., Kawaguchi, K., Tagawa, K. & Hagihara, B. (1969). Hemoproteins in anaerobically grown yeast cells. *Journal of Biochemistry*, **65**, 375–83.

Lowry, O. H., Rosebrough, N. J., Farr, A. L. & Randall, R. J. (1951). Protein measurement with the Folin phenol reagent. *Journal of Biological Chemistry*, **193**, 265–75.

Marriott, M. S. (1980). Inhibition of sterol biosynthesis in *Candida albicans* by imidazole-containing antifungals. *Journal of General Microbiology*, **117**, 253–5.

Mitropoulos, K. A., Gibbons, G. F., Connell, C. M. & Woods, R. A. (1976). Effect of triarimol on cholesterol biosynthesis in rat-liver subcellular fractions. *Biochimica Biophysica Research Communications*, **71**, 892–900.

Omura, T. & Sato, R. (1964). The carbon monoxide-binding pigment of liver microsomes. I. Evidence for its hemoprotein nature. *Journal of Biological Chemistry*, **239**, 2370–8.

Roberts, R. L., Bowers, B., Slater, M. L. & Cabib, E. (1983). Chitin synthesis and localization in cell division cycle mutants of *Saccharomyces cerevisiae*. *Molecular and Cellular Biology*, **3**, 922–30.

Rogers, H. J., Perkins, H. R. & Ward, J. B. (1980). Biosynthesis of wall components in yeast and filamentous fungi. In *Microbial Cell Walls and Membranes*, pp. 478–507. London: Chapman & Hall.

Sato, R. & Omura, T. (1979). *Cytochrome P-450*. Tokyo: Kodansha Ltd.

Siegel, M. R. & Ragsdale, N. N. (1978). Antifungal mode of action of imazalil. *Pesticide Biochemistry and Physiology*, **9**, 48–56.

Van Cutsem, J., Van Gerven, F., Zaman, R., Heeres, J. & Janssen, P. A. J. (1983). Pharmacological and preclinical results with a new oral and topical broad-spectrum antifungal, R 51 211. *Proceedings 13th International Congress of Chemotherapy* (Vienna, 1983), ed. K. H. Spitzy & K. Karrer, SS4.8/1–11. Vienna: Verlag H. Egermann.

Vanden Bossche, H., Lauwers, W., Willemsens, G., Marichal, P., Cornelissen, F. & Cools, W. (1984). Molecular basis for the antimycotic and antibacterial activity of *N*-substituted imidazoles and triazoles: inhibition of isoprenoid biosynthesis. *Pesticide Science*, **15**, 188–98.

Vanden Bossche, H., Marichal, P., Lauwers, W. F. & Willemsens, G. (1983). Biochemical differences between yeast and mycelia. Do they determine the antimycotic activity of ketoconazole? In *Proceedings 13th International Congress of Chemotherapy* (Vienna, 1983), ed. K. H. Spitzy & K. Karrer, PS4.8/3–9. Vienna: Verlag H. Egermann.

Vanden Bossche, H., Ruysschaert, J. M., Defrise-Quertain, F., Willemsens, G., Cornelissen, F., Marichal, P., Cools W. & Van Cutsem, J. (1982). The interaction of miconazole and ketoconazole with lipids. *Biochemical Pharmacology*, **31**, 2609–17.

Vanden Bossche, H., Willemsens, G., Cools, W., Cornelissen, F., Lauwers, W. F. & Van Cutsem, J. M. (1980). *In vitro* and *in vivo* effects of the antimycotic drug ketoconazole on sterol synthesis. *Antimicrobial Agents and Chemotherapy*, **17**, 922–8.

Vanden Bossche, H., Willemsens, G., Cools, W. & Lauwers, W. F. (1981). Effect of miconazole on the fatty acid pattern in *Candida albicans*. *Archives Internationales de Physiologie et de Biochimie*, **89**, B134.

Vanden Bossche, H., Willemsens, G., Cools, W., Lauwers, W. F. J. & Le Jeune, L. (1978). Biochemical effects of miconazole on fungi. II. Inhibition of ergosterol biosynthesis in *Candida albicans*. *Chemico-Biological Interactions*, **21**, 59–78.

Vanden Bossche, H., Willemsens, G. & Van Cutsem, J. M. (1975). The action of miconazole on the growth of *C. albicans. Sabouraudia*, **13**, 63–73.

White, R. E. & Coon, M. J. (1980). Oxygen activation by cytochrome P-450. *Annual Review of Biochemistry*, **49**, 315–56.

Wilkinson, C. F., Hetnarski, K. & Yellin, T. O. (1972). Imidazole derivatives – a new class of microsomal enzyme inhibitors. *Biochemical Pharmacology*, **21**, 3187–92.

Willemsens, G., Cools, W. & Vanden Bossche, H. (1980). Effects of miconazole and ketoconazole on sterol synthesis in a subcellular fraction of yeast and mammalian cells. In *The Host Invader Interplay*, ed. H. Vanden Bossche, pp. 691–4. Amsterdam: Elsevier Biomedical Press.

Yoshida, Y., Aoyama, Y., Kumaoka, H. & Kubota, S. (1977). A highly purified preparation of cytochrome P-450 from microsomes of anaerobically grown yeast. *Biochemical and Biophysical Research Communications*, **78**, 1005–10.

16

The polyene macrolide antibiotics and 5-fluorocytosine: molecular actions and interactions

D. KERRIDGE and W. L. WHELAN

Department of Biochemistry, Tennis Court Road, Cambridge CB2 1QW, UK

Introduction

Nystatin was the first polyene macrolide antibiotic to be isolated (Hazen & Brown, 1950) and heralded a major breakthrough in the treatment of fungal infections. Other polyenes were subsequently found and amphotericin B, the only one sufficiently non-toxic to be used in the treatment of systemic fungal infections, was isolated by Gold *et al.* in 1956. The clinically important polyenes are shown in Table 16.1. Amphotericin B is still used to treat patients suffering from a variety of systemic fungal infections, even though it has serious side effects, ranging from nausea and vomiting to kidney damage. As it is not absorbed from the gut, it has to be administered intravenously. The role of amphotericin B in the therapy of human infections is complex and not

Table 16.1. *The clinical use of 5-fluorocytosine and the polyene macrolide antibiotics*

Therapy	Disease
Amphotericin B	Systemic candidosis Cryptococcosis Blastomycosis Coccidioidomycosis Histoplasmosis
5-Fluorocytosine	Systemic candidosis Cryptococcosis
5-Fluorocytosine plus Amphotericin B	Systemic candidosis Cryptococcosis
Candicidin, Natamycin and Nystatin	Topical and gastrointestinal fungal infections

Table 16.2. *The antifungal activity of 5-Fluorocytosine*

Organism	Minimum growth inhibitory concentration (μg ml^{-1}) of 5-fluorocytosine (median value)
Candida spp.	0.1–2
Torulopsis glabrata	0.1–25
Cryptococcus neoformans	0.1–25
Aspergillus fumigatus	1.0–12
Sporothrix schenkii	100–1000
Blastomyces dermatitidis	100–1000
Madurella mycetomi	100–1000
(Acremonium) spp.	100–1000
Coccidioides immitis	Greater than 1000
Histoplasma capsulatum	Greater than 1000
Dermatophytic fungi	Greater than 1000

restricted to an inhibitory effect on the fungal pathogen. There have been a number of reports of a stimulatory effect of the drug at low concentrations on macrophages, and there is evidence for both a potent humoral immunostimulant effect and an augmentation of cell-mediated tissue response (reviewed by Medoff, Brajtburg & Kobayashi, 1983). These effects lie outside the scope of this chapter but are clearly important in understanding the therapeutic effects of this drug *in vivo*.

The discovery of amphotericin B was rapidly followed by the synthesis (Duschinsky, Pleven & Heidelberger, 1957) and subsequent use (Grunberg, Prince & Utz, 1967) of 5-fluorocytosine (5-FC) as an oral antifungal agent in the treatment of candidosis and cryptococcosis. Unlike amphotericin B, 5-FC has a limited antimicrobial spectrum (Table 16.2), being most effective against *Candida*, *Torulopsis* and *Cryptococcus* spp, but ineffective against many important fungal pathogens including *Coccidioides immitis* and *Histoplasma capsulatum* (Scholer, 1970). Unlike the polyene macrolide antibiotics, 5-FC is rapidly absorbed after oral administration and the concentration of drug in both cerebrospinal fluid and serum can reach up to 70 μg ml^{-1} during recommended therapy (Scholer, 1980). In general, tolerance to 5-FC is good and large doses can be given over prolonged periods without significant side effects; the world record is 10.7 kg given over a period of 3 years for a case of cryptococcosis (Zylstra, 1974)! Many of the side effects observed during therapy may result from the metabolism of the drug to 5-fluorouracil (Diasio, Lakings & Bennett, 1978), presumably by the intestinal flora. It is unfortunate that such a well-tolerated drug should suffer from two drawbacks, first the limited range of sensitive pathogenic fungi and second, the high frequency with which variants resistant to it appear during therapy (Scholer, 1980).

A synergistic effect of amphotericin B and 5-FC was first observed by Medoff, Comfort & Kobayashi (1971); this has subsequently been confirmed both *in vitro* and *in vivo* and has provided the basis for a combined therapy in which the dose of each individual drug is reduced, so minimising toxic effects. This may well be the treatment of choice for cryptococcal meningitis and disseminated cryptococcosis (Utz *et al.*, 1975; Hay, 1980). So we have two disparate antifungal drugs which are linked by the fact that, for certain serious systemic infections, a combination of the two is the preferred treatment. There have been a number of excellent articles recently on antimycotic drugs (Speller, 1980; Gale *et al.*, 1981; Ryley *et al.*, 1981; Medoff *et al.*, 1983) which we do not intend to duplicate in this chapter. We will concentrate on the molecular bases of action and the mechanisms by which fungi can become resistant to these drugs. We will endeavour to provide an explanation for the efficacy of the combined therapy.

Chemical and biological properties
The polyene macrolide antibiotics

During the past 30 years some 200 different polyenes isolated from a variety of species of *Streptomyces* have been reported (Hamilton-Miller, 1973; Ryley *et al.*, 1981). These compounds are characterised by a macrolide ring of carbon atoms closed by lactonisation, containing both a system of conjugated double bonds all in the trans position, and a hydrophilic region characterised by a number of hydroxyl groups. The structural formulae of certain of the clinically important polyene macrolide antibiotics are shown in Fig. 16.1. The polyenes differ in the number of carbon atoms within the ring (range 12–37), the number of hydroxyl groups (range 6–14), and in the presence or absence of a carbohydrate moiety connected to the macrolide ring through a glycosidic bond. In all instances this carbohydrate is 3-amino-3,6-dideoxy-mannose (mycosamine). In addition carboxyl, aromatic or aliphatic groups may be attached at specific positions to the macrolide ring (Hamilton-Miller, 1973; Kobayashi & Medoff, 1977; Gale *et al.*, 1981; Ryley *et al.*, 1981). The complete three-dimensional structure of amphotericin B has been elucidated by X-ray crystallography (Mechlinski *et al.*, 1970). It is a rigid rod-shaped molecule with opposing hydrophobic and hydrophilic faces (Fig. 16.2). The mycosamine moiety is present at one end of the molecule and there is a single hydroxyl group at the other end of the rod. The overall length of molecule is 2.1 nm and this is roughly equivalent to the length of a phospholipid molecule, a feature of some significance in its mode of action.

The polyenes are effective against eukaryotic organisms and those prokaryotic organisms (*Mycoplasmas* and the *Acholeplasmas*) containing sterols in their plasma membranes. The importance of membrane sterols is supported by the finding of Gottleib *et al.* (1958) that fungi can

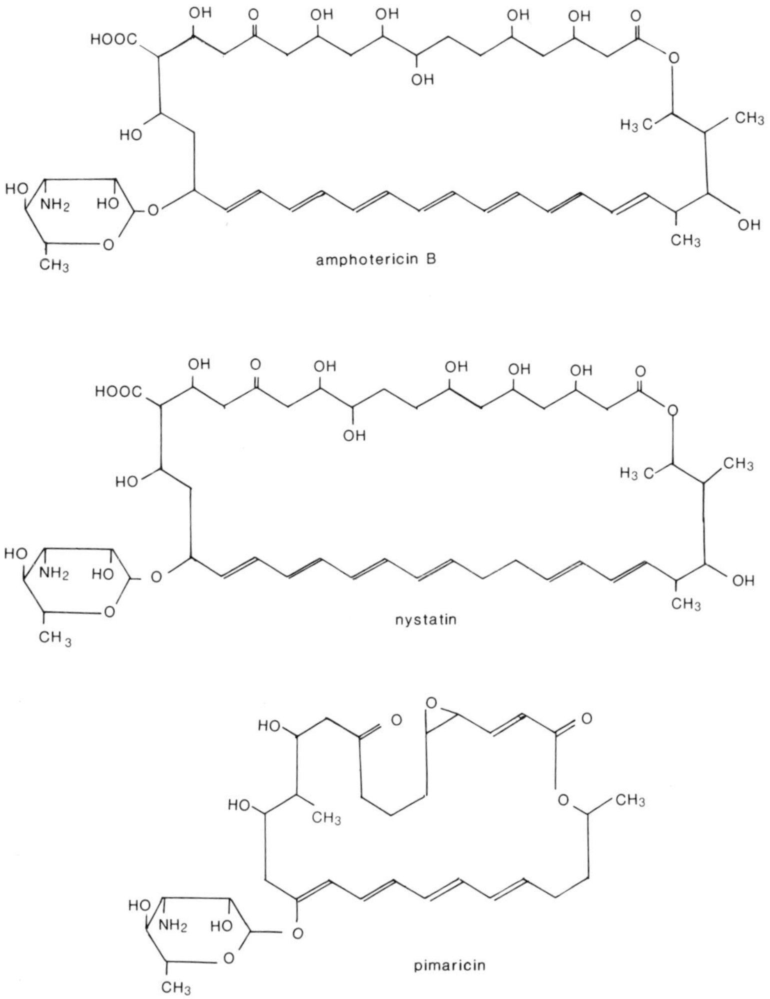

amphotericin B

nystatin

pimaricin

Fig. 16.1. The structural formulae of certain clinically important polyene macrolide antibiotics.

be protected from the inhibitory action of certain polyenes by the addition of sterols to the growth medium. This protective effect was later shown by Lampen, Arnov & Safferman (1960) to result from a hydrophobic interaction between the antibiotic and the sterol in the growth medium with consequent reduction in the effective drug concentration. Other hydrophobic compounds, for example fatty acids (Iannitelli & Ikawa, 1980), also protect sensitive organisms against the action of these compounds, presumably in a similar manner. Sterols can also have the opposite effect; Mas & Pina (1980) found that addition of ergosterol to cultures of nystatin-resistant strains of *Candida albicans* which lack ergosterol resulted in the organism becoming sensitive to the antibiotic. This was a phenotypic change, and on subculture in the absence of the sterol these strains rapidly regained resistance. The minimum growth-inhibitory concentrations are not only affected by the nature of the polyene antibiotic but also by the composition of the growth medium, the growth temperature and the inoculum size (Kitahara *et al.*, 1976). It is also possible to distinguish two groups of polyenes, those where it is possible to distinguish between fungistatic and fungicidal effects and those where this distinction cannot be made (Kolter-Brajtburg *et al.*, 1979).

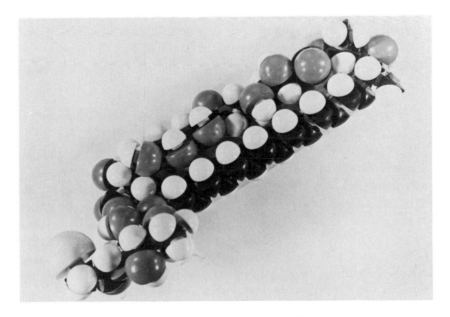

Fig. 16.2. Space-filling model of amphotericin B.

5-Fluorocytosine

In contrast to the polyene macrolide antibiotics, 5-FC (Fig. 16.3) has a simple chemical structure. It was first synthesised by Duschinsky *et al.* in 1957 as a potential cytostatic drug and, although it was subsequently shown to be ineffective (Heidelberger *et al.*, 1958), it was found to have antifungal properties *in vitro* (Berger & Duschinsky, 1962). The first successful use of 5-FC in the treatment of human fungal infections was reported by Grunberg *et al.* (1967). This drug exerts both fungistatic and fungicidal effects, with the fungicidal effects requiring a higher concentration of drug and a longer exposure to it (Scholer, 1970). The antifungal activity of 5-FC is strongly antagonised by complex nitrogen sources, in particular purines and pyrimidines, in the growth medium (Scholer, 1968).

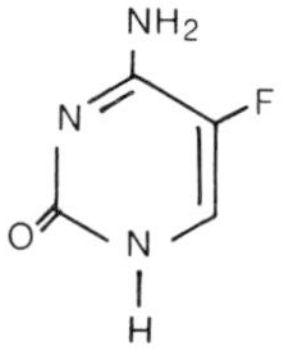

Fig. 16.3. 5-Fluorocytosine.

The mechanism of action of 5-FC is quite distinct from that of the polyene macrolide antibiotics and will be discussed separately in the next section; however, the differences in their mode of action provide a justification for a combined chemotherapy and this will be discussed in the final section of this chapter.

Molecular basis of drug action
The polyene macrolide antibiotics

The addition of a polyene macrolide antibiotic at the growth-inhibitory concentration to cultures of sensitive organisms results in a rapid cessation of growth, due to an interaction of the antibiotic with the plasma membrane and a resultant impairment of barrier function and a consequent loss of cellular constituents. The specificity of interaction and the extent of the damage caused depends both on the nature of the antibiotic and the lipid constituents of the membrane (Norman, Spielvogel & Wong, 1976). At one extreme there is filipin, a compound far too toxic to be used clinically, which causes gross membrane damage in sensitive cells, releasing both low molecular weight material and small proteins (Lampen, 1966; Norman *et al.*, 1972); at the other extreme, there are two antibiotics, *N*-succinyl-perimycin (Borowski & Cybulska, 1967) which induces the release of potassium ions only, and pimaricin found by de Kruijff *et al.* (1974) to be without effect on the permeability properties of egg lecithin liposomes and the cytoplasmic membranes of *Acholeplasma laidlawii*.

One of the earliest detectable effects of a polyene macrolide antibiotic is the release of potassium ions from sensitive cells; in 1966, Lampen postulated that the primary effect of the polyene was to render the membrane permeable to potassium ions. This loss of potassium ions is associated with an uptake of protons, and it is the resultant acidification of the cell contents which Lampen considered to be responsible for the fungicidal effects. This interpretation was supported by the finding that addition of potassium and magnesium ions at relatively high concentrations (85 mM and 45 mM respectively) to cultures of either *Saccharomyces cerevisiae* or *Candida albicans* (Liras & Lampen, 1974; Kerridge, Koh & Johnson, 1976a) protects the organisms against the inhibitory effects of candicidin and amphotericin B methyl ester. Under the experimental conditions used, these antibiotics are still able to interact with the plasma membrane causing an impairment of barrier function, but the protecting ions maintain the intracellular ionic concentration.

The hypothesis that death results from acidification associated with potassium loss was generally accepted until 1978, when it was modified by Palacios & Serrano as a result of their studies on the effects of amphotericin B and nystatin on glucose and maltose fermentation in *S. cerevisiae*. Maltose fermentation in *S. cerevisiae* is more sensitive to inhibition by nystatin than glucose fermentation and, unlike the latter, is not protected by the addition of potassium ions. (Table 16.3). The transport of maltose into yeast is an active process coupled to proton translocation (Serrano, 1977) and, unlike glucose transport, is inhibited by both nystatin and amphotericin B at low concentrations. In addition, direct measurements of the effects of these antibiotics on cells depleted of their internal ATP demonstrated a drug-induced increase proton permeability of the plasma membrane. The authors considered that the primary effect of these polyenes on the plasma membrane of *S. cerevisiae* is to increase its permeability to protons. In yeasts, the proton gradient plays an important role in the functioning of the plasma membrane, energising the transport of amino acids and other nutrients (Foury & Goffeau, 1975) and the maintenance of the potassium pool (Pena, 1975). The rapid release of potassium induced by polyenes, and

Table 16.3. *The inhibitory effects of polyenes on glucose and maltose fermentation in Saccharomyces cerevisiae*

| | | Fermentation rate (nmol CO_2/min^{-1} (mg dry wt cells)$^{-1}$) | | | |
| | | Glucose | | Maltose | |
Antibiotic	Concentration ($\mu g\,mg^{-1}$ yeast)	$-KCl$	$+KCl$	$-KCl$	$+KCl$
Control	—	53	53	44	43
Nystatin	0.05	44	48	9	9
Nystatin	0.13	32	48	8	6
Nystatin	0.54	3	38	2	2
Amphotericin B	0.54	3	30	2	2

Data taken from Palacios & Serrano, 1978.

used by a number of workers to monitor their interaction with the protoplast membrane (Gale, 1974; Hammond, Lambert & Kliger, 1974), can now be considered a secondary effect resulting from an increased permeability to protons.

Although the presence of exogenous potassium and magnesium ions protects against loss of viability, the importance of potassium loss in the fungicidal action of amphotericin B has been questioned by Hsu-Chen & Feingold (1974). In a study of the correlation between potassium loss and viability in polyene-resistant strains of *C. albicans*, they found that (in one class of mutant) polyene-induced potassium loss occurrred at concentrations lower than those required to kill the cells. Subsequent studies on the correlation between potassium loss and cell viability for a variety of polyene macrolide antibiotics provided evidence that, for certain polyenes at least, the loss of potassium does not result in cell death. Malewicz, Jenkin & Borowski (1981) questioned this apparent dissociation of potassium leakage from loss of viability and considered that it could result from the reversibility of interaction of the drug with the membrane; this allowed the cells to recover, even though they had lost an appreciable proportion of their internal potassium. The finding that this interaction with the plasma membrane is reversible has been used by a number of workers to alter the permeability of the plasma membrane of animal cells to a variety of compounds ranging from ions (Cass & Dalmark, 1973) to DNA (Kumar *et al.*, 1974), and of yeast to other inhibitors including 5-FC and rifampicin (Medoff *et al.*, 1972).

The specificity of action of the polyene macrolide antibiotics can be readily correlated with the presence or absence of sterols in cell membranes. The structural requirements of the sterol necessary to render *Acholeplasma* spp. sensitive to polyene macrolide antibiotics have been studied by De Kruijff *et al.* (1974). This organism can incorporate a wide variety of exogenous sterols into its plasma membrane; the authors concluded that the membrane sterol must have a 3-β-hydroxyl group, a planar ring structure, and a hydrophobic side chain attached to C17 to make the cells sensitive to the permeabilising effects of these compounds. Using potassium leakage as a measure of relative sensitivity to polyenes, Gale (1974) found that ergosterol-containing *Candida albicans* is more sensitive to amphotericin B than the cholesterol-containing mouse LS cells. Similar results were obtained by Chen, Chou & Feingold (1978). When ergosterol is present in the lipid bilayer, the membranes are more sensitive to disruption by amphotericin B than when cholesterol is present; the converse is true for filipin (De Kruijff *et al.*, 1974; Archer & Gale, 1975; Archer, 1976). It would appear that the greater affinity of amphotericin B for the ergosterol-containing fungal membranes is responsible for its clinical effectiveness.

Physical interaction between membrane lipids and the polyenes can be monitored spectrophotometrically, and these techniques have been used to study both the stoichiometry of the interaction and the conformational changes which occur in polyene molecules and in associated lipids. Polyene–sterol complexes have been detected in both water and in water–alcohol mixes and, although this does not prove conclusively that such complexes exist in the plasma membranes of sensitive cells, it does provide circumstantial evidence for their existence (Medoff *et al.*, 1983). Amphotericin B molecules form molecular aggregates in water at concentrations as low as 5×10^{-7} M (Mazerski, Bolard & Borowski, 1982) and, in the formation of a polyene–sterol complex *in vitro*, the sterols will have to compete with these aggregation forces. Addition of ethyl alcohol affects both polyene–polyene and polyene–sterol interactions, and there is a narrow range of concentrations where the polyene–sterol interaction can be successfully monitored; under these conditions Gruda *et al.* (1980) demonstrated a differential selectivity of amphotericin B towards ergosterol over cholesterol. Readio & Bittman (1982) overcame the problem of self association of polyene molecules in water by studying their interaction with sterols incorporated into liposomal vesicles; they estimated that one polyene molecule associated with one membrane sterol molecule. There is a significant difference between the binding constants of amphotericin B to ergosterol and cholesterol, which is consistent with its selectivity *in vivo*.

The spectrophotometric method most sensitive to changes in the conformation of the drug molecule is circular dichroism and, by using this technique, Bolard, Seigneuret & Boudet (1980) demonstrated that the drug exists in a number of conformational states in association with lipid bilayers. These conformational states depend upon the physical properties of the membrane, its sterol content and the time elapsed after the addition of the antibiotic. It is likely that one, or at the most two, of these are responsible for the drug-induced permeability changes (Vertut-Croquin et al., 1983). The active conformers are the same when either ergosterol or cholesterol is present in the plasma membrane, but the concentration of amphotericin B required to achieve this conformational state is greater when cholesterol is present than when ergosterol is the membrane sterol.

Evidence that the presence of sterols in the membrane is not essential for the disruptive effects of the polyene antibiotics came from studies of glucose release from liposomal vesicles (Hsu-Chen & Feingold, 1973). Liposomes prepared from lecithin with dipalmitoyl or distearoyl side chains are sensitive to polyenes, but become resistant if a sterol is incorporated into the lipid bilayer. However, similar results were obtained by Archer (1976) using *Mycoplasma mycoides* subsp. *capri* in which the composition of the membrane had been modified by growth in the presence of specific lipid supplements. The interaction of amphotericin B with liposomal vesicles is readily reversible and, *in vitro*, the exchange of the drug between vesicles is rapid with a half time of 30 s (Bolard et al., 1981); it has been possible using this technique to measure, by circular dichroism and proton efflux, the influence of the physical state of the membrane on drug–membrane interaction. The binding of amphotericin B to cholesterol-containing liposomal vesicles in the gel state is 200 times greater than to those where the lipids are in the crystalline state. It would appear that both the membrane sterol and the physical state of the membrane are important in determining the interaction of the polyene antibiotic and that, for a disruptive interaction to occur, the lipids in the membrane must be in the ordered state.

The effect of polyene macrolide antibiotics on membrane morphology

There have been a number of electron-microscopic studies of drug–membrane interactions using both negatively stained and freeze–etched preparations (Tillack & Kinski, 1973; Verkleij *et al.*, 1973; De Kruijff & Demel, 1974; Nozawa *et al.*, 1974; Kitajima, Sekiya & Nozawa, 1976; Pesti *et al.*, 1981; Sekiya, Yano & Nozawa, 1982). The most pronounced changes that have been observed are associated with a redistribution and clustering of the intramembranous particles. This is considered by Pesti *et al.* (1981) to be a specific effect of the polyene since it does not occur on addition of nystatin to a nystatin-resistant strain of *Candida albicans*. There were no changes in the intracellular membranes, a finding consistent with the absence of polyene translocation across the plasma membrane. Similar findings were obtained by Sekiya *et al.* (1982). These authors observed differences between the effects of amphotericin B and its methyl ester on the plasma membrane of *C. albicans*, with the methyl ester giving rise to elevated particle-free domains towards the outside of the cell. There was also a marked reduction in the density of the intramembranous particles. Both groups considered that this redistribution of the intramembranous particles results from localised changes in the physical state of the membrane, induced by the dissociation of the membrane sterol from the acyl side chains of the phospholipids, as a result of its interaction with the polyene molecule.

Electrical-conductance measurements can provide evidence for the presence of antibiotic-induced aqueous pores in membranes (for references see Gale *et al.*, 1981). Discrete current fluctuations, diagnostic of an aqueous pore, have been detected in black lipid films treated with amphotericin B at concentrations in the region of 10^{-8}–10^{-9} M (Ermishkin, Kasumov & Potzeluyev, 1976, 1977; Kasumov *et al.*, 1979). Cholesterol was essential for the polyene-induced fluctuation and there was evidence for preferential anion selectivity of a single channel. These channels undergo a large number of transitions between the open and closed states during a life time of several minutes (Ermishkin *et al.*, 1977). Conflicting results were obtained by Romine *et al.* (1977), who examined the mechanism of nystatin-induced conductance increases in planar bilayer membranes, formed from lecithin and cholesterol, by spectral analysis of kinetic fluctuations. The Lorentzian components of the power spectrum which are characteristic of channel activity were not observed for nystatin-doped membranes. Their findings were in better agreement with a carrier mechanism involving the transport of single charges in an uncorrelated fashion. Discrete current fluctuations can also be observed in the absence of added polyene in lipid bilayers formed from distearoyl phosphorylcholine at the phase-transition temperature and, in this instance, it was suggested that the ionic channels resulted from an interaction of different lipid domains within the bilayer (Antonov *et al.*, 1980). It is unfortunate that conductance measurements, although providing evidence for the existence of ionic channels in the presence of amphotericin B, have not provided conclusive proof for the existence of an aqueous pore bounded by polyene molecules.

Models of polyene action

A number of molecular models have been proposed to explain the interaction of polyenes with lipid bilayers and the resultant impairment of membrane function (De Kruijff & Demel, 1974; Marty & Finkelstein, 1975; Van Hoogevest & De Kruijff, 1978). Essentially, the models are similar in that the hydrophobic face of the rigid polyene molecule interacts with either sterol molecules or the acyl side chains of the membrane phospholipids. These complexes then move laterally in the membrane and aggregate to produce an annulus comprising eight polyene molecules in which the hydrophobic face of the polyene molecule is directed inwards, so forming an aqueous pore of internal diameter 0.8 nm. In the first model, proposed by Van Deenen's group at Utrecht (De Kruijff & Demel, 1974), two opposing annuli are required to span the lipid bilayer. In this model, the mycosamine moiety of the polyene is positioned at the membrane/water interface and the single hydroxyl group at the other end of the rigid rod is positioned within the bilayer. It is easy to see how such a double annulus would form on addition of the antibiotic to both sides of a lipid bilayer, but the antibiotic is added *in vivo* to one side of the membrane only; to form such a pore, antibiotic molecules would need to be translocated across the bilayer. De Kruijff & Demel (1974) suggested that a polyene-induced distortion within the bilayer might, in fact, aid the translocation of the antibiotic across the membrane. However, since Aracava *et al.* (1981) were unable to detect any movement of amphotericin B across the membranes of multilamellar vesicles and Pesti *et al.* (1981) could not observe any effect of mystatin on the morphology of the intracellular membranes of *Candida albicans*, it is unlikely that such a translocation occurs.

A modified model of polyene–membrane interaction was proposed by Marty & Finkelstein (1975) to take this objection into account. In this model, either a double or a single annulus could span the lipid bilayer and, in each case, produce a distortion of the bilayer in the vicinity of the pore (Fig. 16.4). Further support for the single annulus spanning the lipid bilayer came from studies of the effect of bilayer thickness on pore formation (Van Hoogevest & de Kruijff, 1978). Addition of amphotericin B to one side of vesicles, prepared from phosphatidylcholine which contained mainly oleic acid, resulted in potassium leakage but, when the vesicles were prepared from didocosanoyl phosphatidylcholine, it was necessary to add the antibiotic to both sides of the bilayer for leakage to occur. The length of the polyene molecule (2.1 nm) is less than the thickness of the hydrophobic core of the lipid bilayer formed from egg phosphatidyl choline (3.5 nm), and the authors proposed that the flexibility of the acyl side chains allowed meniscus formation to occur in the vicinity of the pore. There is also the possibility that the polyene annulus can move vertically within the bilayer providing some mobile characteristics for the pore. The hydrophobic core of the lipid bilayer formed from didocosanoyl phosphatidylcholine is too thick to allow a single annulus of polyene molecules to span it.

There are differences in the ion selectivity of the pore when the polyenes are added to either one or both sides of the membrane. Addition to one side only results in the membrane becoming permeable to cations whereas, when the antibiotic is added to both sides, the membrane is permeable to anions (Marty & Finkelstein, 1975). This difference in ion selectivity is assumed to result from differences in the

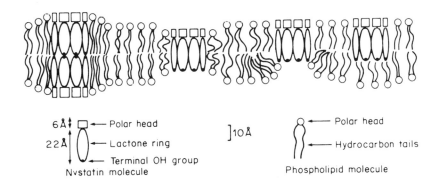

6 Å Polar head
22 Å Lactone ring
Terminal OH group
Nystatin molecule

]10 Å

Polar head
Hydrocarbon tails
Phospholipid molecule

Fig. 16.4. A model for the interaction of nystatin with lipid bilayers (from Marty & Finklestein, 1975).

structure of the single and double annulus. When the polyene is added to one side only, the ring of hydroxyl groups at the hydrophobic end of the annulus will impart a negative potential and hence provide a cation-selective gate at the entrance to the pore. In contrast to this, on addition to both sides of the membrane, the hydroxyl groups within the double annulus will impart a positive charge to the interior of the pore relative to the bulk of medium; this will be anion selective.

Do these models provide an adequate explanation for amphotericin-induced permeability? The answer is: probably not. At concentrations below those required to inhibit growth, the polyenes enhance the penetration of substances as diverse as ions (Cass & Dalmark, 1973) and DNA (Kumar *et al.*, 1974) into animal cells, and rifampicin and 5-FC into yeast cells (Medoff *et al.*, 1972). The penetration of large molecules into the cell without an enhanced leakage of cellular contents cannot be readily explained by the production of an aqueous pore of internal diameter 0.8 nm in the plasma membrane. The evidence for an aqueous pore comprising eight polyene molecules is not conclusive, although clearly if such pores form they will alter the permeability properties of the membrane, allowing leakage of ions and low molecular weight material from the cell. Membrane sterols play a major role in controlling the fluidity of the lipid bilayer of the plasma membrane; interaction of the sterol molecules with polyene molecules will prevent their normal interaction with the acyl side chains of membrane phospholipids and, as a result, affect membrane fluidity. The importance of membrane fluidity in the movement of ions across the membrane has already been demonstrated by Antonov *et al.* (1980) who suggested that this could result from an interaction of lipid domains within the lipid bilayer; it may be that the enhanced permeability to protons results not from the production of nonstatic aqueous pores bounded by polyene molecules within the membrane, but indirectly from the localised effect of the antibiotic on the physical state of the lipids within the bilayer. At high concentrations of the antibiotic, there is clearly a gross disorganisation of the membrane structure with an associated loss of both low and high molecular weight cellular constituents.

5-Fluorocytosine

In contrast to the polyene macrolide antibiotics, the molecular basis of the inhibitory effects of this drug is well understood (Polak & Scholer, 1980). 5-FC is metabolised by sensitive organisms to a number of metabolic derivatives which then exert their growth-inhibiting effects by interfering with the synthesis of cellular macromolecules (Fig. 16.5). The analogue is transported into both *Saccharomyces cerevisiae* and *Candida albicans* by cytosine permease, which is usually responsible for the uptake of cytosine, adenine, guanine and hypoxanthine (Polak & Grenson, 1973). The affinity of each substrate for the permease varies with K_m values ranging from 6 μM for adenine to 60 μM for 5-FC. The uptake of each compound is competitively inhibited by each of the others and the K_i values for this inhibition are the same as the K_m values for their uptake. The uptake of these bases is an energy-dependent process coupled to a proton gradient with a stoichiometry of one proton per substrate molecule transported (Forêt, Schmidt & Reichert, 1978). Once inside the cell, 5-FC is deaminated to 5-fluorouracil by a cytosine deaminase. This is the key enzyme in determining the clinical use of this drug; since it is absent or present only in very low amounts in mammalian cells, the drug cannot be converted by host cells to 5-fluorouracil and thence to a variety of inhibitory metabolic derivatives. In *C. albicans*, 5-fluorouracil is metabolised to 5-fluorouridylate by uridine monophosphate pyrophosphorylase and is then incorporated into RNA. The extent to which 5-fluorouracil will replace uracil in cellular RNA is considerable, and up to 50% replacement has been reported for yeast ribosomal and transfer RNA (Polak & Scholer, 1975).

There is a good correlation between the incorporation of radioactivity from 5-F(^{14}C)-cytosine into cellular nucleic acids and the sensitivity of a number of strains of *Candida albicans* to the drug. Sensitive strains

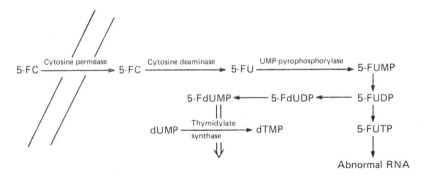

Fig. 16.5. Metabolism of 5-fluorocytosine in sensitive fungi. 5-FC: 5-fluorocytosine; 5-FU: 5-fluorouracil; 5-FUMP: 5-fluorouridylate; 5-FUTP: 5-fluorouridine triphosphate; 5-FdUMP: 5-fluorodeoxy-uridylate; 5-FdUDP: 5-fluorodeoxyuridine diphosphate; dUMP: deoxyuridylate: dTMP: deoxythymidylate.

incorporated up to 75 ng per 10^6 cells whereas, for resistant strains, the values were only 2 ng per 10^6 cells. It is to be expected that such a massive replacement of uracil by 5-fluorouracil in the RNA of sensitive cells will affect its functioning and there is evidence for this occurring in yeast. Incubation of *Saccharomyces cerevisiae* in the presence of 5-FC results in disturbances in protein synthesis and associated changes in the relative proportions of free amino acids within the pool fraction (Polak, 1974). Transfer RNA isolated from *S. cerevisiae* after incubation in the presence of 5-FC is still capable of amino acylation, but Giégé & Weil (1970) observed changes in the relative proportions of the individual amino acids compared to tRNA isolated from control cells. Extensive studies have been carried out on the effects of 5-fluorouracil and its derivatives on tissue culture cells. In these not only is the rate of RNA synthesis reduced in the presence of the analogues, but also the post-transcriptional processing of the RNA is affected by the replacement of uracil in the RNA by 5-fluorouracil (Wilkinson, Tisty & Hanar, 1975; Tseng, Medina & Randerath, 1978; Glazer & Hartman, 1980). Although it is probable that similar effects on post-transcriptional modification will occur in yeasts, these have not yet been reported.

In 1957 Heidelberger *et al.* predicted that 5-fluorouracil would not only be incorporated into cellular RNA but would also inhibit DNA synthesis. This prediction was soon confirmed for tumour cells (Danenberg, Montag & Heidelberger, 1958). The interest in 5-fluorouracil and its derivatives as potential anticancer and antiviral agents stems from their conversion to 5-fluoro-2-deoxyuridylate. This compound was found by Danenberg *et al.* (1958) to inhibit the incorporation of ^{14}C from ^{14}C-formate (a precursor of the methyl group of thymine) into DNA in tumour cells, and to induce thymineless death in *Escherichia coli* (Cohen *et al.*, 1958). The inhibition of DNA synthesis results from an inhibition of thymidylate synthase, a key enzyme in the synthesis of DNA since it is responsible for the *de novo* synthesis of thymidine within the cell.

5-Fluoro-2′-deoxyuridylate has played an important role in elucidating the enzyme action of thymidylate synthase (reviewed by Danenberg, 1977), yet it was not until 1977 that Polak & Wain reported that DNA synthesis in *Candida albicans* ceased immediately on addition of 5-FC; in the following year it was established that the drug was in fact metabolised to 5-fluoro-2′-deoxyuridylate in *C. albicans* and that this compound inhibited thymidylate synthase (Diasio, Bennett & Myers, 1978). Subsequently, Wagner & Shadomy (1979) demonstrated a similar interconversion in *Aspergillus fumigatus*. The mechanism by which 5-fluoro-2'-deoxyuridylate inhibits thymidylate synthase has been the subject of extensive studies using purified enzymes derived from *Streptococcus faecalis*, phage-infected *Escherichia coli*, and methotrexate-resistant *Lactobacillus casei* (Danenberg, 1977). In the absence of the cofactor 5,10-methylene tetrahydrofolate, 5-fluoro-2'-deoxyuridylate associates reversibly with thymidylate synthase but, in the presence of the cofactor, it is covalently bound and the complex is stable to treatment with urea and is not dissociated on precipitation with trichloroacetic acid or by SDS gel electrophoresis (Lagenbach, Danenberg & Heidelberger, 1972; Santi & McHenry, 1972). In the covalent complex (Fig. 16.6) the 6 position of the analogue is covalently bound to the nucleophilic catalyst at the active site via a thioether bond, and the 5 position is linked to the one carbon of the 5,10-methylene tetrahydrofolate. This complex is analogous to the steady-state intermediate of the normal reaction and, although slowly broken down, is responsible for the inhibition of the enzyme (Danenberg & Danenberg, 1978).

So, unlike the polyene macrolide antibiotics which physically interact with the plasma membrane and as a result impair its barrier function, 5-FC has two distinct effects: a production of aberrant RNA and an inhibition of thymidylate synthase (and hence DNA synthesis). Clearly the fungistatic and fungicidal effects could result from either or both of these, and the first indication that there might be differences in their relative importance came from studies on a resistant isolate of

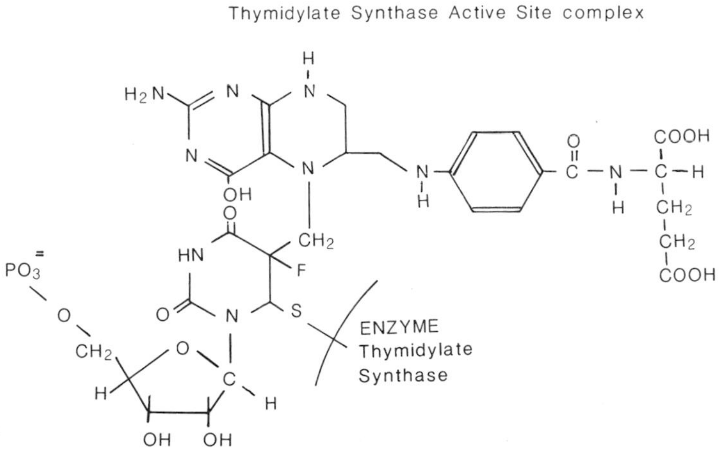

Thymidylate Synthase Active Site complex

Fig. 16.6. The inhibitory complex at the active site of thymidylate synthase.

Cryptococcus neoformans where the incorporation of radioactivity into RNA from 5F-(^{14}C)-cytosine was similar in both the resistant and sensitive strain. This was interpreted by Diasio, Bennett & Myers (1978) as evidence for a resistance mechanism involving thymidylate synthase and that, for this strain at least, the inhibition of DNA synthesis is more important than the production of aberrant RNA in mediating the effects of the drug. Waldorf & Polak (1983) carried out a more detailed assessment of the relative importance of the two inhibitory actions of this drug. They analysed a large number of isolates of *Candida albicans* and demonstrated that, for the majority of them, it was not possible to distinguish between the effects on DNA and RNA synthesis in their contribution to growth inhibition. In a number of the remaining strains, either RNA or DNA synthesis occurred at the control rate while the synthesis of the other nucleic acid was inhibited. These data would suggest that either effect can be responsible for growth inhibition in *C. albicans*.

Morphological effects of 5-fluorocytosine

The morphological and ultrastructural changes observed on incubation of *Candida albicans* and *Aspergillus fumigatus* in the presence of 5-FC are consistent with its effect on the synthesis of DNA (Arai *et al.*, 1977; Wain & Polak, 1979; Wain, Polak & Florio, 1981). The yeast form of *C. albicans* increases in size and the mycelial form continues to extend in the presence of the drug, a result consistent with the continued synthesis of macromolecules other than DNA. Ultrastructural changes occur in both the nucleus and cell wall of this yeast; after 2 h incubation, the nucleus is larger and after 12 h incubation the nucleus is further enlarged and translucent with filamentous components appearing in it while the cell wall becomes progressively thinner (Arai *et al.*, 1977). These changes are analogous to those observed by Cohen *et al.* (1958) in their studies on the effect of inhibition of DNA synthesis on the morphology of both bacterial and mammalian cells; they provide further evidence for the importance of the inhibition of thymidylate synthase in the antifungal activity of this drug.

Drug resistance

The maximum safe serum level of amphotericin B is taken to be $2-3 \mu g \, ml^{-1}$ (Hamilton-Miller, 1972*a*) while 5-FC is tolerated at levels as high as $100 \mu g \, ml^{-1}$. These levels are well in excess of the minimum concentrations necessary to abolish growth of typical sensitive strains under laboratory conditions. But resistance to these drugs at these levels has been repeatedly observed and is of two types: phenotypic and genotypic.

Resistance to amphotericin B

Phenotypic resistance to amphotericin B has been repeatedly observed in batch cultures of *Candida albicans* after the cessation of growth (Gale, 1974; Hammond *et al.*, 1974). In these experiments, stationary-phase cells were found to be resistant to polyene-induced potassium leakage while exponential-phase cells were sensitive by this criterion. This resistance is associated with changes in the cell wall since spheroplasts derived from phenotypically resistant cells are sensitive (Gale *et al.*, 1975; Kerridge *et al.*, 1976*b*). It is perhaps not surprising that changes in the cell wall of *C. albicans* can affect its sensitivity to amphotericin B. Certainly the addition of amphotericin B to a suspension of protoplasts causes an immediate potassium leakage whereas there is an appreciable time lag before potassium leakage can be detected from intact cells. Changes in the ultrastructure of the cell wall of *C. albicans* occur after the cessation of growth in batch culture (Cassone, Kerridge & Gale, 1979); the cell wall loses its characteristic layered appearance and is thicker (211 ± 58 nm) than the cell wall in exponentially growing cells (143 ± 22 nm). Although there are significant changes in cell wall morphology after the cessation of growth, the porosity of these structures to polyethylene glycols does not alter (Cope, 1980) and it is unlikely that a simple increase in the barrier function of the cell wall can explain the enhanced resistance.

The resistance of stationary phase *Candida albicans* to amphotericin B methyl ester can be modified in a number of ways, for example by treatment with —SH agents (Gale *et al.*, 1975), by changing the oxygen tension or pH value of the suspending medium (Gale, Johnson & Kerridge, 1977) or by treating the cells with exogenous glucanases (Gale *et al.*, 1980). These factors, although diverse, have one underlying feature in that they all affect the activity of the endogenous cell wall glucanases; those factors which increase glucanase activity decrease phenotypic resistance and those which decrease glucanase activity enhance resistance (Table 16.4; Notario *et al.*, 1982). The cell wall of *C. albicans* is not a static structure and its molecular organisation will presumably result from a balance between the synthesis and degradation of the structural β-1,3-and β-1,6-glucans within the wall. The polyenes are large rigid molecules, and it is not surprising therefore that

Table 16.4. *Relationship between β-glucanase activity and resistance to amphotericin B methyl ester in Candida albicans*

Condition or treatment	β-glucanase activity	Polyene resistance
Stationary phase cultures	Decrease	Increase
pH value of the medium increased to 7	Increase	Decrease
pH value of the medium maintained at 8.0	Inactivation	Irreversible increase
Growth medium supplemented with glutamate (stationary phase cells)	Increase	Decrease
Oxygen content of the medium		
Decreased	Increase	Decrease
Increased	Decrease	Increase
Addition of trichodermin	Decrease	Increase
Addition of 2-mercaptoethanol	Increase	Decrease
Addition of *N*-ethylmaleimide	Inhibition	Irreversible increase
Incubation with chitinase	Increase	Decrease
Incubation with β-glucanase	Increase	Decrease

After Table 5 in Notario *et al.* (1982).

even minor changes in the organisation of the cell wall, not detected by porosity measurements with polyethylene glycols, can have such pronounced effects on their penetration to the plasma membrane. Are these phenotypic changes in sensitivity important clinically? Obviously in lesions the population of yeast will be heterogenous, with both growing and nongrowing cells; if these nongrowing cells are resistant to amphotericin B at the concentration present in the lesion, then there will be a reservoir of infection which can develop after the cessation of therapy.

Genotypic resistance to amphotericin B is not generally considered a clinical problem. In a study of some 2015 clinical isolates of *Candida albicans*, Athar & Winner (1971) found none resistant to either nystatin or amphotericin B. There is, however, some evidence that monitoring for resistance to the polyene antibiotics might be advisable for certain groups of patients at risk from fungal infections (Dick, Merz & Saral, 1980). These authors surveyed isolates of *C. albicans* from patients undergoing extensive therapy for acute leukaemia and bone marrow transplantation and found a significant incidence of polyene-resistant isolates.

Production of polyene-resistant mutants in the laboratory is relatively easy and, in general, these can be distinguished from the parental strain by changes in the lipid composition of the cell; from the data available, however, it is not possible to derive a unified hypothesis for the biochemical basis of resistance (cf. 5-FC). Athar & Winner (1971) selected mutants by continuous subculture in the presence of amphotericin B and found resistance associated with a decreased level of ergosterol. In contrast to this finding, strains of *C. albicans* selected for resistance after mutagenesis with *N*-methyl-*N'*-nitro-*N*-nitrosoguanidine had an increased level of ergosterol within the cell (Hamilton-Miller, 1972*b*).

Detailed analyses of the lipid composition of polyene-resistant strains of *C. albicans* have been performed by Subden *et al.* (1977) and Pierce *et al.* (1978), but these have thrown little light on the mechanism of resistance and it is clear that alteration in the lipid content alone may not be sufficient to explain the resistance in all strains of *C. albicans*. The data on the lipid composition of the resistant mutants was obtained by analysis of intact cells and not of the isolated and purified plasma membrane; no account was taken of possible phenotypic variation in the lipid content of these organisms. In general, analysis of the lipid composition of polyene-resistant mutants has been more useful in providing information on the pathways of sterol biosynthesis in these organisms than on the molecular basis of drug resistance.

5-Fluorocytosine

Heritable resistance to 5-FC is much more common among *Candida albicans* isolates than is heritable resistance to amphotericin B, and it is this significantly frequent occurrence of resistant strains which poses the main problem in 5-FC therapy. Various workers using different methods to assess resistance have reported a wide range of frequencies at which resistant strains occur (Scholer, 1980). Drouhet, Mercier-Soucy & Montplaisir (1974) observed that the frequency of *C. albicans* resistant to 5-FC at concentrations greater than 25 μg ml^{-1} was greater among strains of serotype B than serotype A. Other workers have reported similar findings (Auger, Dumas & Joly, 1979; Stiller *et al.*, 1982) but the biochemical basis of this association is unknown.

Stiller *et al.* (1982) made a thorough study of the factors which affect the laboratory assessment of resistance to 5-FC and reported the existence of three classes of strains: (i) sensitive; (ii) intermediate in resistance; and (iii) highly resistant. The existence of strains of intermediate resistance was originally reported by Normark & Schönebeck (1972) and later by Defever *et al.* (1982) and by Whelan *et al.* (1981). In the latter study, it was shown that three typical partially resistant strains were heterozygous for resistance; the partially resistant strains gave rise to sensitive variants and highly resistant variants. The resistant and sensitive variants arose by segregation from the heterozygous state $(F/f \rightarrow F/F + f/f)$ rather than by *de novo* mutation. That study showed that the genetic determinants for resistance, represented here as the recessive (f) allele of the gene which determined sensitivity, preexists in the general *Candida albicans* population and that homozygosity (f/f) resulted in a high level of resistance. In the heterozygous state (F/f), the recessive allele is expressed at a low but readily detectable level, resulting in the observed partial resistance.

It is useful to consider together the two studies Stiller *et al.* (1982) and Defever *et al.* (1982) which reported partially resistant strains using different methods to assess resistance (Table 16.5). The results of these two independent studies indicate that partially resistant strains constitute a significant fraction of a random sample of strains in the USA. All partially resistant strains encountered in the study of Defever *et al.* (1982) gave rise to highly resistant variants at a readily detectable frequency and resembled the isolates previously shown to be hetero-

Table 16.5. *Susceptibility of Candida albicans to 5-fluorocytosine: two US studies*[a]

	Susceptibility group		
	Sensitive	Partially resistant	Highly resistant
Stiller *et al.*	0.60	0.36	0.04
Defever *et al.*	0.57	0.37	0.06

[a]The frequencies of isolates in three susceptibility groups are shown. Stiller *et al.* (1982) examined 402 isolates and classified them according to minimum inhibitory concentrations; their groups II and III are here taken together as one partially resistant group. Defever *et al.* (1982) examined 137 isolates and classified them according to ability to grow on a defined minimal agar medium containing 5-FC ($50 \, \mu g \, ml^{-1}$).

zygous. Our present working hypothesis is that all or most of the partially resistant strains encounted in these studies (Table 16.5) were heterozygous for resistance and that heterozygotes constitute the main obstacle to successful therapy with 5-FC.

A simple method has been described (Whelan *et al.*, 1981) by which strains of *Candida albicans* may be classified as either homozygous sensitive, partially resistant heterozygous or homozygous resistant. The homozygous sensitive strains constitute the majority in both studies of strains isolated in the USA (Table 16.5); treatment with 5-FC is appropriate in these cases. Homozygous sensitive strains also constitute the majority in a sample of isolates studied by us in Britain (W. L. Whelan & D. Kerridge, unpublished observations). Whether infections due to heterozygous strains should be treated with 5-FC alone remains to be determined. It seems probable that treatment of an infection caused by a heterozygous (F/f) strain will result in selection for the highly resistant variants which are likely to exist in any sizeable population (i.e. greater than 10^5 cells) of heterozygous cells, and that this selection is responsible for treatment failure. It is, however, possible that treatment would be successful in most cases since the resistance of the heterozygote is slight and there is a low frequency of resistant variants. It should be noted that Schönebeck & Ånséhn (1973) considered that infections due to partially resistant strains are amenable to 5-FC therapy. This question is not resolved at present and clearly animal studies using well characterised heterozygous strains would be useful, as would studies of the heterozygote to homozygote change during human infections. It does seem likely, however, that combined therapy with 5-FC and amphotericin B is most useful when the infection strain is heterozygous for 5-FC resistance, since the polyene antibiotic should reduce the likelihood of emergence of a 5-FC-resistant segregant.

The efficacy of 5-FC as an antifungal agent depends on its uptake into the cell and subsequent interconversion by enzymes already present within the cell to derivatives affecting nucleic acid synthesis. Provided that the mutation is not lethal, resistance can result from a block at any of these enzymatic steps. Jund & Lacroute (1970,1974) distinguished six types of resistant mutants in *Saccharomyces cerevisiae*. They proposed the following five mechanisms for resistance:

   (i) a deficiency in cytosine permease

  (ii) a deficiency in cytosine deaminase

 (iii) a deficiency in uridylic acid pyrophosphorylase

 (iv) a loss of the feed back regulation of aspartate transcarbamylase by ATP giving increased *de novo* synthesis of pyrimidines

  (v) an increased *de novo* synthesis of pyrimidines resulting from a stimulation of orotidylate pyrophosphorylase and orotidylate decarboxylase.

It is likely that similar considerations apply to *Candida albicans*, although the absence of a complete sexual genetic system has precluded a comprehensive study of all mechanisms by which this species may become resistant. Jund & Lacroute (1970) showed that the pattern of

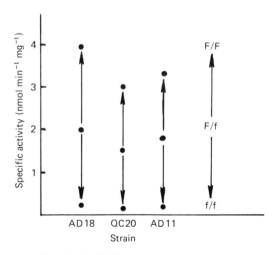

Fig. 16.7. UMP-pyrophosphorylase activity in partially resistant strains of *Candida albicans* and sensitive and resistant segregants derived from them.

resistance to 5-FC, 5-fluorouracil, 5-fluorouridine was consistent with the demonstrated position of mutational blocks in their resistant mutants. For example, a cytosine-permease-deficient mutant is resistant to 5-FC but remains sensitive to the other compounds. Polak & Scholer (1975) employed this method in a study of 29 resistant and partially resistant isolates of *C. albicans*. They suggested that resistance in two isolates was due to a deficiency in cytosine deaminase, and the resistance in each of the remaining isolates might be due to a deficiency in UMP-pyrophosphorylase activity or to increased *de novo* pyrimidine synthesis. Normark & Schönebeck (1972) described a mutant obtained in the laboratory in which resistance to 5-FC is associated with a marked decrease in UMP-pyrophosphorylase activity.

We have investigated the enzymic basis of resistance to 5-FC in highly resistant clinical isolates and in resistant segregants derived from certain heterozygous strains. These studies will be reported in detail elsewhere but we shall discuss the essential points here. The most common enzyme deficiency associated with resistance is a deficiency in the activity of UMP pyrophosphorylase; five out of six highly resistant isolates obtained in the USA possess markedly less activity than do sensitive or partially resistant isolates. However, the partially resistant isolates typically possess less UMP-pyrophosphorylase activity than do homozygous sensitive isolates, which suggests that the partially resistant isolates are heterozygous at a gene which determines this activity. That result is consistent with a simple gene-dosage hypothesis that the specific activity of UMP-pyrophosphorylase is directly proportional to the number of functional (F) alleles in the series $F/F:F/f:f/f$. In certain heterozygotes, we have shown that the sensitive segregant (F/F) does possess about twice the activity of the heterozygote while the resistant (f/f) segregant possesses markedly reduced activity (Fig. 16.7). We consider that a deficiency in this enzyme is the most frequent cause of resistance in highly resistant clinical isolates, in partially resistant clinical isolates and in highly resistant segregants obtained from the heterozygotes.

Combined therapy

In 1971, Medoff *et al.* reported that, in combination with amphotericin B, the antifungal activity of 5-FC was enhanced against *Cryptococcus neoformans* and various species of *Candida*. Since then there have been a number of reports of additive or synergistic effects of these antibiotics both *in vitro* and in experimental infections with *Candida albicans* (Scholer, 1980); in fact combination therapy with these two antifungal drugs is recommended for the treatment of cryptococcal meningitis and disseminated cryptococcosis (Hay, 1980).

There have been a number of studies on the biochemical basis of these effects, but the evidence is conflicting and the mechanism not understood. There are two important reasons for this: the first is that the strains of *Candida albicans* used in these studies have not been characterised either biochemically or genetically with regard to their resistance to 5-FC, and the second is that in animal studies no account has been taken of the effects of the drugs on the host's normal defence system. It has generally been assumed that the limiting factor in 5-FC resistance is the uptake of drug into the cell, and that the addition of amphotericin B at sub-growth-inhibitory concentrations allows the unrestricted uptake of 5-FC into the cell. Resistance associated with the loss of cytosine permease is not common in clinical isolates of *C. albicans* (Polak & Scholer, 1975; W. L. Whelan & D. Kerridge, unpublished observations) nor is there any evidence that cytosine permease is the rate-limiting step in the metabolism of the drug in normal sensitive strains of *C. albicans*. Medoff *et al.* (1972) demonstrated that amphotericin B at sub-growth-inhibitory concentrations enhanced incorporation of ^{14}C from 5-F-(^{14}C) cytosine into the acid-precipitable fraction of *C. albicans*. They interpreted these data as indicating an amphotericin-B-enhanced uptake of 5-FC. Supporting evidence was obtained by Polak (1978) for an enhanced uptake into *C. albicans*, but surprisingly no stimulation of uptake was demonstrated for *Cryptococcus neoformans* (although synergism has been reported in this organism). Conflicting results were reported by Beggs & Sarosi (1982) who obtained evidence for a sequential effect of the drugs on *C. albicans*, with amphotericin B first killing the majority of the cells in the population and 5-FC subsequently acting to prevent the growth of the surviving cells once the amphotericin B had broken down to a non-inhibitory level. During the first 3 h of incubation, amphotericin B strongly inhibited the uptake of radioactive 5-FC into the cells. This inhibition was reversed after about 3–6 h incubation, correlating with the breakdown of amphotericin B present in the medium.

In the combined therapy of fungal infections the situation is clearly more complex, since amphotericin B affects not only the fungus but also the host's defence mechanisms. A sequential mechanism could provide an explanation for the efficacy *in vivo*, with amphotericin B reducing the population of fungi to a level such that the occurrence of a 5-FC-resistant variant would not be expected. Combined therapy is clearly effective and, in order to develop its full potential, there is an obvious need for an understanding of the biochemical basis for the additive/synergistic effects of these two drugs both *in vitro* and *in vivo*. With the recent advances in understanding the biochemical and genetic basis of drug resistance it should prove possible to solve this problem, and hopefully to improve the therapy of fungal infections.

Acknowledgements. The authors wish to thank the Medical Research Council and the Nuffield Foundation for financial support.

References

Antonov, V. F., Petrov, V. V., Molnar, A. A., Predvoditelev, D. A. & Ivanov, A. S. (1980). The appearance of single-ion channels in unmodified lipid bilayer membranes at the phase transition temperature. *Nature*, **283**, 585–6.

Aracava, Y, Schreier, S., Phadke, R., Deslauriers, R. & Smith, I. P. C. (1981). Effects of amphotericin B on membrane permeability-kinetics of spin probe reduction. *Biophysical Chemistry*, **14**, 325–32.

Arai, T., Nikani, Y., Yokoyama, T. & Masuda, K. (1977). Morphological changes in yeasts as a result of the action of 5-fluorocytosine. *Antimicrobial Agents and Chemotherapy*, **12**, 355–60.

Archer, D. B. (1976). Effect of the lipid composition of *Mycoplasma mycoides* subspecies *Capri* and phosphatidylcholine vesicles upon the action of polyene antibiotics. *Biochimica et Biophysica Acta*, **436**, 68–76.

Archer, D. B. & Gale, E. F. (1975). Antagonism by sterols of the action of amphotericin B and filipin on the release of potassium ions from *Candida albicans* and *Mycoplasma mycoides* subsp. *Capri. Journal of General Microbiology*, **90**, 187–90.

Athar, M. A. & Winner, H. L. (1971). The development of resistance by *Candida albicans* to polyene antibiotics *in vitro*. *Journal of Medical Microbiology*, **4**, 505–17.

Auger, P., Dumas, C. & Joly, J. (1979). A study of 666 strains of *Candida albicans*: correlation between serotype and susceptibility to 5-fluorocytosine. *Journal of Infectious Diseases*, **139**, 590–4.

Beggs, W. H. & Sarosi, G. A. (1982). Further evidence for sequential action of amphotericin B and 5-fluorocytosine against *Candida albicans*. *Chemotherapy (Basel)*, **28**, 341–4.

Berger, J. & Duschinsky, R. (1962). Controlling fungi with 5-fluorocytosine. *U.S. Patent Application Ser.*, No. 181, p. 822.

Bolard, J., Seigneuret, M. & Boudet, G. (1980). Interaction between phospholipid bilayer membranes and the polyene antibiotic amphotericin B. Lipid state and cholesterol content dependence. *Biochemica et Biophysica Acta*, **599**, 280–93.

Bolard, J., Vertut-Croquin, A., Cybulska, B. E. & Gary-Bobo, C. M. (1981). Transfer of the polyene antibiotic amphotericin B between single-walled vesicles of dipalmitoylphosphatidylcholine and egg yolk phosphatidylcholine. *Biochimica et Biophysica Acta*, **647**, 241–8.

Borowski, E. & Cybulska, B. (1967). Potassium-less death of *Saccharomyces cerevisiae* cells treated with *N*-succinyl perimycin and the reversal of fungicidal effect by potassium ions. *Nature*, **213**, 1034–5.

Cass, A. & Dalmark, H. (1973). Equilibrium dialysis of ions in nystatin-treated red cells. *Nature*, **244**, 47–9.

Cassone, A., Kerridge, D. & Gale, E. F. (1979). Ultrastructural changes in the cell wall of *Candida albicans* following cessation of growth and their possible relationship to the development of polyene resistance. *Journal of General Microbiology*, **110**, 339–49.

Chen, W. C., Chou, D.-L. & Feingold, D. S. (1978). Dissociation between ion permeability and the lethal action of the polyene antibiotics in *Candida albicans*. *Antimicrobial Agents and Chemotherapy*, **13**, 914–17.

Cohen, S. S., Flaks, J. G., Barner, H. D., Loeb, M. R. & Lichtenstein, J. (1958). The mode of action of 5-fluorocytosine and its derivatives. *Proceedings of the National Academy of Sciences USA*, **44**, 1004–12.

Cope, J. E. (1980). The porosity of the cell wall of *Candida albicans*. *Journal of General Microbiology*, **119**, 253–5.

Danenberg, P. V. (1977). Thymidylate synthetase – a target enzyme in cancer chemotherapy. *Biochimica et Biophysica Acta*, **473**, 73–92.

Danenberg, P. V. & Danenberg, K. D. (1978). Effect of 5,10-methylenetetrahydrofolate on the dissociation of 5′-fluoro-2′-deoxyuridylate from thymidylate synthetase: evidence for an ordered mechanism. *Biochemistry*, **17**, 4018–26.

Danenberg, P. V., Montag, B. J. & Heidelberger, C. (1958). Studies on fluorinated pyrimidines. Effects on nucleic acid metabolism. *Cancer Research*, **18**, 329–34.

Defever, K. S., Whelan, W. L., Rogers, A. L., Beneke, E. S., Veselenak, J. M. & Soll, D. R. (1982). *Candida albicans* resistance to 5-fluorocytosine: frequency of partially resistant strains among clinical isolates. *Antimicrobial Agents and Chemotherapy*, **22**, 810–15.

De Kruijff, B. & Demel, R. A. (1974). Polyene antibiotic–sterol interactions in membranes of *Acholeplasma laidlawii* cells and lecithin liposomes. III. Molecular structure of the polyene antibiotic–cholesterol complexes. *Biochimica et Biophysica Acta*, **339**, 57–70.

De Kruijff, B., Gerritsen, V. J., Oerlemans, A., Van Dijck, P. W. M., Demel, R. A. & Van Deenen, L. L. M. (1974). Polyene antibiotic sterol interactions in membranes of *Acholeplasma laidlawii* cells and lecithin liposomes. II. Temperature dependence of the polyene antibiotic-sterol complex formation. *Biochimica et Biophysica Acta*, **339**, 44–56.

Diasio, R. B., Bennett, J. E. & Myers, C. E. (1978). The mode of action of 5-fluorocytosine. *Biochemical Pharmacology*, **27**, 703–7.

Diasio, R. B., Lakings, D. E. & Bennett, J. E. (1978). Evidence for conversion of 5-fluorocytosine to 5-fluorouracil in humans: possible factor in 5-fluorocytosine clinical toxicity. *Antimicrobial Agents and Chemotherapy*, **14**, 903–8.

Dick, J. D., Merz, W. G. & Saral, R. (1980). Incidence of polyene-resistant yeasts recovered from clinical specimens. *Antimicrobial Agents and Chemotherapy*, **18**, 158–63.

Drouhet, E., Mercier-Soucy, L. & Montplaisir, S. (1974). Sensibilité et résistance des *Candida* à la 5-fluorocytosine au 5-fluorouracile et à la 5-fluorouridine. Relation avec l'ecologie et le sérotype de *C. Albicans*. *Compte Rendus de l'Académie des Sciences, Paris*, **278**, 605–8.

Duschinsky, R., Pleven, E. & Heidelberger, C. (1975). The synthesis of 5-fluoropyrimidines. *Journal of the American Chemical Society*, **79**, 4559–60.

Ermishkin, L. N., Kasumov, Kh. M. & Potzeluyev, V. M. (1976). Single ionic channels induced in lipid bilayers by polyene antibiotics amphotericin B and nystatin. *Nature*, **262**, 698–9.

Ermishkin, L. N., Kasumov, Kh. M. & Potzeluyev, V. M. (1977). Properties of amphotericin B channels in a lipid bilayer. *Biochimica et Biophysica Acta*, **470**, 357–67.

Forêt, M., Schmidt, R. & Reichert, U. (1978). On the mechanism of substrate binding to the purine-transport system of *Saccharomyces cerevisiae*. *European Journal of Biochemistry*, **82**, 33–43.

Foury, F. & Goffeau, A. (1975). Stimulation of active uptake of nucleosides and amino acids by cyclic adenosine 3′,5′-monophosphate in the yeast *Schizosaccharomyces pombe*. *Journal of Biological Chemistry*, **250**, 2354–62.

Gale, E. F. (1974). The release of potassium ions from *Candida albicans* in the presence of polyene antibiotics. *Journal of General Microbiology*, **80**, 451–65.

Gale, E. F., Johnson, A. M., Kerridge, D. & Koh, T. Y. (1975). Factors affecting the changes in amphotericin sensitivity of *Candida albicans* during growth. *Journal of General Microbiology*, **87**, 20–36.

Gale, E. F., Johnson, A. M. & Kerridge, D. (1977). The effect of aeration and metabolic inhibitors on resistance to amphotericin in starved cultures of *Candida albicans*. *Journal of General Microbiology*, **99**, 77–84.

Gale, E. F., Ingram, J., Kerridge, D., Notario, V. & Wayman, F. (1980). Reduction of amphotericin resistance in stationary phase cultures of *Candida albicans* by treatment with enzymes. *Journal of General Microbiology*, **117**, 383–91.

Gale, E. F., Cundliffe, E., Reynolds, P. E., Richmond, M. H. & Waring, M. J. (1981). *The Molecular Basis of Antibiotic Action*, 2nd edn. New York: John Wiley & Sons.

Giégé, R. & Weil, J. H. (1970). Etude des tRNA de levure ayant incorporé du 5-fluoruracile provenant de la désamination *in vivo* de la 5-fluorocytosine. *Bulletin de la Société de Chimie Biologique*, **52**, 135–44.

Glazer, R. & Hartman, K. U. (1980). The effect of 5-fluorouracil on the synthesis and methylation of low molecular weight nucleic acid in L1210 cells. *Molecular Pharmacology*, **17**, 245–9.

Gold, W., Stout, H. A., Pagano, J. F. & Donovick, R. (1956). Amphotericins A and B, antifungal antibiotics produced by a streptomycete. I. *In vitro* studies. *Antibiotics Annual*, 1956, 576–86.

Gottleib, D., Carter, H. E., Sloneker, J. H. & Immann, A. (1958). Protection of fungi against polyene antibiotics by sterols. *Science*, **128**, 361.

Gruda, I., Nadeau, P., Brajtburg, J. & Medoff, G. (1980). Application of differential spectra in the ultraviolet-visible region to study the formation of amphotericin B-sterol complexes. *Biochimica et Biophysica Acta*, **602**, 260–8.

Grunberg, E., Prince, H. N. & Utz, J. P. (1967). Observations on the activity of two newer antifungal agents, saramycetin (X-5079C) and 5-fluorocytosine. *Proceedings of the 5th International Congress of Chemotherapy*. Vienna, **4**, pp. 69–76.

Hamilton-Miller, J. M. T. (1972a). A comparative *in vitro* study of amphotericin B, clotrimazole and 5-fluorocytosine against clinically isolated yeasts. *Sabauroudia*, **10**, 276–83.

Hamilton-Miller, J. M. T. (1972b). Sterols from polyene resistant mutants of *Candida albicans*. *Journal of General Microbiology*, **73**, 201–3.

Hamilton-Miller, J. M. T. (1973). Chemistry and biology of the polyene macrolide antibiotics. *Bacteriological Reviews*, **37**, 166–96.

Hammond, S. M., Lambert, R. A. & Kliger, B. N. (1974). The action of polyene antibiotics; induced potassium leakage in *Candida albicans*. *Journal of General Microbiology*, **81**, 325–30.

Hay, R. J. (1980). The treatment of systemic mycoses caused by specific pathogenic fungi. In *Antifungal Chemotherapy*, ed. D. C. E. Speller, pp. 285–331. Chichester, John Wiley & Sons.

Hazen, E. L. & Brown, R. (1950). Two antifungal agents produced by a soil actinomycete. *Science*, **112**, 423.

Heidelberger, C., Chaudhuri, N. K., Danneberg, P., Mooren, D., Griesbach, L., Duschinsky, R., Schnitzer, R. J., Pleven, E. & Scheiner, T. (1957). Fluorinated pyrimidines, a new class of tumour inhibitory compounds. *Nature*, **179**, 663–6.

Heidelberger, C., Griesbach, L., Montag, B. J., Mooren, D., Cruz, O., Schnitzer, R. J. & Grunberg, E. (1958). Studies on fluorinated pyrimidines. II. Effects on transplanted tumours. *Cancer Research*, **18**, 305–17.

Hsu-Chen, C.-C. & Feingold, D. S. (1973). Polyene antibiotic action on lecithin liposomes: effect of cholesterol and fatty acyl chains. *Biochemical and Biophysical Research Communications*, **51**, 972–8.

Hsu-Chen, C.-C. & Feingold, D. S. (1974). Two types of resistance to antibiotics in *Candida albicans*. *Nature*, **251**, 656–9.

Iannitelli, R. C. & Ikawa, M. (1980). Effect of fatty acids on action of polyene antibiotics. *Antimicrobial Agents and Chemotherapy*, **17**, 861–4.

Jund, R. & Lacroute, F. (1970). Genetic and physiological aspects of resistance to 5-fluoropyrimidines in *Saccharomyces cerevisiae*. *Journal of Bacteriology*, **102**, 607–15.

Jund, R. & Lacroute, F. (1974). Génétique et physiologie de la résistance aux 5-fluoropyrimidines chez *Saccharomyces cerevisiae*. *Bulletin de la Société Française Mycologie Médicale*, **3**, 5–8.

Kasumov, Kh. M., Borisova, M. P., Ermishkin, L. N., Potzeluyev, V. M., Silberstein, A. Ya. & Vainshtein, V. A. (1979). How do ionic channel properties depend on the structure of polyene antibiotic molecules? *Biochimica et Biophysica Acta*, **551**, 229–37.

Kerridge, D., Koh, T. Y. & Johnson, A. M. (1976a). The interaction of amphotericin B methyl ester with protoplasts of *Candida albicans*. *Journal of General Microbiology*, **96**, 117–23.

Kerridge, D., Koh, T. Y., Marriott, M. S. & Gale, E. F. (1976b). The production and properties of protoplasts from the dimorphic yeast *Candida albicans*. In *Microbial and Plant Protoplasts*, ed. J. F. Peberdy, A. H. Rose, H. J. Rogers & E. C. Cocking, pp. 23–38. London: Academic Press.

Kitahara, M., Seth, V. K., Medoff, G. & Kobayashi, G. S. (1976). Antimicrobial sensitivity testing of six clinical isolates of *Aspergillus*. *Antimicrobial Agents and Chemotherapy*, **9**, 908–14.

Kitajima, Y., Sekiya, T. & Nozawa, Y. (1976). Freeze–fracture ultrastructural alterations induced by filipin, pimaricin, nystatin and amphotericin B in the plasma membrane of *Epidermophyton*, *Saccharomyces* and red blood cells: a proposal of models for polyene ergosterol complex-induced membrane lesions. *Biochimica et Biophysica Acta*, **455**, 452–65.

Kobayashi, G. S. & Medoff, G. (1977). Antifungal agents: recent developments. *Annual Review of Microbiology*, **31**, 291–308.

Kolter-Brajtburg, J., Medoff, G., Kobayashi, G. S., Boggs, S., Schlessinger, D., Panday, R. C. & Rinehart, K. L. (1979). Classification of polyene antibiotics according to chemical structure and biological effects. *Antimicrobial Agents and Chemotherapy*, **15**, 716–22.

Kumar, B. V., Medoff, G., Kobayashi, G. S. & Schlessinger, D. (1974). Uptake of *Escherichia coli* DNA into HeLa cells enhanced by amphotericin B. *Nature*, **250**, 323–5.

Lampen, J. O. (1966). Interference by polyene antibiotics (especially nystatin and filipin) with specific membrane functions. In *Biochemical Studies of Antimicrobial Drugs*, ed. B. A. Newton & P. E. Reynolds, pp. 111–30. Cambridge University Press.

Lampen, J. O., Arnov, P. M. & Safferman, R. S. (1960). Mechanism of protection by sterols against polyene antibiotics. *Journal of Bacteriology*, **80**, 200–6.

Langenbach, R. J., Danenberg, P. V. & Heidelberger, C. (1972). Thymidydate synthetase: mechanism of inhibition by 5-fluoro-2'-deoxyuridylate. *Biochemical and Biophysical Research Communications*, **48**, 1565–71.

Liras, P. & Lampen, J. O. (1974). Protection by K^+ and Mg^{++} of growth and macromolecular synthesis in candicidin treated yeast. *Biochimica et Biophysica Acta*, **374**, 159–63.

Malewicz, B., Jenkin, H. M. & Borowski, E. (1981). Repair of membrane alterations induced in baby hamster kidney cells by polyene macrolide antibiotics. *Antimicrobial Agents and Chemotherapy*, **19**, 238–47.

Marty, A. & Finkelstein, A. (1975). Pores formed in lipid bilayers by nystatin, differences in one sided and two sided action. *Journal of General Physiology*, **65**, 515–26.

Mas, J. & Pina, E. (1980). Disappearance of nystatin resistance in *Candida* mediated by ergosterol. *Journal of General Microbiology*, **117**, 249–52.

Mazerski, J., Bolard, J. & Borowski, E. (1982). Self-association of some polyene macrolide antibiotics in aqueous media. *Biochemica et Biophysica Acta*, **719**, 11–17.

Mechlinski, W., Schaffner, C. B., Ganis, P. & Avitabile, G. (1970). Structure and absolute configuration of the polyene macrolide antibiotic amphotericin B. *Tetrahedron Letters*, **44**, 3873–6.

Medoff, G., Brajtburg, J. & Kobayashi, G. S. (1983). Antifungal agents useful in therapy of systemic fungal infections. *Annual Reviews of Pharmacology and Toxicology*, **23**, 303–30.

Medoff, G., Comfort, M. & Kobayashi, G. S. (1971). Synergistic action of amphotericin B and 5-fluorocytosine against yeast-like organisms. *Proceedings of the Society of Experimental Biology and Medicine*, **138**, 571–4.

Medoff, G., Kobayashi, G. S., Kwan, C. N., Schlessinger, D. & Venkov, P. (1972). Potentiation of rifampicin and 5-fluorocytosine as antifungal antibiotics by amphotericin B. *Proceedings of the National Academy of Sciences USA*, **69**, 196–9.

Norman, A. W., Demel, R. A., De Kruijff, B., Geurts van Kessel, W. S. M. & Van Deenen, L. L. M. (1972). Studies on the biological properties of polyene antibiotics: comparison of other polyenes with filipin in their ability to interact specifically with sterol. *Biochimica et Biophysica Acta*, **290**, 1–14.

Norman, A. W., Spielvogel, A. M. & Wong, R. G. (1976). Polyene antibiotic–sterol interactions. *Advances in Lipid Research*, **14**, 127–70.

Normark, S. & Schönebeck, J. (1972). *In vitro* studies of 5-fluorocytosine resistance in *Candida albicans* and *Torulopsis glabrata*. *Antimicrobial Agents and Chemotherapy*, **2**, 114–21.

Notario, V., Gale, E. F., Kerridge, D. & Wayman, F. (1982). Phenotypic resistance to amphotericin B in *Candida albicans*: relationship to glucan metabolism. *Journal of General Microbiology*, **128**, 761–77.

Nozawa, Y., Kitajima, Y., Sekiya, T. & Ino, Y. (1974). Ultrastructural alterations induced by amphotericin B in the plasma membrane of *Epidermophyton floccosum* as revealed by freeze–etch electron microscopy. *Biochimica et Biophysica Acta*, **367**, 32–8.

Palacios, J. & Serrano, R. (1978). Proton permeability induced by polyene antibiotics: a plausible mechanism for their inhibition of maltose fermentation in yeast. *FEBS Letters*, **91**, 198–201.

Pena, A. (1975). Studies on the mechanism of K^+ transport in yeast. *Archives of Biochemistry and Biophysics*, **167**, 397–419.

Pesti, M., Novak, E. K., Ferenczy, L. & Svoboda, A. (1981). Freeze fracture electron microscopical investigation of *Candida albicans* cells sensitive and resistant to nystatin. *Sabouraudia*, **19**, 17–26.

Pierce, A. M., Pierce, H. D., Unrau, A. M. & Oehlschlager, A. C. (1978). Lipid composition and polyene antibiotic resistance of *Candida albicans* mutants. *Canadian Journal of Biochemistry*, **56**, 135–42.

Polak, A. (1974). Effects of 5-fluorocytosine on protein synthesis and amino acid pool in *Candida albicans*. *Sabouraudia*, **12**, 309–19.

Polak, A. (1978). Synergism of polyene antibiotics with 5-fluorocytosine. *Chemotherapy (Basel)*, **24**, 2–16.

Polak, A. & Grenson, M. (1973). Evidence for a common transport system for cytosine, adenine and hypoxanthine in *Saccharomyces cerevisiae* and *Candida albicans*. *European Journal of Biochemistry*, **32**, 276–82.

Polak, A. & Scholer, H. J. (1975). Mode of action of 5-fluorocytosine and mechanisms of resistance. *Chemotherapy (Basel)*, **21**, 113–30.

Polak, A. & Scholer, H. J. (1980). Mode of action of 5-fluorocytosine. *Revue de l'Institut Pasteur de Lyon*, **13**, 233–44.

Polak, A. & Wain, W. H. (1977). The inhibition by 5-fluorocytosine of nucleic acid synthesis in *Candida albicans, Cryptococcus neoformans* and *Aspergillus fumigatus. Chemotherapy (Basel)*, **23**, 243–59.

Readio, J. D. & Bittman, R. (1982). Equilibrium binding of amphotericin B and its methyl ester and borate complex to sterols. *Biochimica et Biophysica Acta*, **685**, 219–24.

Romine, W. O., Sherette, G. B., Brown, G. B. & Bradley, R. J. (1977). Evidence that nystatin may not form channels in thin lipid membranes. *Biophysical Journal*, **17**, 269–74.

Ryley, J. F., Wilson, R. G., Gravestock, M. B. & Poyser, J. P. (1981). Experimental approaches to antifungal chemotherapy. *Advances in Pharmacology and Chemotherapy*, **18**, 49–176.

Santi, D. V. & McHenry, C. S. (1972). 5-Fluoro-2'-deoxyuridylate: covalent complex with thymidylate synthetase. *Proceedings of the National Academy of Sciences USA*, **69**, 1855–7.

Scholer, H. J. (1968). Chemotherapie von Mykosen innerer Organe. *Schweizerische Medizinische Wochenschrift*, **98**, 602–11.

Scholer, H. J. (1970). Antimykoticum 5-Fluorocytosin. *Mykosen*, **13**, 179–88.

Scholer, H. J. (1980). Flucytosine. In *Antifungal Chemotherapy*, ed. D. C. E. Spller, pp. 35–106. Chichester: John Wiley & Sons.

Schönebeck, J. & Ånséhn, S. (1973). 5-Fluorocytosine resistance in *Candida* spp. and *Torulopsis glabrata. Sabouraudia*, **11**, 10–20.

Sekiya, T., Yano, K. & Nozawa, Y. (1982). Effects of amphotericin B and its methyl ester on plasma membranes of *Candida albicans* and erythrocytes as examined by freeze–fracture electron microscopy. *Sabouraudia*, **20**, 303–11.

Serrano, R. (1977). Energy requirements for maltose transport in yeast. *European Journal of Biochemistry*, **80**, 97–102.

Speller, D. C. E. (ed.) (1980). *Antifungal Chemotherapy*. Chichester: John Wiley & Sons.

Stiller, R. L., Bennett, J. E., Scholer, H. J., Hall, M., Polak, A. & Stevens, D. A. (1982). Susceptibility to 5-fluorocytosine and prevalence of serotype in 402 *Candida albicans* isolates from the United States. *Antimicrobial Agents and Chemotherapy*, **22**, 482–7.

Subden, R. E., Safe, L., Morris, D. C., Brown, R. G. & Safe, S. (1977). Eburicol, lichesterol ergosterol and obtusifoliol from polyene resistant mutants of *Candida albicans*. *Canadian Journal of Microbiology*, **23**, 751–4.

Tillack, L. T. W. & Kinski, S. C. (1973). A freeze–etch study of the effects of filipin on liposomes and human erythrocyte membranes. *Biochimica et Biophysica Acta*, **323**, 43–54.

Tseng, W., Medina, D. & Randerath, K. (1978). Specific inhibition of transfer RNA methylation and modification in mice treated with 5-fluorouracil. *Cancer Research*, **38**, 1250–7.

Utz, J. P., Garriques, J. L., Sande, M. A., Warner, J. F., McGehee, R. F. & Shadomy, S. (1975). Combined flucytosine–amphotericin B treatment of cryptococcal infections. *Journal of Infectious Diseases*, **132**, 368–73.

Van Hoogevest, P. & De Kruijff, B. (1978). Effect of amphotericin B on cholesterol containing liposomes of egg phosphatidylcholine and didocosanoyl phosphatidylcholine: a refinement of the model for the formation of pores by amphotericin B in membranes. *Biochimica et Biophysica Acta*, **511**, 397–407.

Verkleij, A. J., De Kruijff, B., Gerritsen, W. F., Demel, R. A., Van Deenen, L. L. M. & Ververgaert, P. H. J. (1973). Freeze–etch electron microscopy of erythrocytes, *Acholeplasma laidlawii* cells and liposomal membranes after the action of filipin and amphotericin B. *Biochimica et Biophysica Acta*, **291**, 577–81.

Vertut-Croquin, A., Bolard, J., Chabbert, M. & Gary-Bobo, C. (1983). Differences in the interaction of the polyene antibiotic amphotericin B with cholesterol- or ergosterol-containing phospholipid vesicles: a circular dichroism and permeability study. *Biochemistry*, **22**, 2939–44.

Wagner, G. E. & Shadomy, S. (1979). Studies on the mode of action of 5-fluorocytosine in *Aspergillus* species. *Chemotherapy (Basel)*, **25**, 61–9.

Wain, W. H. & Polak, A. (1979). The effect of flucytosine on the germination of *Candida albicans*. *Postgraduate Medical Journal*, **55**, 671–3.

Wain, W. H., Polak, A. & Florio, R. A. (1981). The effect of 5-fluorocytosine on the germination of *Aspergillus fumigatus* and on subsequent nuclear division. *Sabouraudia*, **19**, 147–53.

Waldorf, A. R. & Polak, A. (1983). Mechanisms of action of 5-fluorocytosine. *Antimicrobial Agents and Chemotherapy*, **23**, 79–85.

Whelan, W. L., Beneke, E. S., Rogers, A. L. & Soll, D. R. (1981). Segregation of 5-fluorocytosine-resistant variants by *Candida albicans*. *Antimicrobial Agents and Chemotherapy*, **19**, 1078–81.

Wilkinson, D. S., Tisty, T. D. & Hanar, R. J. (1975). The inhibition of ribosomal RNA synthesis and maturation in Novikoff hepatoma cells by 5-fluorouridine. *Cancer Research*, **35**, 3014–26.

Zylstra, W. (1974). Cryptococcosis and 5-fluorocytosine. *Australian and New Zealand Journal of Medicine*, **4**, 296–9.

17

Enhancement of efficacy of antifungal agents by entrapment inside liposomes

M. L. CHANCE and R. R. C. NEW

Department of Parasitology, Liverpool School of Tropical Medicine,
Pembroke Place, Liverpool L3 5QA, UK

Introduction

Liposomes are small vesicles composed of membrane bilayers which form spontaneously when phospholipids are dispersed in aqueous media. They comprise an aqueous compartment entrapped in the centre of the vesicle, and a lipid compartment consisting of the fatty acid chains of phospholipids; this may be quite sizeable in the case of multilamellar vesicles (MLVs), which are composed of several concentric membranes.

A variety of different types of liposome may be manufactured (Ryman & Tyrrell, 1979), but perhaps the simplest to prepare are multilammelar vesicles. These are made by dissolving the lipid components in organic solvent (usually chloroform or methanol), and evaporating to dryness to give a thin film which is resuspended by shaking with the aqueous phase. The large multilamellar vesicles formed may be ultrasonicated briefly to reduce their size below 1 μm in diameter. A drug may be incorporated into the liposome by including it in either the organic or the aqueous phase according to its solubility, unentrapped material being removed if necessary by dialysis or centrifugation.

Administration of drugs entrapped inside liposomes is an effective way of bringing those drugs into close contact with pathogenic organisms residing in the reticulo-endothelial cells of the body (Gregoriadis, 1970; Ryman & Tyrrel, 1979). Liposomes, like all small particles, are taken up by macrophages, especially in the liver and spleen, and are rapidly digested in the secondary lysosomes to release their contents into the vacuole and surrounding cytosol. Since liposomes are rapidly cleared from the bloodstream, the toxicity of the drug to the host may often be reduced. In this chapter, we discuss how liposomes have widened the scope of therapy using antifungal agents, both in terms of increased efficacy of the drugs and also in the treatment of non-fungal infections.

Amphotericin B in liposomes
Mode of action and toxicity

Because of its amphipathic nature, and its ability to integrate into biological lipid membranes, amphotericin B is an ideal candidate for incorporation into liposomes; it has been studied in model membrane systems by a variety of workers (Chen & Bittman, 1977; Van Hoogevest & de Kruijff, 1978; Kasumov *et al.*, 1979; Bolard, Seigneuret & Boudet, 1980). From such studies, it is thought that the amphotericin molecules inserting in the membrane spontaneously assemble to form a pore with a central channel large enough to permit passage of low molecular weight ions and water (de Kruijff & Demel, 1974). In this model, a pore structure consisting of eight amphotericin units is stabilised by sterols, in particular 3-β-hydroxy-5,7-dienes such as ergosterol (Verkleij *et al.*, 1973; Kotler-Brajtburg *et al.*, 1974; Nozawa *et al.*, 1974; Feigin *et al.*, 1979). The strong interaction with ergosterol explains the selectivity of amphotericin for fungal membranes, whose integrity is compromised as ions pass freely from one side to the other, resulting in the death of the cell. The lesser, but not insignificant, interaction of amphotericin with cholesterol in membranes is probably responsible for the well-known host-directed toxicity observed in clinical practice.

An important feature of amphotericin B liposomes is the reduction in toxicity when given intravenously, observed in comparison with the proprietary deoxycholate stabilised formulation Fungizone®. Fig. 17.1 shows the results of an acute toxicity test in mice in which the influence of lipid composition was demonstrated (New, Chance & Heath, 1981*a*). The presence of sterol in the neutral liposome membrane markedly increased the LD_{50} from 3.2 mg kg^{-1} for liposomes containing egg yolk lecithin only, to 14 mg kg^{-1}, i.e. it decreased the toxicity. Liposomes containing ergosterol instead of cholesterol are less toxic still.

The degree of saturation (i.e. hydrogenation) of the phospholipids is also important. Amphotericin B contained within hydrogenated lecithin liposomes was non-toxic at the highest dose (20 mg kg^{-1}) that could be administered without using excessively large volumes. Free amphotericin, when given as a particulate suspension produced by sonicating pure drug, has a relatively low toxicity, with an LD_{50} of 125 mg kg^{-1}, when compared to Fungizone®. An increased degree of saturation of the phospholipids, and the presence of sterols in membranes would be expected to increase their order and stability, and thus decrease the flux of molecules across the membranes. Some support for this hypothesis is found in the studies of Chen & Bittman (1977), who found that the rate of uptake of amphotericin into liposomes was decreased by the presence of hydrogenated lecithin or sterols in the liposome membrane. Thus if the same factors control the movement of amphotericin in the reverse direction, the decreased toxicity associated with amphotericin-containing liposomes composed of either hydrogenated lecithin or containing sterols may possibly be explained on the basis of the controlled release of the liposome contents. Some contradictory conclusions on the relationship between liposome composition and amphotericin have, however, been reached by Bolard *et al.* (1980).

Antifungal activity

Although the toxicity of intravenous amphotericin B is reduced by entrapment in liposomes, its efficacy in experimental models of fungal infection is retained. Taylor *et al.* (1982) have shown that the median survival time of nude mice carrying a fatal infection of *Histoplasma capsulatum* is extended from 13 d (untreated) to 19 d by administration of the maximum non-toxic dose of Fungizone® (10 μg dose^{-1}) on three successive days. In contrast, treatment with the maximum non-toxic dose of positively charged ergosterol-containing amphotericin liposomes (90 μg dose^{-1}) resulted in an 80% survival of these mice at

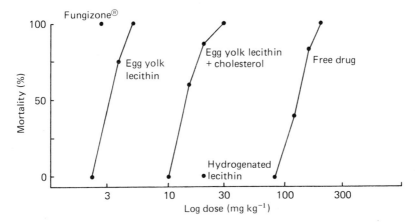

Fig. 17.1. Effects of the liposome composition on the toxicity of entrapped amphotericin B administered intravenously to uninfected mice.

35 d, when all animals from the other groups had died. The surviving mice, however, were shown to contain morphologically intact yeast forms in their tissues.

In the same laboratories, Graybill *et al.* (1982) obtained similar results in the treatment of murine cryptococcosis. In a model where 10^6 colony forming units of *Cryptococcus neoformans* was given intraperitoneally, followed by treatment intravenously on three separate days using levels of drug close to the maximum tolerated dose, the median survival time was prolonged from 22 d (control or Fungizone® 10 μg dose^{-1}) to 36 d with 48 μg dose^{-1} amphotericin B liposomes. No treatment regime, however, was found to be superior when using equivalent amounts of drug. These authors also report that liposomes conferred longer survival on mice infected intracerebrally.

Recently, Juliano *et al.* (1983) have shown that the therapeutic index of amphotericin B (defined as ratio LD_{50} to minimum effective dose) in murine candidosis can also be increased, by entrapment in negative liposomes, from 2 to between 4 and 12. It is interesting to note that no abnormalities in blood chemistry or histology were noted in the animals treated with liposomes, thus allaying fears that organ toxicity might result from the marked tropism of liposomes for the liver and spleen. These workers have also demonstrated that administering amphotericin-containing liposomes 2 d before inoculating with *Candida albicans* has a marked prophylactic effect (Lopez-Berenstein *et al.*, 1983).

In all of the systems described here, it would appear that the efficacy of amphotericin is unchanged by incorporation into liposomes, and that the advantage gained results from the reduced toxicity of the liposomal preparation, allowing administration of larger quantities of drug than are possible in the free form.

Antileishmanial activity

In addition to treating infections of fungal origin, antifungal agents are being successfully used clinically in the treatment of leishmaniasis (Manson-Bahr, 1982), a disease caused by parasitic protozoa of the genus *Leishmania*. In the mammalian host, these parasites are almost invariably found within macrophages of the reticulo-endothelial system. The leishmanial organisms survive within parasitophorous vacuoles even though lysosomal fusion is not inhibited (Lewis & Peters, 1977). These authors showed that two separate phagocytic vacuoles can undergo fusion, which suggests that chemotherapy with lysosomatrophic agents such as liposome-entrapped drugs would be successful.

Table 17.1. *Clearance of amastigotes from liver after treatment with amphotericin B, either in dispersions or liposomes*

Treatment	Amastigotes per 200 liver cell nuclei ($\pm$SD)[a]	
	Dispersion	PC liposomes
Amphotericin	501 ± 39	—
Amphotericin/cholesterol	173 ± 80	4 ± 6
Amphotericin/dehydrocholesterol	201 ± 44	1 ± 2
Amphotericin/ergosterol	354 ± 64	1 ± 2

[a]All animals received three doses of $10\,\text{mg}\,\text{kg}^{-1}\text{d}^{-1}$ of amphotericin B; PC: phosphatidylcholine. The amastigote count in untreated animals was 508 ± 76.

Thus, liposomes contained within one phagocytic vacuole may fuse with another vacuole containing parasites, resulting in drug being delivered directly into the vicinity of the parasite.

Intravenously administered liposomes are preferentially taken up into macrophages of the liver and spleen – which are the tissues in which the leishmanial amastigotes are present in visceral leishmaniasis. The expectation of successful chemotherapy of visceral infection using liposome-entrapped drugs has been realised in experimental studies with the antimonial drugs that are in clinical use (Black, Watson & Ward, 1977; Alving *et al.*, 1978; New *et al.*, 1978).

Similarly, intravenous administration of amphotericin-containing liposomes was seen to reduce the number of amastigotes within the liver of experimentally infected mice (New *et al.*, 1981*a*). A sonicated dispersion of amphotericin was without effect, though mixtures of amphotericin and sterol had some effect (Table 17.1). This effect was, however, small when compared to that seen with amphotericin within liposomes, when the parasite load was virtually eliminated at a dose at which free amphotericin was ineffective. Empty liposomes are known to be without effect.

Experiments with lower levels of amphotericin which allowed Fungizone® to be included (Table 17.2) showed that, in contrast to the models of fungal disease discussed above, amphotericin-containing liposomes performed better than Fungizone® at roughly equivalent doses, provided that ergosterol was also present.

In the other main clinical manifestation of leishmaniasis, the parasites are not present in visceral organs, but are restricted to macrophages

Table 17.2. *Clearance of amastigotes from liver after treatment with three doses of amphotericin B liposomes containing different sterols*

Treatment	Dose $(mg\,kg^{-1}d^{-1})$	Parasites per 200 liver cell nuclei $(\pm SD)$
None	0	1321 ± 111
Amp	1	1417 ± 51
Fungizone®	0.75	1229 ± 72
Amp/PC	0.5	1434 ± 35
Amp/PC/cholesterol	1	1248 ± 113
Amp/PC/ergosterol	1	720 ± 171

within cutaneous lesions. From what has been outlined of the behaviour of liposomes *in vivo*, the prospects of successful liposome-mediated chemotherapy of cutaneous leishmaniasis would not appear to be good. Effective treatment of cutaneous lesions in mice with liposome-entrapped antimonial drugs has, however, been achieved (New *et al.*, 1981*b*). It has also been shown that amphotericin-containing liposomes, of the same composition as those that proved effective in treating visceral infections, suppress the development of cutaneous lesions (Fig. 17.2). Free amphotericin and Fungizone® were not effective, though because of its toxicity, the amount of amphotericin in the Fungizone® treatment was at a much lower level.

Griseofulvin and 5-fluorocytosine in liposomes

Two other well-known antifungal agents have been tested in our mouse model for visceral leishmaniasis. Neither griseofulvin nor 5-fluorocytosine have been shown, either experimentally or clinically, to have any significant antileishmanial activity. This was confirmed in our experiments (Table 17.3) in which free drug was without effect, whereas liposome-entrapped drug had a marked effect on the parasite load.

Discussion

The leishmanicidal activity of a number of antifungal agents with different modes of action suggests the existence of possible biochemical similarities between protozoa and fungi, which could be exploited further. It is interesting to note that it has recently been reported that promastigotes of *Leishmania* spp., like fungi, have large amounts of ergosterol and related sterols in their membrane (Goad, Holz & Beach, 1984), which by analogy with fungi explains the high susceptibility of these organisms to amphotericin.

Table 17.3. *Clearance of liver amastigotes by three doses of griseofulvin or 5-fluorocytosine: liposome enhancement of activity*

	Parasites per 200 nuclei ($\pm$SD)	
	Griseofulvin (25 mg kg^{-1})	5-fluorocytosine (37 mg kg^{-1})
Control	644 $\pm$ 82	1318 $\pm$ 87
Free drug	685 $\pm$ 53	1222 $\pm$ 74
Drug/liposomes	154 $\pm$ 32	681 $\pm$ 104

It is quite clear from the results with experimental cutaneous leishmaniasis that liposome-entrapped drugs can exert an effect at peripheral sites remote from the centres of liposome concentration. Thus the treatment of non-visceral infections may be possible with liposome-entrapped drugs, though only when the infecting organism is associated with macrophages.

Though liposome-entrapped drugs have their greatest potential in treating infections associated with the reticulo-endothelial system, the

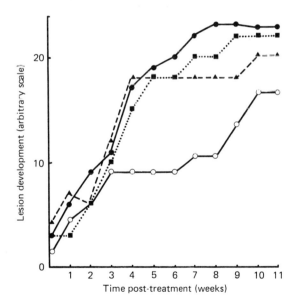

Fig. 17.2. Effect of liposome-entrapped and free amphotericin B on development of lesions in mice due to *Leishmania mexicana amazonensis*. Control: filled circles; free amphotericin (10 mg kg^{-1} day^{-1} ($\times$3)): squares; entrapped amphotericin: open circles; Fungizone® (0.75 mg kg^{-1} day^{-1} ($\times$3)): triangles.

preferential distribution of liposomes to the liver and spleen may result in a sufficient concentration of drug in these organs to exert an effect on visceral infections, even though the target organisms are not located specifically within macrophages.

We have shown that liposome-mediated therapy holds great promise for the treatment of leishmaniasis. It may also have great relevance to fungal infections, particularly histoplasmosis, where the parasite is present in cells of the reticulo-endothelial system.

References

Alving, C. R., Steck, E. A., Chapman, W. L., Waits, V. B., Hendricks, L., Swartz, G. H. & Hanson, W. L. (1978). Therapy of leishmaniasis: superior efficacies of liposome encapsulated drugs. *Proceedings of the National Academy of Sciences, USA*, **75**, 2959–63.

Black, C. D. V., Watson, G. J. & Ward, R. J. (1977). The use of pentostam liposomes in the chemotherapy of experimental leishmaniasis. *Transactions of the Royal Society of Tropical Medicine and Hygiene*, **71**, 550–2.

Bolard, J., Seigneuret, M. & Boudet, G. (1980). Interaction between phospholipid bilayer membranes and the polyene antibiotic amphotericin B: lipid state and cholesterol dependence. *Biochimica et Biophysica Acta*, **599**, 280–93.

Chen, W. C. & Bittman, R. (1977). Kinetics of association of amphotericin B with vesicles. *Biochemistry*, **16**, 4145–9.

de Kruijff, B. & Demel, R. A. (1974). Polyene antibiotic–sterol interactions in membranes of *Acholeplasma laidlawii* cells and lecithin liposomes. III. Molecular structure of the polyene antibiotic–cholesterol complexes. *Biochimica et Biophysica Acta*, **339**, 57–70.

Feigin, A., Byelousova, I., Yakhimovich, R., Vasilyevskaya, V. & Tereshin, I. (1979). Role of the number and position of double bonds in the tetracyclic nucleus of sterols in the interaction with polyene antibiotics. *Biofizica*, **24**, 330–1.

Goad, L. J., Holtz, G. G. & Beach, D. H. (1984). Sterols of *Leishmania* spp.: nature and biosynthesis. *Molecular and Biochemical Parasitology*, **10**, 161–70.

Graybill, J. R., Craven, P. C., Taylor, R. L., Williams, D. M. & Magee, W. E. (1982). Treatment of murine cryoptococcocis with liposome associated amphotericin B. *Journal of Infectious Diseases*, **145**, 748–52.

Gregoriadis, G. (1970). The liposome drug-carrier concept: its development and future. In *Liposomes in Biological Systems*, ed. G. Gregoriadis & A. C. Allison, pp. 25–86. New York: John Wiley & Sons.

Juliano, R. L., Lopez-Berenstein, G., Mehta, R., Hopfer, R., Mehta, K. & Kasi, L. (1983). Pharmacokinetics and therapeutic consequences of liposomal drug delivery: fluoreodeoxyuridine and amphotericin B as examples. *Biology of the Cell*, **47**, 39–46.

Kasumov, K. M., Borisova, M. P., Ermishkin, L. N., Potseluyev, V. M., Silberstein, A. Y. & Vainstein, V. A. (1979). How do ionic channel properties depend on the structure of polyene antibiotic molecules? *Biochimica et Biophysica Acta*, **551**, 229–37.

Kotler-Brajtburg, J., Price, H. D., Medoff, G., Schlessinger, D. & Kobayashi, G. S.

(1974). Molecular basis for the selective toxicity of amphotericin B for yeast and filipin for animal cells. *Antimicrobial Agents and Chemotherapy*, **5**, 383–7.

Lewis, D. H. & Peters, W. (1977). The resistance of intracellular *Leishmania* parasites to digestion by lysosomal enzymes. *Annals of Tropical Medicine and Parasitology*, **71**, 295–310.

Lopez-Berenstein, G., Mehta, R., Hopfer, R. L., Mills, K., Kasi, L., Mehta, K., Fainstein, V., Luna, M., Hersh, E. M. & Juliano, R. (1983). Treatment and prophylaxis of disseminated infection due to *Candida albicans* in mice with liposome-encapsulated amphotericin B. *Journal of Infectious Diseases*, **147**, 939–45.

Manson-Bahr, P. E. C. (1982). Leishmaniasis. In *Manson's Tropical Diseases*, 18th edn., 1982, ed. P. E. C. Manson-Bahr & F. I. C. Apted, pp. 93–115. London: Baillière Tindall.

New, R. R. C., Chance, M. L. & Heath, S. (1981*a*). Antileishmanial activity of amphotericin and other antifungal agents entrapped in liposomes. *Journal of Antimicrobial Chemotherapy*, **8**, 371–81.

New, R. R. C., Chance, M. L. & Heath, S. (1981*b*). The treatment of experimental cutaneous leishmaniasis with liposome-entrapped Pentostam. *Parasitology*, **83**, 519–27.

New, R. R. C., Chance, M. L., Thomas, S. C. & Peters, W. (1978). The antileishmanial activity of antimonials entrapped in liposomes. *Nature*, **272**, 55–6.

Nozawa, U., Kitajima, Y., Sekiya, T. & Ito, Y. (1974). Ultrastructural alterations introduced by amphotericin B in the plasma membrane of *Epidermophyton floccosum* as revealed by freeze–etch electron microscopy. *Biochimica et Biophysica Acta*, **367**, 32–8.

Ryman, B. E. & Tyrrell, D. A. (1979). Liposomes – methodology and applications. In *Lysosomes in Biology and Pathology*, vol. 6, 1979, ed. J. T. Dingle & H. B. Fell, pp. 549–74. Amsterdam: North Holland.

Taylor, R. L., Williams, D. M., Craven, P. C., Graybill, J. R., Drutz, D. J. & Magee, W. E. (1982). Amphotericin B in liposomes: a novel therapy for histoplasmosis. *American Review of Respiratory Diseases*, **125**, 610–11.

Van Hoogevest, P. & de Kruijff, B. (1978). Effect of amphotericin B on cholesterol-containing liposomes of egg phosphatidyl choline and didocosenoyl phosphatidyl choline. A refinement of the model for the formation of pore by amphotericin B in membranes. *Biochimica et Biophysica Acta*, **511**, 397–407.

Verkleij, A. J., de Kruijff, B., Gerritsen, W. F., Demel, R. A., van Deenen, L. L. M. & Ververgaert, P. H. J. T. (1973). Freeze–etch electron microscopy of erythrocytes, *Acholeplasma laidlawii* cells and liposomal membranes after the action of filipin and amphotericin B. *Biochimica et Biophysica Acta*, **291**, 577–81.

Species index

Italic page numbers refer to diagrams

Subject index

Italic page numbers refer to diagrams